AF717330

WORKS ISSUED BY
THE HAKLUYT SOCIETY

Series Editors
Janet Hartley
Joyce Lorimer
Maurice Raraty

Series editor for this volume: Janet Hartley

TAKING NEWTON ON TOUR:
THE GRAND TOUR TRAVEL DIARY OF MARTIN FOLKES
(1690–1754)

THIRD SERIES
NO. 45

THE HAKLUYT SOCIETY

Trustees of the Hakluyt Society 2024–2025

PRESIDENT

Dr Gloria Clifton

VICE-PRESIDENTS

Professor Felipe Fernández-Armesto
Captain Michael Barritt RN
Professor Nandini Das
Bruce Hunter
Dr Sarah Tyacke CB

COUNCIL

(with date of election)

Dr Lubaaba Al-Azami (2023)
Dr Katie Bank (2021)
Dr Jack Benson (co-opted)
Professor Zoltán Biedermann (2024)
Professor Daniel Carey MRIA (2020)
Dr Matthew Day (2022)
Dr Natalya Din-Kariuki (2022)
Dr Derek L. Elliott (2021)
Dr Bertie Mandelblatt (2021)
Dr Richard Marsh (2022)
Dr Guido van Meersbergen (2020)
Dr Elaine Murphy (2022)
Professor Ladan Niayesh (2023)
Professor Joan-Pau Rubies (2023)
Professor Suzanne Schwarz (2024)
Dr Emily Stevenson (2022)
Dr Mark Williams (2024)

HONORARY TREASURER

Alastair Bridges

HONORARY JOINT SERIES EDITORS

Professor Janet Hartley
Professor Joyce Lorimer
Dr Maurice Raraty

HONORARY EDITOR, ONLINE JOURNAL OF THE HAKLUYT SOCIETY

Raymond Howgego

HONORARY ADMINISTRATIVE EDITOR

Dr Katherine Parker

(Dr Parker is also Royal Geographical Representative on the Council)

HONORARY ADVISER FOR CONTRACTS

Stephen Edwards

HONORARY ARCHIVIST

Dr Margaret Makepeace

ADMINISTRATION

Email: office@hakluyt.com

Postal address only

The Hakluyt Society, c/o Map Library, The British Library, 96 Euston Road, London NW1 2DB

Telephone: +44 (0)7568 468066

Website: www.hakluyt.com

Registered Charity No. 313168

VAT No. GB 233 4481 77

INTERNATIONAL REPRESENTATIVES OF THE HAKLUYT SOCIETY

Australia	John A. Greig, 18 Kiandra Close, The Gap, Brisbane, QLD 4061, Australia
Canada	Professor Cheryl A. Fury, Department of History and Politics, University of New Brunswick, Saint John, NB, E2L 4L5
Central America	Dr Stewart D. Redwood, P.O. Box 0832-0757, World Trade Center, Panama, Republic of Panama
China	Professor May Bo Ching, Department of Chinese and History, City University of Hong Kong, Kowloon Tong, Hong Kong
France	Dr Sandhya Patel, Université Paul-Valéry Montpellier 3, Département d'Études Anglophones, Route de Mende, Montpellier 34199 Cedex 5
Germany	Monika Knaden, Lichtenradenstrasse 40, 12049 Berlin
Iceland	Anna Agnarsdóttir, Professor Emerita, Department of History, University of Iceland, Arnagardur Vid Sudurgotu, Reykjavík 101
India	Professor (Emerita) Supriya Chaudhuri, Department of English, Jadavpur University, Kolkata
Ireland	Professor Jim McAdam OBE, Queen's University of Belfast and Royal Society of Antiquaries of Ireland, c/o 16B Dirnan Road, Cookstown, Co Tyrone BT80 9XL
Japan	Derek Massarella, Professor Emeritus, Chuo University, Higashinakano 742–1, Hachioji-shi, Tokyo 192–03
Netherlands	Dr Anita van Dissel, Room number 2.66a, Johan Huizingagebouw, Doezensteeg 16, 2311 VL Leiden
New Zealand	Dr John Holmes, PO Box 6307, Dunedin North, Dunedin 9059
Portugal	Dr Manuel Ramos, Av. Elias Garcia 187, 3Dt, 1050 Lisbon
Russia	Professor Alexei V. Postnikov, Institute of the History of Science and Technology, Russian Academy of Sciences, 1/5 Staropanskii per., Moscow 103012
South America	Sr Sergio Zagier, Cuenca 4435, 1419 Buenos Aires, Argentina
South East Europe	Dr Derek L. Elliott, Ljubljana International School, Rakovniška ulica 5, 1000 Ljubljana, Slovenia
Spain	Ambassador Dámaso de Lario, Embajador de España, Calle Los Borja 3, 3ª, 46003 València
Switzerland	Dr Tanja Bührer, Universität Bern, Historisches Institut, Unitobler, Länggasstrasse 49, 3000 Bern 9
United States	Professor Mary C. Fuller, Literature Section, 14N-405, Massachusetts Institute of Technology, 77 Massachusetts Avenue, Cambridge, MA, 02139-4307

Jonathan Richardson the Elder, *Martin Folkes*. Oil on canvas, 76 × 63 cm.
Society of Antiquaries of London. Bridgeman Images.

TAKING NEWTON ON TOUR: THE GRAND TOUR TRAVEL DIARY OF MARTIN FOLKES (1690–1754)

Edited by
ANNA MARIE ROOS

Published by
The Boydell Press
for
THE HAKLUYT SOCIETY
LONDON
2025

© The Hakluyt Society 2025

All Rights Reserved. Except as permitted under current legislation no part of this work may be photocopied, stored in a retrieval system, published, performed in public, adapted, broadcast, transmitted or reproduced in any form or by any means, without the prior permission of the copyright owner

The right of Anna Marie Roos to be identified as the author of this work has been asserted in accordance with sections 77 and 78 of the Copyright, Designs and Patents Act 1988

First published 2025

A publication of The Hakluyt Society
published by The Boydell Press
an imprint of Boydell & Brewer Ltd
Bridge Farm Business Park, Top Street, Martlesham, IP12 4RB
and of Boydell & Brewer Inc.
668 Mt Hope Avenue, Rochester, NY 14620, USA
website: www.boydellandbrewer.com
website: www.hakluyt.com

ISBN 9781916931190

The Boydell Press is an imprint of Boydell & Brewer Ltd
PO Box 9, Woodbridge, Suffolk IP12 3DF, UK
and of Boydell & Brewer Inc.
668 Mt Hope Avenue, Rochester, NY 14620–2731 , USA
website: www.boydellandbrewer.com

Boydell & Brewer's Authorised Representative for product safety in the EU is Easy Access System Europe – Mustamäe tee 50, 10621 Tallinn, Estonia, *gpsr.requests@easproject.com*

A CIP catalogue record for this book is available from the British Library

The publisher has no responsibility for the continued existence or accuracy of URLs for external or third-party internet websites referred to in this book, and does not guarantee that any content on such websites is, or will remain, accurate or appropriate

Typeset in Garamond Premier Pro
by Waveney Typesetters, Wymondham, Norfolk

To Ian Benton and in memory of Jim Bennett

CONTENTS

LIST OF ILLUSTRATIONS

Map

Colour Plates

Figures

PREFACE AND ACKNOWLEDGEMENTS

This volume provides a critical edition of an exceptional example of the 'Scientific Grand Tour' taken by Martin Folkes. Folkes (1690–1754) was Newton's protégé, antiquary, mathematician, and the only simultaneous President of the Royal Society and the Society of Antiquaries. In 1733–5, he went on a Grand Tour as a scientific ambassador for the Royal Society, demonstrating Newtonian optics to Italian virtuosi. He also measured ancient and Renaissance buildings to understand past architectural engineering and design. His 97-page illustrated diary (in the Bodleian Library at the University of Oxford, shelfmark MS Eng. misc.c.444) also challenges the long-standing, mistaken impression among scholars that the Royal Society was in decline in the eighteenth century. Analysing Folkes's activities abroad and creating an edition from this source tracing his Italian route provide a novel reading of Newtonianism and the purpose of the Grand Tour as a vehicle for scientific research and statesmanship. There is related correspondence in London and Italian archives, and extant objects which he collected on his tour (such as a plaster cast of the Farnese Globe now in the British Museum) which enriches the annotation and illustration of the volume.

The diary reveals that in Venice Folkes recreated Newton's experiments in optics for an invited and sceptical audience at the Palazzo Giustiniani, primarily to counter the claims of the Paduan natural philosopher Giovanni Rizzetti (1675–1751) that he could not reproduce the experiments in the refrangibility of light in Newton's *Opticks*. These experiments were recreated in Florence with naturalist Antonio Cocchi and fellow Freemasons. Other contextual material reveals that Folkes also discussed ancient and Renaissance use of tricks of optical illusion and architectural perspective with Francesco Algarotti, part of his interest in probing the relationship between seeing and believing in scientific observation and aesthetics. When in Venice, Folkes also contacted Anders Celsius (1701–44) and arranged for Celsius to obtain precision scientific instruments used in his expedition with Pierre Louis Maupertuis (1698–1759) to Lapland to measure the shape of the earth by triangulation to prove Newton's theory that the earth was a flattened sphere. At the same time, the diary reflected Folkes's antiquarian interests; not surprising as the Royal Society at this time was also involved in projects that integrated natural philosophy ('science') and antiquarianism (history/analysis of past material culture, a form of nascent archaeology) particularly before the re-establishment of the Society of Antiquaries of London in 1717.

My work in editing Folkes's diary has a long history. Analysing his diary was the subject of an All Souls Visiting Fellowship at the University of Oxford in Trinity Term in 2017, resulting in a published article about his journal in the *British Journal of the History of Science* the same year.[1] A major biography of Folkes followed in 2021 with Oxford University Press thanks to the good offices of acquisitions editor Dan Taber.[2]

[1] Roos, 'Taking Newton on Tour', pp. 569–601.

[2] Roos, *Martin Folkes (1690–1754): Newtonian, Antiquary, Connoisseur.*

I have many people to thank who facilitated my own scholarly journey. I would first like to thank Sir Robert ffolkes who kindly gave me permission to prepare Folkes's diary for publication. The Bodleian Library staff should be thanked, particularly Richard Ovenden for his assistance with questions of copyright and provenance. The Royal Society Library has been exceptionally kind in providing digital materials and research assistance for this work, with a special mention of Rupert Baker, Ellen Embleton, Louisiane Ferlier, Anne McLaughlin, and Keith Moore. All Souls College at the University of Oxford provided the most amenable place for me to think and write about Folkes, natural philosophy, and antiquarianism, for which I am most grateful. The Huntington Museum and Art Gallery also gave me a visiting fellowship to examine their collection of theatrical materials, important in understanding Folkes's wife, the Drury Lane actress Lucretia Bradshaw, and her experience of her husband's scientific Grand Tour. The staff at Northwestern University who gave me access to and scans of the correspondence of Antonio Cocchi have my deep gratitude, with a special mention of Peter B. Burtch, D. J. Hoek, Geoffrey Welles Morse, Kurt Munson, and Sarah M. Pritchard. My colleagues at the University of Lincoln are also to be thanked for their encouragement and collegiality. I am very grateful that the University provided me with a pump priming grant for archival research to finish this edition.

For fruitful discussions on Folkes and his journey, I am equally indebted to Christian Dekesel, William Eisler, Patricia Fara, Moti Feingold, Rob Iliffe, Sir Noel Malcolm, Cesare Pastorino, Cornelis Schilt, Benjamin Wardhaugh, and Alexander Wragge-Morley. My particular thanks are due to the Hakluyt Society for accepting the present work, as well as their wonderful editors. Jim Bennett, the past president, President Gloria Clifton, Janet Hartley, Katie Parker, and Will Ryan have been unfailingly helpful throughout the process. Janet was exceptionally helpful in her suggestions for improving the volume and, with Mary Murphy, copyedited the edition thoroughly. I would also like to thank Barrie Fairhead for his expertise in typesetting this volume. All remaining errors are my own.

My ultimate debt of gratitude is to my husband Ian Benton who has stood lovingly by my side and to whom I dedicate this work. I also dedicate this work to Professor Jim Bennett who sadly passed away when this edition was being completed; he was a remarkable scholar, museum curator, mentor, and my friend. Rest in peace, Jim.

INTRODUCTION

1. Introduction and Provenance of the Diary

Martin Folkes (1690–1754) was the scion of a wealthy Norfolk family.[1] His father, also named Martin, was the realm's indispensable solicitor-general who served as estate manager for Catherine of Braganza, the long-suffering Queen of Charles II, and, among other duties, discreetly arranged the deeds of ownership of the house of Nell Gwyn, Charles's mistress. The elder Folkes built a considerable estate through shrewd purchases and marriage to a local knight's co-heir. His seat was Hillington Hall close to the present Sandringham Estate; the Hall later became known for its landscape park.[2]

His son, however, spent most of his childhood not in Norfolk, but in London. Educated by the Huguenot and probability theorist Abraham de Moivre (1667–1754), and subsequently at Clare Hall, Cambridge, Folkes demonstrated precocious giftedness in mathematics and was elected to the Royal Society at the age of twenty-three. In addition to more serious research of the aurora borealis, the rakish young man spent part of his time in the Royal Society calculating gambling odds for his friends and debating if eating asparagus gives a distinctive smell to various bodily fluids. Despite these peccadillos, he subsequently became Sir Isaac Newton's protégé (who did not routinely gamble but may have eaten asparagus), as well as an English antiquary, mathematician, numismatist and astronomer.

Folkes would in the latter part of his career become simultaneously President of the Royal Society and of the Society of Antiquaries, the only individual to have that distinction. A clubbable man, he was also a member of the Spalding Gentlemen's Society and the Académie royale des sciences, becoming a friend of Montesquieu and Voltaire, who called him their 'brother'. Whilst on the Continent, Folkes was a participant in French salon culture and a correspondent of one of its doyennes, Madame Marie Thérèse Rodet Geoffrin (1699–1777); they debated if animals had souls, and she worried about his ever increasing girth due to his avid membership in the Royal Society Dining Club.[3] He had other administrative roles in his varied career, serving on the Board of Longitude and as one of the first governors of Coram's Foundling Hospital.

Folkes also was a patron of the arts, having his official portrait as Royal Society President done by his friend William Hogarth (1697–1764). Dressed in a luxe brown

[1] Some of the material in this introduction has been taken from my journal article about his diary, 'Taking Newton on Tour: The Scientific Travels of Martin Folkes, 1733–1735', *British Journal of the History of Science*, 50, 4, 2017, pp. 569–601, Copyright © British Society for the History of Science 2017, 'Reprinted with permission'. Other material is also from my biography of Folkes: *Martin Folkes (1690–1754): Newtonian, Antiquary, Connoisseur* (Oxford University Press, 2021).

[2] Williamson, *Archaeology of the Landscape Park*, pp. 129–30.

[3] Brown, 'Madam Geoffrin and Martin Folkes', pp. 215–41.

velvet coat with gold buttons, the bewigged Folkes extends his right hand and index finger as if to appear in active debate.[1] He also was a frequent attendee of the London theatre, one of the first members of the gentry to marry a Drury Lane actress, Lucretia Bradshaw (d. *c.*1755), incurring significant family disapproval for transcending boundaries of class and propriety. Antiquary William Stukeley (1687–1765) related that Folkes's mother, Dorothy Hovell, 'grieved at it so much that she threw her self out of a window and broke her arm'.[2] Folkes was also a deist and deputy master of the Grand Masonic Lodge of London and active in the short-lived Egyptian Society (1741–3), helping to dissect a mummy at Richmond House, the London seat of his friend Charles Lennox, 2nd Duke of Richmond (1701–50), early cricket's greatest patron. Folkes was possibly the best-connected, self-determined and most versatile natural philosopher and antiquary of his age, an epitome of Enlightenment sociability, yet he is today a surprisingly neglected figure. His was an intellectually vibrant world in which the long shadow of Newton – Folkes's patron and hero – has tended to obscure those who followed him.

Luckily for the historian, Folkes kept a travel diary of his Grand Tour, his departure from London on 24 March 1732/3 and his return on 3 September 1735, recording the Italian part of his peregrination from Padua to Rome.[3] The diary reveals that during his journey he served as a scientific ambassador for the Royal Society, demonstrating Newtonian optics to Italian virtuosi. He also measured ancient and Renaissance buildings to understand past architectural engineering and design. The manuscript of his journey evinces Folkes's intellectual verve and challenges the received wisdom among scholars that the Royal Society was in a nadir in the first half of the eighteenth century. Analysing Folkes's activities abroad and creating an edition from this source not only helps rescue this remarkable man from obscurity, but also tracing his Italian route provides a novel reading of Newtonianism and the purpose of the Grand Tour as a vehicle for scientific diplomacy.

Folkes's diary is in the Bodleian, bound in vellum. In 1966, the Bodleian bought it from bookseller H. W. Edwards, who had acquired it from Sotheby's on 28 June 1965 at an auction of the estate of Sir Thomas Phillipps.[4] Phillipps (1792–1872) was one of the most pre-eminent English collectors of books and manuscripts, acquiring over 60,000 manuscripts kept on his estate of Middle Hill at Broadway, Worcestershire.[5] His stated ambition was 'to have one copy of every book in the world', though the disarray of his collection led Sir Francis Madden to describe Phillipps's residence as a 'literary charnel house'.[6] Phillipps was also an obsessive manuscript collector, treating his friends to a 'dessert of manuscripts'; after dinner, there was no trifle but as one French visitor recounted, instead 'a choice treat of the most previous manuscripts of Middlehill, and

[1] Hogarth, 'Portrait of Martin Folkes', *c.*1740, Royal Society, London, image RS.9283, https://pictures.royalsociety.org/image-rs-9283 [Accessed 1 Jan 2024].

[2] Stukeley, *The Family Memoirs of the Rev. William Stukeley*, I, p. 99.

[3] MS Oxford, Bodleian Library, Martin Folkes, 'Journey from Venice to Rome', MS Eng. misc. c. 444 (hereafter Folkes, 'Journey from Venice to Rome' or to page numbers in the text). There is a supplemental account book for his journey: Norfolk Record Office, Martin Folkes Esq., 'Memoranda in case he should go abroad again, respecting his Estates, foreighn [*sic*] money and Places', NRS 20658, f. 18r (hereafter NRS 20658).

[4] 'Journal of Martin Folkes in Venice and Rome', https://archives.bodleian.ox.ac.uk/repositories/2/resources/3931 [Accessed 11 May 2023]; Bibliotheca Phillippica New Series, lot 165.

[5] For a short biography and authority file for Phillipps, see SNAC (Social Networks and Archival Context), https://snaccooperative.org/view/3040193 [Accessed 15 Oct 2020]. Edwards, *Cat. 114*, p. 45.

[6] Blom, *To Have and to Hold*, p. 150; Munby, *Portrait of an Obsession*, p. 267.

we were able, at will, to pass them from one to the other until all hours of the night'.[1] After Phillipps failed to sell his collection to the nation, it was subsequently bequeathed to Katharine Fenwick, his daughter. It was later inherited by Thomas FitzRoy Fenwick (1856–1938), his grandson, who oversaw sales of the Phillipps collection, auctioned from 1886 by Sotheby's. The manuscript of the diary has in pencil on the flyleaf a reference to Phillipps's published catalogue of his books: *Catalogus librorum manuscriptorum in bibliotheca D. Thomae Phillipps, Bart.* This work was self-published by Phillipps in 1837, and the diary is listed on p. 33, no. 2930, as 'Journal of an Englishman at Venice'.

How did Phillipps acquire the diary? After Martin Folkes's only son died in a riding accident in 1740, Folkes bequeathed his estate at Hillington Hall and other landed properties to his younger brother, William (d. 1773). However, Folkes left his daughter Lucretia his house in Queen Square, London. In the townhouse was his library, which, as the newspapers reported, was 'said to equal, if not exceed' the library of his friend bibliophile Richard Mead (1673–1754). (Mead was Newton's physician and a renowned collector.) Lucretia also received her father's manuscripts, plate, coin and instrument collection, together with a sum of £12,000 as well as £200 in stocks of South Sea Annuities. She sold most of the library, art collection, and significant manuscripts at auction shortly after Folkes's death. Lucretia then married Richard Betenson in 1758 and died shortly after; they had no children. Richard Betenson subsequently died intestate, and his estate was administered in June 1786. There is reason to believe that a good portion of Folkes's private manuscripts, perhaps including the diary, were in Betenson's hands. Betenson, for example, possessed two of Folkes's ledgers indexing the papers presented at weekly meetings of the Royal Society. These ledgers were bought by the Royal Society in 1859 at an auction in Brighton.[2] Betenson also possessed an account in Folkes's hand about the Royal Society from its first institution which the Royal Society also purchased.[3] At this point, the diary disappears until Phillipps himself bought it, but he did not identify it as written by Folkes, but only as by an 'Englishman'.

During the 1965 Sotheby's sale of the Phillipps collection, art historian Sir Richard Brinsley Ford (1908–99) contacted Sotheby's about the diary. Brinsley Ford was compiling material for his project on English participants in the Grand Tour, his notes later used by John Ingamells for his *Dictionary of British and Irish Travellers in Italy, 1701–1800* (1997). Brinsley Ford noted that one 'Mr Hobson' had given his secretary permission to examine and take notes from the diary on his behalf before the auction took place.[4]

[1] Munby, *Portrait of an Obsession*, p. 136.

[2] RS/MS/702, An Account of the Royal Society from its First Institution; RS/MS/704, Index of Papers Read Before the Royal Society, 1716–38; RS/MS/213, Table of Contents of Journal Book, 1741–8.

[3] Folkes's manuscripts and books however were not preserved in one place. While it is impossible to know how many of his own papers or correspondence were destroyed on his instruction, we do know a good proportion of them were scattered due to the sale of the Macclesfield Library at Shirburn Castle in 2004. In 1872, the Royal Commission on Historical Manuscripts documented the papers of Folkes, consisting of four portfolios of letters and five folios of scientific correspondence. The papers survived at Hillington Hall in Norfolk for another 60 years, at which time they were sold, dispersed, and in some cases presumed lost.

[4] Letter of Sir Richard Brinsley Ford, 9 June 1965, The Grand Tour, Adams Matthew Digital, p. 9, https://www.grandtour.amdigital.co.uk/Documents/Detail/folkes-martin/22269671?item=22280448 [Accessed 19 Sept 2023]. Amongst other resources, the 'Grand Tour' collection is a digitization of Brinsley Ford's paper archive which is in the Paul Mellon Centre for Studies in British Art in London. McKitterick, 'Anthony Robert Alwyn Hobson', p. 56.

'Mr Hobson' was Anthony Robert Alywn Hobson (1921–2014), who oversaw the book department at Sotheby's and spent most of his professional career cataloguing Phillipps's collection. From his examination of notes taken from the diary, Brinsley Ford subsequently realized the diary was by Martin Folkes, and the manuscript was subsequently sold via H. W. Edwards to the Bodleian Library.[1] (It is unclear why the Royal Society Library did not purchase it.) As we will see below, the diary ends rather abruptly, and Folkes's journey continued in Italy for some months after the last entry. This gives us the tantalizing possibility that more of his journal is extant, yet to be discovered. Perhaps it exists in the collection of another obsessive bibliophile like Phillipps.

Folkes's Grand Tour keyed to the list of places he indicated he visited in his travel account book. Norfolk Record Office. See Appendix A.

[1] McKitterick, 'Anthony Robert Alwyn Hobson', p. 57–8.

2. The Historiography of Folkes's Tour

Other than vellomania, it is not clear why Phillipps had wished to possess the diary, as in some ways Folkes's journey was typical in its route, if not necessarily in its travel mishaps or its purpose. Folkes did not take his tour as a young man to see the world, but as a married man, with Lucretia and their three children, a parrot, a fashionable 'Dutch dog' (a pug), and a monkey.[1] The family journeyed from Ferrara to Ravenna and then to Loreto where he unkindly deplored the beggars – 'walking a hundred yards in the street it goes beyond all human imagination, and is irksom beyond expression'.[2] There are intimate details of family life, as Folkes recorded buying large purple grapes for his children from an itinerant merchant when they were travelling from Ferrara, and he rose early 'before the rest of the family was up' to go with his adolescent son Martin to Ancona to show him the sights.[3] There were also mishaps. Both Folkes and his wife Lucretia suffered from prickly heat rash, or 'scotture', whilst in Venice, and Folkes found the temperature there unbearable, buying cooling drinks at a 'fresco shop', the early modern equivalent of the lemonade stand. Folkes also reported on their way to Narni:

> I had the bad luck to lose my dutch Dog, of which I have unable to hear no more tho I sent two servants back and offerd a very great reward, as it has really given me far more uneasiness than I think such a sort of thing should but I have that weakness and am really fretted and vext about it beyond measure.[4]

Happily, the pug was recovered, having wandered back towards Venice in search of Folkes, and was 'almost starvd', Folkes commenting that 'the Dog moans still and seems to attempt at telling his misfortunes'.[5]

Figure 1. Folkes owned a 'Dutch Dog' or English Pug, a breed fashionable in the eighteenth century, engraved here by Andrew Bell. 'Pug D.', Plate CXIX, *Encyclopaedia Britannica C–L*, ed. A. Bell and C. MacFarquhar, Edinburgh, 1771, digitized at the National Library of Scotland, Creative Commons Attribution 4.0 International License.

His travel diary, like Folkes himself, is also understudied, which led to later misunderstandings about the purpose of the journey. According to William Stukeley, Folkes went on his journey because he lost the election for the Royal Society presidency

[1] *The Family Memoirs of the Reverend William Stukeley*, p. 99. Stukeley also claims they took a cat.
[2] See below, p. 151.
[3] See below, p. 142.
[4] See below, p. 163.
[5] See below, p. 175.

to Sir Hans Sloane, collector, physician and founder of the British Museum and Chelsea Physic Garden.[1]

Despite Stukeley's claims, Folkes's Grand Tour was *five years* after his loss of the Royal Society election. Stukeley's opinion about the purpose of Folkes's journey was also partisan. As a devout Anglican vicar, Stukeley was hostile to Folkes's irreligion, claiming that Folkes was 'an errant infidel', who 'set up an infidel Club at his house on Sunday evenings where ... others of the heathen stamp, assembled'.[2] He then went on to claim that Folkes thought 'there are no differences between us and animals; but what is owing to the different structure of our brain, as between man and man', and that he called himself 'the Godfather of all Monkeys'. Folkes's infidel club in London included Charles Lennox, Duke of Richmond (1701–50) and John, 2nd Duke of Montagu (1690–1749), the first aristocratic Grand Master of the Premier Grand Lodge of England.[3] Folkes certainly entertained materialist philosophies before and during his tour of Italy. The Mancunian poet John Byrom (1692–1763) recorded their conversation when at Folkes's house on Queen Square in London for dinner on 26 February 1725/6, to see his 'books and rarities'. They talked with the surgeon William Cheselden (1688–1752) and the watchmaker George Graham (1673/5–1751) about Mrs Folkes's very comical pet monkey: they 'disputed whether he had reason or no', concluding that 'he was a man without reason'.[4]

To give due to Stukeley, Folkes's travels to Italy from 1733 to 1735 certainly confirmed his freethinking beliefs. He noted in his diary that he considered Catholicism and Protestantism 'as only a tool of the state to keep the vulgar in awe', noting that Venetian intellectuals were 'fond of any sort of freethinking book, and with what eagerness they speak on the subject and enquire after the ways of thinking of a nation more used to liberty of thought than themselves'.[5] That said, Folkes continued:

> The bible is in no ways known and I have met with stupendous instances of mistakes about the things contained in it from men otherwise very well acquainted with the common points of Litterature ... and it has been with the greatest surprize several have heard me speake of the measure of learning of all sorts that very antient to say no more book contains.[6]

The diary also records his visit to the Vatican Library to see the Codex Vaticanus. The 4th-century AD Uncial Codex Vaticanus is one of the most famous authoritative biblical texts, one of two of the oldest Biblical manuscripts in parchment that are still extant, and a cornerstone of the Septuagint and the New Testament. Folkes was particularly interested in examining a passage, the Comma Johanneum, which provides the clearest support for the doctrine of the Trinity, but which is considered in modern scholarship as an interpolation. Folkes was a deist, and his mentor Newton an anti-Trinitarian, so this also may have been a motivating factor for his investigations. However, Folkes had some more specific reasons for inspecting the passage. Jonathan Richardson the Younger (the son of Folkes's neighbour, the artist Jonathan Richardson the Elder), asserted that

[1] Stukeley, *The Family Memoirs of the Reverend William Stukeley*, I, p. 99.

[2] Ibid., pp. 99–100.

[3] Péter, 'Admissions and Lodge Meetings', p. 2.

[4] Byrom, *Private Journal*, I, pp. 209–10.

[5] See below, p. 71.

[6] See below, p. 71.

he indeed saw the Comma Johanneum in the Codex Vaticanus, and Folkes wished to put paid to such false assertions to avoid being publicly accused of anti-Trinitarianism.

When the diary has been assessed more recently by scholars, it is usually done so to provide examples of historical topography, or to show the importance of the Grand Tour. John Ingamells's *Dictionary of British and Irish Travellers in Italy, 1701–1800*, is typical in its summary of his activities, mentioning the people Folkes met and his character assessments.[1] And, certainly in some ways, Folkes *was* a typical traveller. He clearly read the Italian travel guides of Joseph Addison (1705), Maximilien Misson (1714), and Edward Wright (1730), to shape his itinerary, in some cases agreeing with their aesthetic judgments when assessing paintings and sculpture, setting these within the framework of an informed connoisseur.[2] Folkes's initial perceptions about Venice were also representative of many Grand Tourists. Rosemary Sweet has noted:

> Folkes complained that there were no other squares of any size, nor even any gardens or yards. Whilst his observation was certainly an exaggeration, what is important is the perception that he and others had formed that Venice occupied a uniquely densely inhabited space. The maps and plans produced for tourists would only have served to reinforce this impression of crowded in-filling. Venice appears in these prints as a mass of blocks of tightly packed buildings, intersected by narrow canals – unlike the plans of cities such as Rome or Turin which celebrated the grand open spaces and rectilinear streets so highly prized at the time. Folkes also found Venice confusing and difficult to navigate: it was disorientating due to the absence of a regular street plan and the visitor was confronted with a labyrinthine network of canals and *calle*. There were no perspective views, no straight lines, no open vistas, except on the Grand Canal. It was only after he had purchased a plan of Venice that he was able to make any sense of the city spatially.[3]

Folkes has also been studied for his interests in numismatics and medals, analyses of this grand tour accomplished through this lens. Folkes published the first comprehensive treatise of English silver and gold coinage (posthumously published by the Society of Antiquaries).[4] He was also a collector of coins and medals, the twenty-six-page auction catalogue of his estate indicating that it took nearly a week to disperse his numismatic specimens, as well as his English, Swiss, Roman, papal, and Danish medals.[5] Folkes was in fact later responsible for the establishment of the Copley Medal in the Royal Society, as Rebekah Higgitt elucidated in her case study of its creation.[6] William Eisler also discussed the medal Folkes had cast with his visage and iconography whilst in Rome to indicate his allegiance to ideals of Freemasonry and noted that he moved in Jacobite circles, though it was unlikely Folkes was a Jacobite himself.[7] Folkes's medal, which he had struck at Rome with Egyptian and Freemasonic motifs, featured the pyramid of Cestius near the Porta San Paolo. On the reverse, the medallion had symbols 'taken directly from the seal of the Neapolitan "Perfetta Unione" [Masonic] lodge: the pyramid

[1] Ingamells, *Dictionary*, pp. 355–6.

[2] Hurd, ed., 'Remarks on several parts of Italy', *Works of Joseph Addison*; Wright, *Some Observations Made in Travelling through France, Italy, etc*; Misson, *A New Voyage to Italy*.

[3] Sweet, *Cities and the Grand Tour*, p. 210.

[4] Folkes, *Tables of English silver and gold coins*; Burnett, *Hidden Treasures*, II, pp. 917–19, 1105–14.

[5] *A Catalogue of the Genuine, Entire and Choice Collection of Coins, Medals, and Medallions*, titlepage. The auction catalogue indicates no author, though the auctioneer was Samuel Langford.

[6] Higgitt, '"In the Society's Strong Box"', pp. 284–316.

[7] Eisler, 'The Construction of the Image of Martin Folkes', Parts I and II.

(but here transformed into the pyramid of Cestius at Rome), pillars of the porch of the Temple of Solomon, sphinx with crescent moon, acacia branch, and sunburst'.[1] This iconography indicates that Folkes had established ties with the Masonic lodges during his journey there, as well as with the lodge in Florence when he visited his friend and fellow Mason Antonio Cocchi. The phrase on the medal, *sua sidera norunt* – His own constellations have acknowledged him – is from the sixth book of the *Aeneid,* where Virgil described an encounter in Elysium between Aeneas and his father Anchises. The inscription also indicates that the medal may have been struck at Rome to commemorate Folkes's successful journey there to spread Newtonian and Masonic principles, as well as 'to show the high esteem in which Folkes was held in the city of antiquities ... about the time that he was elected a member of the French Academy' of Sciences.[2]

Figure 2a. Ottone Hamerani, *Martin Folkes* (1742), bronze medal (obverse), 3.7 cm. Author's collection. Author's photograph.

Figure 2b. Ottone Hamerani, *Martin Folkes* (1742), bronze medal (reverse), 3.7 cm. Author's collection. Author's photograph.

The iconography of his medal was detailed, and no doubt the result of the particular instructions of Folkes as a connoisseur. His impressions of artworks that he saw in his diary are equally specific. It is important to consider Folkes's highly empirical observations of art collections and the content of libraries in Venice in his diary which were written before the fall of the Republic to Napoleon on 12 May 1797. The religious orders were suppressed and there was unbridled plunder, despoliation and loss of works of art from churches and monasteries. Vast quantities of Venetian art were taken to Paris or auctioned elsewhere, and sculptures of gold and silver were melted down. As Gilks has argued: 'supporters of the seizures argued that it was no dynastic *enlèvement* (abduction) of *butin* (booty) but rather the orderly extraction, confiscation, or *saisie* (seizure) by a legitimate authority in order to advance civilization'.[3] Arguments were also made that appropriating art saved it from the 'negligence' of owners, and it was 'widely believed –

[1] Small and Small, 'South Italy', p. 335.

[2] *Martin Folkes Medal,* object M.8466, British Museum, London. https://www.britishmuseum.org/collection/object/C_M-8466 [Accessed 1 Nov 2019].

[3] Gilks, 'Attitudes to the Displacement of Cultural Property', p. 118.

in France and elsewhere – that possessing masterpieces increased national wealth by inspiring manufacturers, providing instruction, and attracting tourists'.[1] Folkes's diary is thus useful to ascertain placement and provenance of these lost works before this period of unbridled Napoleonic aggression.[2]

3. Taking Newton on Tour

Whilst these previous assessments are valuable, there has been little consideration of to what extent in his journey Folkes melded both aesthetic, antiquarian and natural-philosophical (scientific) interests. His tour was one of not only connoisseurship, but also of scientific peregrination, or 'science on the move', an activity that was prevalent from 1650 to 1750.[3] The origins of the scientific peregrination were in the '*peregrinatio academica* of early modern aspiring scholars', and/or the *peregrinatio medica*, as medicine was the subject for which foreign travel was most valuable. Medical students brought back new techniques, knowledge, and *materia medica* to their homeland and to attain accomplishments to become gentlemen of quality, with a cosmopolitanism to add polish to their English education. But, by the time of Folkes's journey, scientific tours apart from the reasons of *politesse* and medical pedagogy had long developed as a separate enterprise. In 1698, physician and naturalist, Martin Lister (1639–1712), wrote a guidebook of his *Journey to Paris*, intending it specifically to appeal to fellow natural philosophers. Lister directed the reader to his interests in natural history using his mature judgment and own eyes, offering 'clean Matter of Fact, and some short notes of an unprejudiced Observer'.[4]

Folkes had similar sentiments reflecting his specialism in mathematics and optics and interest in architectural measurement; his journey was designed to answer particular questions of natural philosophy. His diary reflects 'antiquarian science', a form of peregrination in which he used metrology to understand not only the aesthetics but the scientific principles of antique buildings and artefacts, as well as their context and place in the Italian landscape. The Royal Society at this time was also involved in projects that integrated natural philosophy ('science') and antiquarianism (history/analysis of past material culture, a form of nascent archaeology), particularly before the re-establishment

[1] Ibid., p. 119.

[2] The catalogues of Antonio Maria Alessandro Zanetti (1706–78) are referred to often in this volume. Zanetti was a noted Venetian historian and scholar of antiquities, architecture, and sculpture. With his homonymous older cousin Antonio Maria Zanetti, he wrote a series of works about important collections in Venice and served as the custodian of the Biblioteca Marciana. A work dedicated to Venetian paintings in public spaces, the *Descrizione di tutte le pubbliche pitture della città di Venezia e isole circonvicine* (1733) was highly successful. This work integrated a 17th-century catalogue by Mario Boschini, *Le Minere della pittura* (1664). Zanetti also wrote a large-scale account of the Venetian school of painting, *Della pittura veneziana e delle opere pubbliche de' veneziani maestri* (1771), as well as a two-volume work illustrating classical sculptures preserved in Venetian public collections, Zanetti providing many of the drawings himself. His work is still utilized today to track provenance and placement of early modern Venetian paintings and sculptures before the Napoleonic conquest of the Republic in 1797, and the subsequent suppression and despoliation of Venetian churches and monasteries. It is hoped Folkes's diary, like Zanetti's works, will also assist in these future investigations. Sorensen, 'Zanetti, Antonio Maria Alessandro', *Dictionary of Art Historians*, https://arthistorians.info/zanetti [Accessed 21 Sept 2023].

[3] Greengrass et al., 'Science on the Move', p. 145.

[4] Stearns, *Martin Lister, A Journey to Paris,* p. 3.

of the Society of Antiquaries of London in 1717. During Folkes's dual presidencies of the Royal Society and the Society of Antiquaries, there were even attempts at unification.[1] In the eighteenth century, the ability to observe nature was thought to make natural philosophers ideal to understand the empirical details of ancient artefacts and how they were created. For example, antiquarianism was 'cognate with natural history at the very least in the sense that the type of activity involved (fieldwork, collection, display, classification) was of a similar character'.[2] These connections were made evident in the Royal Society of the first half of the eighteenth century, particularly during Folkes's presidency, which promoted an admixture of antiquarianism and natural philosophy, disciplines seen as intimately related: the Society of Antiquaries and the Royal Society had many common members and held their meetings on the same day. In the 1740s and 1750s, papers on antiquarian and archaeological subjects comprised between 6 and 7 per cent of articles in the Royal Society's journal, *Philosophical Transactions*, particularly as the Antiquaries' own journal, *Archaeologia,* was not launched until 1770.[3] The antiquarian use of material culture to reconstruct the past was thus practised at both the Royal Society and the Society of Antiquaries in the first half of the eighteenth century, so much so that, as previously mentioned, during Folkes's dual presidencies of these societies there were attempts to unite the two organizations.[4] Attempts at consolidation reflected the simple reality that natural philosophy and antiquarianism were part of the same intellectual enterprise, intellectual currents represented beautifully by Folkes's diary. Folkes's quantitative and empirical approach to antiquarianism was also a riposte to the 'high-flying speculative rationalism' represented by William Stukeley, dubbed the 'arch-druid' by his peers for his naive speculations about ancient British monuments such as Stonehenge.[5]

Folkes's diary also contradicts the long-standing mistaken impression among scholars that the Royal Society was in decline in the eighteenth century. Rather, the eighteenth-century presidents of the Royal Society serving between the two major naturalists, Sir Hans Sloane (1660–1753) and Sir Joseph Banks (1743–1820), are little understood, and few modern biographical studies have been written about them. Fara has noted that the central years of English Science from 1740 to 1770 'remain an under-researched period, with very few studies of social networks of experimenters'.[6] Sorrenson has noted that these unedifying impressions of the Royal Society during this period also stemmed from the belief that the institution failed to match the standard that Newton had set in the *Principia*; its origins also lay in negative opinions voiced by the Society's critics such as John Hill, and from the popularity of Babbage's *Reflection on the Decline of Science in England* (1830).[7] It seems, however, that the Royal Society in the eighteenth century manifested considerable strengths in spreading Newtonianism, and in technological

[1] For attempts at the unification of the Royal Society and the Society of Antiquaries, see Evans, *A History of the Society of Antiquaries*, Rousseau and Haycock, 'Voices Calling for Reform'.

[2] Miller, 'Into the Valley of Darkness', p. 160.

[3] Sorrenson, 'Towards a History of the Royal Society', p. 37; Rousseau and Haycock, 'Voices Calling for Reform', p. 384.

[4] Rousseau and Haycock, 'Voices Calling for Reform'; Roos, *Folkes,* pp. 335–9.

[5] Frischer, 'Ramsey's "Enquiry"', p. 73.

[6] Fara, 'Master of Practical Magnetics', p. 56. Some of this material in this section has come from Roos, *Folkes*, pp. 7–8.

[7] Sorrenson, 'Towards a History of the Royal Society', pp. 29–46.

innovation, setting new standards in experimental precision and refined instrumentation, with Folkes at the centre of such activity.

The diary reveals that, in Venice, Folkes recreated Newton's experiments in optics, cultivating fellow mathematicians and astronomers to gain adherents of the Newtonian research programme in the Royal Society. Although the Venetians were aware of the theories of Newtonian optics, in the 1730s there was a degree of scepticism among natural philosophers there about the validity of the refrangibility of light in the visible spectrum, namely, the degree to which light refracts when passing from one medium into another, or, as Newton wrote, a 'predisposition, which every particular Ray hath to suffer a particular degree of Refraction'.[1] Venice, in particular, was a nexus of Newtonian mathematicians and their intellectual circles, and Folkes went there to convince their virtuosi of the truth of Newton's work in optics. Twenty years previously, Newtonians had fought a campaign against Italian supporters of Leibniz in the calculus dispute over priority of discovery of this mathematical method. In Padua and Venice, after 1710, Jakob Hermann and Nicolas Bernoulli were 'promoters of a fierce campaign against Newton's methods of fluxions in the calculus'.[2] However, there were pro-Newtonians who worked to lessen the prestige of Leibniz through the offices of Abbé Antonio Schinella Conti (1667–1749), who was born and educated in the Veneto; Conti was in London between 1715 and 1718 to become part of Newton's intellectual circle, where he met Folkes. In his travel diary of 1733, Folkes noted: 'I had known the Abb. Very well in England above 16 years ago, I had a very agreeable conversation with him as he is a Gentleman of great knowledg and politeness, and indeed the most knowing man of all the Italians of his side the country ...[3] Conti would also serve as an intermediary between Leibniz and Newton in the calculus dispute.[4]

Before his journey, Folkes had attended a series of Royal Society meetings in which the Society responded to the refutation of Newton's optical work by Giovanni Rizzetti (1675–1751), a Paduan nobleman who belonged to the circle of the mathematician Jacopo Riccati (1676–1754) of Castelfranco Veneto. The Royal Society Journal Book for 27 June 1728, reported that the Society received Rizzetti's, *De luminis affectionibus specimen physico mathematicum*; the tract was referred 'to Dr Desaguliers which he promised to make an extract of'.[5] In *De luminis*, Rizzetti discussed refrangibility. Rizzetti claimed he could not replicate the experiments in Book 1 of the *Opticks*, in which Newton used a candle to illuminate a two-coloured card with black silk threads wrapped around it, demonstrating the principle of chromatic aberration. (Aberration occurs as lenses have a different refractive index for each different wavelength of light.)[6]

[1] Hall et al., *The Correspondence of Isaac Newton*, I, p. 96.

[2] Casini, 'The Reception of Newton's Opticks', p. 217.

[3] See below, p. 90.

[4] Casini, 'The Reception of Newton's Opticks', p. 217.

[5] RS/JBO/14, 21 March 1727, 10 April 1728, 27 June 1728, and 31 Oct 1728, p. 191, p. 235, pp. 249–50.

[6] Newton had placed a glass lens at a distance of six feet from the card and used it to project the light coming from the illuminated card onto a piece of white paper which was at the same distance from the lens on the other side. He moved the piece of white paper back and forth, taking note where and when the red and blue parts of the image were most distinct. The black threads indicated distinctness of the image (when the lines created by the thread were sharpest). Newton found it: 'impossible to focus the image of black silk-thread lines upon a blue background at the same distance as when the same lines are placed against a red background; the distance

The popular lecturer, John Theophilus Desaguliers (1683–1744), had earlier refuted Rizzetti's claims in a series of optical experiments performed at the Royal Society in 1728 in response to Rizzetti's *De luminis*.[1] In his article published in *Philosophical Transactions*, Desaguliers noted: 'we hear indeed in a letter from Sir Thomas Dereham to Sir Hans Sloane, President of the Royal Society, that now Signior Rizzetti alledges, that he was deceived in his experiments by reason of the badness of his prisms which he had from Venice'.[2] In other words, Rizzetti claimed he did not have ideal experimental conditions due to the quality of his prisms made of Venetian glass.

Simon Schaffer has shown that the quality of the prisms in the recreation of Newtonian experiments was key as Francesco Algarotti's attempts in Bologna in 1726 to replicate the Newtonian *experimentum crucis* were described by Rizzetti as failures, as he could not isolate a single red ray of light.[3] Algarotti however eventually realized that his prisms were defective. He subsequently was able to replicate Newton's results when he tried again with some top-quality English prisms. The acceptance or denial of specific experimental apparatus was seminal in disputes over methods in natural philosophy. In this era, we see the beginning of the establishment of standards for professional instrumentation, calibration and measurement.

Folkes had other challenges besides experimental kit in his demonstrations, namely that Venetian natural philosophers had an attitude of 'critical acceptance, if not open aversion' towards Newtonianism.[4] Part of the reason for Folkes's travels in Venice was so he could serve in Desaguliers's stead as the Newtonian demonstrator, and it seemed particularly important that Folkes first prove the truth of Newtonian optics in the city whose glass had been the 'standard against which other glass was to be compared'; Folkes also was there to demonstrate the superiority of English prisms and instrumentation.[5] Folkes's status as a protégé of Newton (his vice-presidential office in the Royal Society was conferred by Sloane six weeks before his travels), his multilingualism, and his renowned affability were also advantages in his scientific campaign. Folkes also came armed with a letter from James Jurin FRS to Giovanni Poleni, the Italian natural philosopher and expert in architecture, hydraulics and classical antiquity (who had also been engaged in the *vis viva* dispute with

between the two positions of sharp focus was as much as one and half inches'. In other words, to get a distinct red image, the paper had to be held 1.5 inches further away than to obtain a distinct blue image. Newton thus concluded that the blue light was refracted more by the lens than the red and was more refrangible. Hall, *All Was Light*, p. 96; see also, Walsh, 'Newton's Epistemic Triad', p. 104. Red and blue light were not strictly homogeneous, and so not all the blue light was more refrangible than all the red light, something Newton admitted. However, these experiments demonstrate a general effect, and Newton described an ideal experiment. The inability to create these ideal conditions was the point upon which Rizzetti seized.

[1] Desaguliers, 'Optical Experiments', pp. 598–9.

[2] Ibid.

[3] Schaffer, 'Glass Works'. 'The *Experimentum crucis* was Newton's proof that white light was composed of coloured light rather than different-coloured light resulted from differing speed through a medium. It was accomplished. 'Newton devised a crucial experiment – he would direct one of the colored rays, say the red one, produced by the first prism, through a second prism. If the ray changed color again, then the prism was effecting the change. But if it stayed red, then the prism was not changing the light, but merely separating the pre-existing colored rays ...'. Ashworth, 'Isaac Newton', Linda Hall Library, https://www.lindahall.org/about/news/scientist-of-the-day/isaac-newton-2/ [Accessed 31 Dec 2023].

[4] Ferrone, *Intellectual Roots*, p. 95.

[5] Schaffer, 'Glass Works', pp. 72, 99.

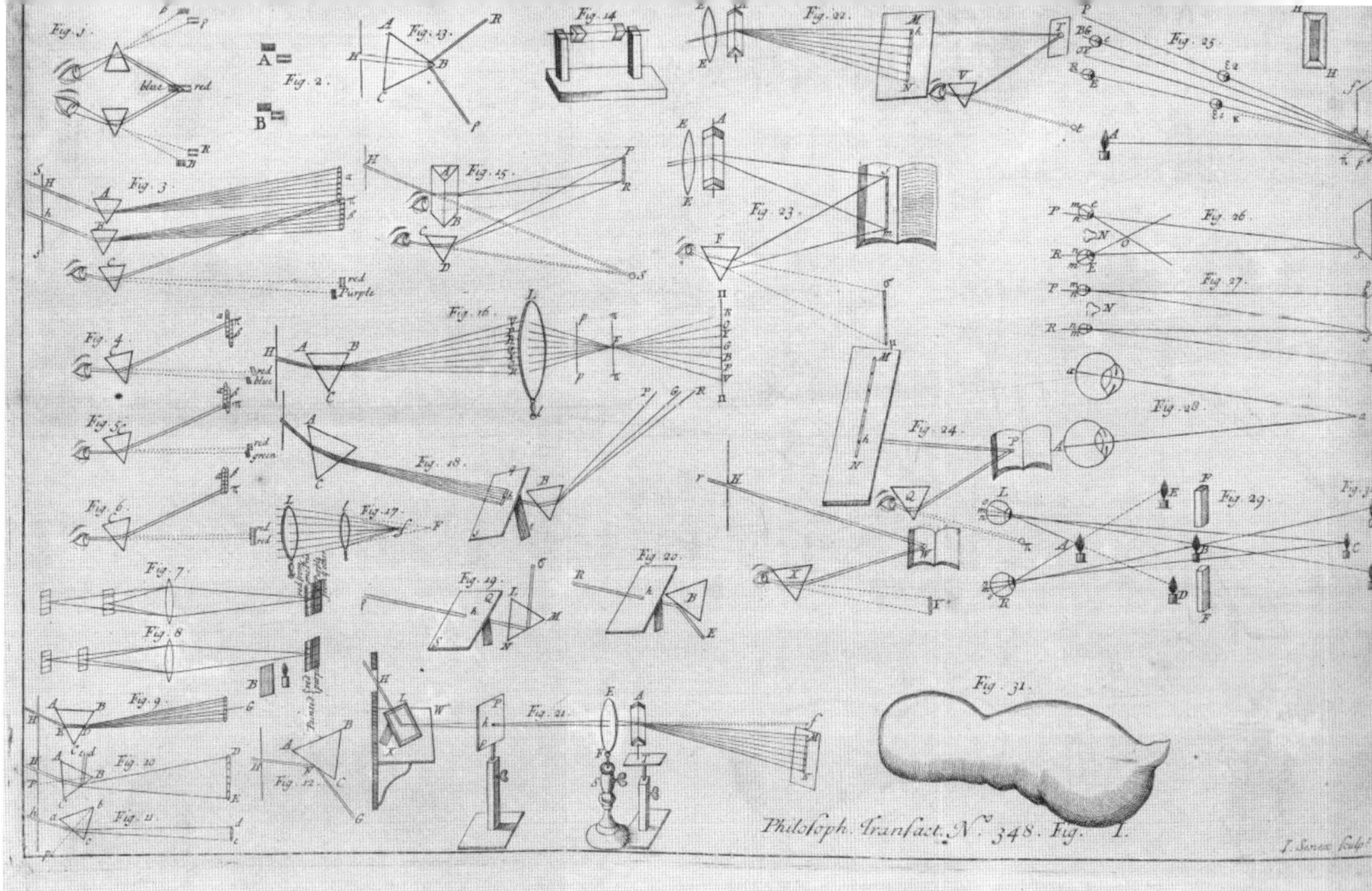

Figure 3. John T. Desaguliers's reconstructions of Newtonian optical experiments in refrangibility, similar to those which Folkes undertook in Venice. The other unrelated picture (labelled fig. 31) is of a gynaecological tumour described in the *Philosophical Transactions*. J. T. Desaguliers, 'An account of some experiments of light and colours, formerly made by Sir Isaac Newton, and mention'd in his opticks, lately repeated before the Royal Society', and 'A plain and easy Experiment to confirm Sir Isaac Newton's Doctrine of the different Refrangibility of the Rays of Light', *Philosophical Transactions*, 29, 348, 1716, pp. 443–52. Courtesy of the Royal Society of London.

Newtonians).[1] Poleni was Professor of Astronomy (1709), Professor of Physics (1715), Professor of Mathematics (1719), and Chair of Experimental Philosophy (1738), all at Padua University, building a remarkable physics laboratory with the involvement of the Venetian Republic. 'His way of lecturing on experimental natural philosophy mainly followed the Anglo-Dutch Newtonians.'[2] Jurin stressed to Poleni that Folkes:

> was an excellent gentleman of high achievement in every kind of literature, especially physics and mathematics; whom Newton valued so highly that, not content to consider him worthy of his own close friendship, he voluntarily chose him to carry out the duties of Vice-President at the formal meetings of the Society in his own absence.[3]

Folkes met Poleni in Venice on 18 July 1733, with whom he discussed meteorology and the state of learning in England, subsequently recommending Poleni to the President, Sir Hans Sloane, for election to the Royal Society.[4]

To counter Rizzetti, Folkes performed a series of optical experiments from Book I of Newton's *Opticks* before an invited audience at the Palazzo Giustiniani which belonged to Giacomo Ascanio Giustiniani (1677–1745), member of the first family of Venice, and a senator. Giustiniani was an erudite and cultured man, friends with Antonio Conti and the musician Benedetto Marcello, and known for his salons for intellectuals and scholars from different social backgrounds. Folkes noted in his travel diary:

> I was at the Palace Justinian in the morning, to have made some of Sir Isaac Newtons optical experiments but the Sun not stirring I could no more than get things in order for another day. The apparatus belonged to Abbe Conti who was there, and he had brought some very good prisms from England.[5]

Clearly on the hunt for apparatus, Folkes had three days earlier indicated: 'I was also at another persons who is a practical Mathematician to see some instruments but nothing was particular what he had best were some English things but none of those neither of the very best sort. There was one good workman here called Bernard Facini ...'.[6] This was Bernardo Facini (1665–1731), a Venetian astronomer, mathematician, and scientific-instrument maker, who was more than a 'good workman'. He was also the inventor of the surveying protractor with magnetic compass (1694), the proportional cyclometer (1695), and (inter alia) a spiral scale slide rule.[7] Folkes also visited during his stay Bernardo Zendrini (1679–1747), the 'public mathematician' of Venice responsible for hydraulics, who Folkes also underestimated and disparaged his 'instruments for levelling etc ... in no ways comparable to what we have in England'.[8] But as he was trying to establish the primacy of English instrumentation in prisms in solving scientific problems, his attitude was understandable, if unfortunate.

[1] The *vis viva* dispute lasted from 1686 until the 1720s and concerned whether the *vis viva*, or momentum, was conserved in the universe. The dispute started between Descartes and Leibniz and resulted over confusion between momentum (mv) and kinetic energy (1/2mv^2). Poleni took the Leibnizian side against Newtonians such as John Theophilus Desaguliers. See Iltis, 'Leibniz and the Vis Viva Controversy', pp. 21–35.

[2] Talas, 'New Light on the Cabinet of Physics of Padua', p. 56.

[3] For the letter of introduction, see Rusnock, ed., *The Correspondence of James Jurin*, p. 391.

[4] See below, p. 71; BL, MS Sloane 4058, f. 60, Folkes to Sloane, 1 Oct 1733.

[5] See below, p. 91.

[6] See below, p. 85.

[7] See Museo Galileo, 'Bernardo Facini', https://catalogue.museogalileo.it/biography/BernardoFacini.html [Accessed 21 June 2021].

[8] See below, p. 85.

Folkes was more impressed by his visit to Domenico Selva's optical workshop near Piazza San Marco. His son Lorenzo documented his father's work in a book, the *Esposizione delle comuni*, among other works on his productions 'which included refracting and reflecting telescopes, microscopes, and surveying levels'. The family also sold helioscopes, opera glasses, magic lanterns and binoculars. Lorenzo became the 'ottico publico', public optician of the Republic of Venice, after his publication of a book on the manufacture of achromatic lenses using flint glass, which enabled him to build superb telescopes. The book was dedicated to the natural philosopher, Francesco Algarotti (1712–64), who was seminal in introducing Newtonian optics to Venice.[1] Folkes recorded he intended to see if he could observe the moons of Jupiter, discovered in 1610 by Galileo, with one of Selva's small reflecting telescopes.

Despite Folkes's unsuccessful visits to Venetian instrument makers to purchase experimental kit, Folkes's optical demonstrations with Conti's English prisms were ultimately successful. On 28 August 1733, Folkes recorded triumphantly: 'I was again at the Giustiniani Palace and made most of the experiments of the first book of Opticks with very good success and to the satisfaction of all that were here which were … men of the first people of Venice.'[2] Folkes was rewarded by Giustiniani with a present of 'spanish Chocolate'.[3] On 2 and 7 September, Folkes then indicated:

> I was again to prepare some of the Experiments of the Opticks at Sig^{r} Giustinianis, our Experiments there have made a good deal of noise and many of the chief Nobility are very desirous to see some if them which I have promised to repeat on ~~saturd~~ tuesday next. tho it is but in a gross manner I can make them want of a convenient apparatus. they are very satisfactory, and carry demonstration to all that are capable of comprehending them … I made the Experiment I had intended at the Justinian Palace I have now shewn all that are considerable in the book of opticks in the first book, and tho my apparatus was none of the best I was sufficiently satisfyd with them and am told it is what was never done with any success here before, there were several of the Chief nobility and some others here present. and every one that has any sense seems thoroughly satisfyed as far as he understands.[4]

Correspondence shows that these experiments were later recreated in Florence with the natural philosopher and physician Antonio Cocchi (1695–1758), and fellow Freemasons. Other contextual material reveals that Folkes also discussed ancient and Renaissance use of tricks of optical illusion and architectural perspective with Francesco Algarotti, part of his interest in probing the relationship between seeing and believing in scientific observation and aesthetics. When in Venice, the diary indicates that Folkes also made contact with Anders Celsius (1701–44), famous for his temperature scale, and later helped arranged for Celsius to obtain precision scientific instruments used in his expedition with Pierre Louis Maupertuis (1698–1759) to Lapland to measure the shape of the earth by triangulation to prove Newton's theory that the earth was a flattened sphere.[5]

[1] 'Domenico Selva', Museo Galileo, https://catalogue.museogalileo.it/biography/DomenicoSelva.html; Buchwald and Fox, eds, *Oxford Handbook of the History of Physics*, pp. 347–8; Lualdi, 'Venetian Makers', pp. 10–13.

[2] See below, p. 91.

[3] See below, p. 91.

[4] See below, p. 93.

[5] Terrall, *The Man Who Flattened the Earth*, p. 137.

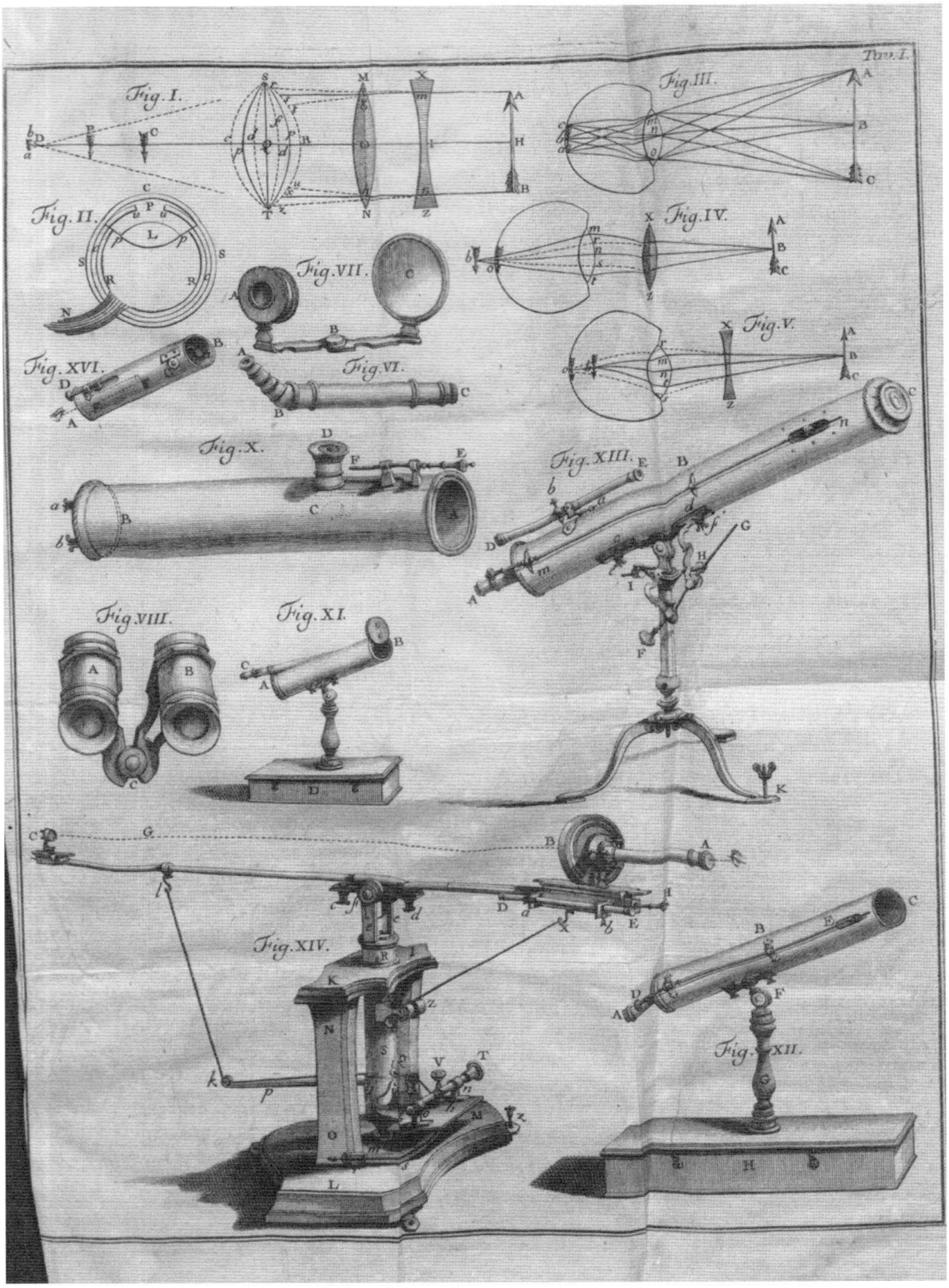

Figure 4. Engraving of Newtonian telescope from Lorenzo Selva's *Esposizione*, Venice, 1761, Plate I, Figures X, XII, p. 99. Here, Selva described 'catodioptric' (mirror plus lenses) of Newtonian and Gregorian patterns. La Fondazione Biblioteca Europea di Informazione e Cultura. Courtesy of La Fondazione Biblioteca Europea di Informazione e Cultura.

In the course of his optical experiments in Venice, Folkes became great friends with Giustiniani and was introduced to the great and good in Venice. Folkes noted in his diary: 'Sig^r Giustiniani is as well bred a good natrd gentleman as I have seen, and is well versd in most parts of Polite learning.'[1] Charles Lennox, Duke of Richmond, also provided introductions to the notables of Venice, Folkes remarking to him in correspondence: 'I have also here had the greatest civility on your Graces account both from the Resident and Mr Smith with whom he being known to you has set me on a very different foot from the generality of travelers that come here'.[2] The Resident was the English ambassador to Venice, Colonel Elizeus Burges, and, 'Mr Smith', Joseph Smith, the art collector and connoisseur, banker to the British community at Venice, and noted host to those on the Grand Tour. 'Joseph Smith passed some seventy years in Venice where he practised as a merchant banker and for some years acted as British consul. His reputation, however, was principally founded on his remarkable (and profitable) activity as a virtuoso and patron of the arts.'[3]

Smith was a patron of Rosalba Carriera (1673–1757), the famous Venetian portrait pastellist, whose godfather was Smith's lawyer. She 'received regular payments from Smith in 1725–8 (some perhaps on behalf of clients) and he eventually owned 38 works by her, including the self-portrait which she gave him in her old age'.[4] Folkes recorded visiting the artist, noting he was: 'extremely well entertained with a great number of fine portraits some of my acquaintance very like ... I saw among others a fine idea representing musick the face was taken I heard from a person now kept by an English Nobleman, but the whole attitude was as fine and well disposed as almost any th[in]g I have seen.'[5] Although Folkes himself did not employ Carriera's services to our knowledge, his younger brother William, who kept track of his brother's accounts, did employ an English pastel painter, William Hoare of Bath, to create his own portrait, the family being au fait with artistic fashion.[6]

Through Giustiniani's good offices, Folkes 'conversed with many' of the Venetian nobility, 'in a manner no Englishmen has done before', and he was invited to several private concerts at the Giustiniani Palace, including an outstanding performance by Giuseppe Tartini (1692–1770), Venetian violinist and composer.[7] Tartini was the first reputed owner of a Stradivarius violin, proficient at the triple stop, where three notes are played simultaneously, usually by sounding the lower notes briefly and allowing them to ring, and then the bow plays the upper notes.[8] Folkes also visited Johann Matthias Reichsgraf von der Schulenburg (1661–1747), a German aristocrat and field marshal, who served in the Venetian army at the beginning of the eighteenth century. Schulenburg defended the Venetian island of Corfu, which was threatened by the Ottoman empire. He retired as a patron of the arts in Venice and had one of the largest private collections

[1] See below, p. 38.

[2] MS Goodwood House, West Sussex Record Office, MS 110, Martin Folkes to Charles Lennox, 8 July 1733.

[3] 'Ingamells, *Dictionary*, pp. 869–71.

[4] 'Dictionary and Archive of Travellers', *The Grand Tour*, Adam Matthews Digital, http://www.grandtour.amdigital.co.uk/Dictionary/TravellerDetails.aspx?documentid=64931 [Accessed 14 Jan 2021].

[5] See below, p. 68.

[6] William Hoare of Bath, *William Folkes of Hillington Hall, Norfolk* (*c*.1740), Hillington Hall, recently sold at Lowell Libson and Jonny Yarker, LTD, https://www.libson-yarker.com/recent-sales/william-folkes-of-hillington-hall-norfolk [Accessed 10 May 2023].

[7] See below, p. 88.

[8] Pollens, *Stradivari*, pp. 123–4.

of the work of sculptor Francesco Bertos (1678–1741). Although his collection was vast, Folkes did not think it had been 'collected with the greatest judgment'; however, due to his passionate interest in cartography, Folkes was more impressed and interested by the Count's collection of books, plaster and papier-mâché reliefs of terrain and fortifications in Corfu and Dalmatia to aid his military campaigns.[1]

4. Cartography, Metrology and Antiquarian Science

Folkes's appreciation of Schulenburg's topographical maps also reflects his interest in the history of cartography, one that also extended into archaeoastronomy. Kristin Lippincott has eruditely analysed the plaster of Paris copy of the Farnese Globe that Folkes had made during his tour.[2] The Globe was the oldest surviving depiction of this set of Western constellations, and Folkes would create a stereographic map of it for Richard Bentley's edition of Manilius's *Astronomia* (1739); Folkes recorded in his travel diary 'Diam of the Farnesian Globe. 2f. 5 inch. 1/10 English Measure', probably indicating that he saw and measured the Globe whilst in Rome.[3] The Globe puts the celestial figures against a grid of circles comprising the celestial equator, the colures, the ecliptic, the tropics, and the Arctic and Antarctic circles. This allows for the constellations to be positioned accurately. 'The declination of the Arctic and Antarctic Circles will correspond to a particular latitude for the observer whose observations were adopted by the sculptor ... a detailed analysis of the globe will reveal the latitude and epoch for the observations incorporated in the Atlas'.[4]

After his return from his Tour, in a presentation he later gave to the Society of Antiquaries on 8 July 1736, Folkes exhibited this model analysing the stellar positions in the constellations carved on his surface to date its creation. It was noted: 'the original Sphere in stone supported by an Atlas being conjectur'd by him upon good ground to have been made in the Year of the Xtian Aera [Christian Era] 112, towards the latter End of the Emperor Trajans's Reign'.[5] It was also noted at the Antiquaries' meeting that 'the colure of this globe passes by those parts of the asterism by which it is said to have past in the days of Hipparchus but the intersection of the Equator and Ecliptic is not at the Colure'.[6] This was a reference to how the 26,000-year precession, or axial wobble, on the Earth that Newton discovered, shifts the positions of the celestial circles over time, so the observed positions on the Farnese Globe can be traced to a particular date in celestial time.

Folkes's diary also shows he had a keen interest in other types of historic maps, particularly the extraordinary *Mappamondo* of Camaldolese monk, Fra Mauro (*c.* 1400–1464), which he encountered in the convent of St Michael of the order of the Camaldolese on the island of Murano.[7] As Matteo Salvadore notes, the map 'resulted from the monk's judicious combination of traditional Ptolemaic and Arab models with

[1] See below, pp. 41–2.
[2] Lippincott, 'A Chapter', pp. 281–99.
[3] Ibid., p. 286.
[4] Schaefer, 'The Epoch of the Constellations', p. 196.
[5] British Library, Add. MSS 4222, f. 24v.
[6] SAL/MS/264 B, vol. 2, pp. 201–2.
[7] See Plate 1.

Figure 5. Giorgio Sommer, *Atlas van Farnese* (*Atlante Museo di Napoli* (*c.* AD 2)) (*c.*1874–9), albumen print, 25 cm × 18.9 cm. Rijksmuseum, Public Domain. Folkes had a reproduction of the Farnese Globe made for him for his archaeoastronomical investigations. Courtesy of the Rijksmuseum, Amsterdam.

modern portolan mapmaking techniques, Chinese cartographical knowledge, and first-hand information. For the latter, the monk was in a perfect location, operating in a city that existed at the crossroad between worlds and attracted both foreign knowledge and distant visitors.'[1] The map is a circular planisphere on parchment of 196 cm in diameter. It is set in a square wooden frame of 223 × 223 cm:

> the globe integrates Ptolemy's *Geography* and marine maps, exhibiting thousands of inscriptions and hundreds of images. The inscriptions derive from written and oral sources that include ancient authors, such as Ptolemy (ca. 100–ca. 178) and Pliny (ca. 23–79), medieval authors, such as Campano de Novara (ca. 1210–1296), Albertus Magnus (ca. 1193–1280) and Thomas Aquinas (ca. 1225–1274), travellers of the previous century, such as Marco Polo (1254–ca. 1324) and Odorico da Pordenone (ca. 1280–1331), contemporary travellers ... Ethiopian monks passing through Venice, and other anonymous testimonies that provided Mauro with detailed information on the main trade routes of the Indian Ocean and on the first navigation of the Portuguese in the Atlantic Ocean.[2]

Folkes dated the map as 1460, which was correct, and a letter he wrote tipped into the back of the diary indicates he may have been planning a published edition or commissioning a copy. The letter noted:

[1] Salvadore, *The African Prester John*, p. 27.

[2] Digital edition of the Fra Mauro Map, Turning the Pages platform exhibit label, Biblioteca Nazionale Marciana, Venice.

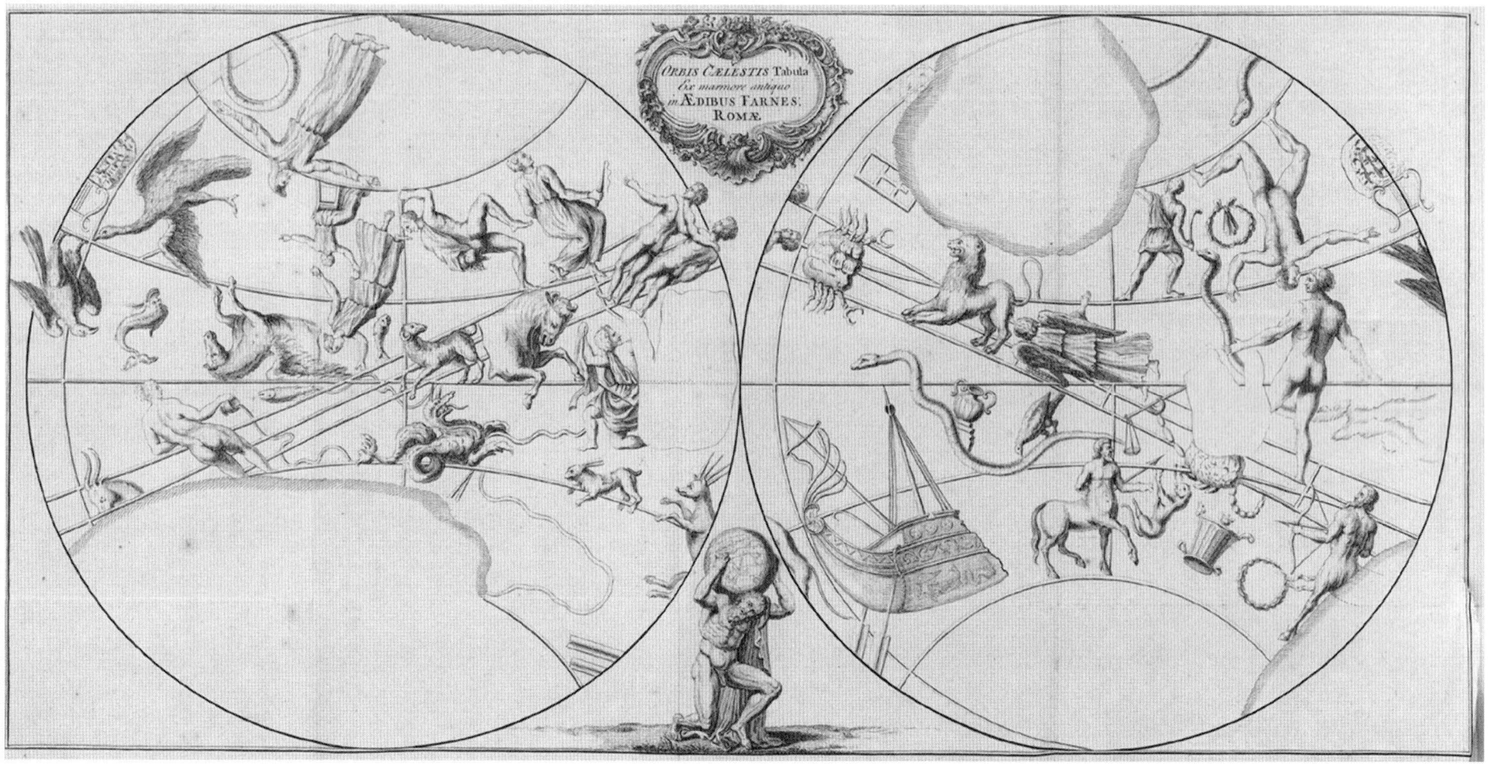

Figure 6. Stereographic projection by Martin Folkes of the Farnese Globe in Richard Bentley, ed., *M. Manilii Astronomicon*, London, 1739, p. xvii. Engraving on folded paper insert, 25.9 × 52.1 cm. The Library, All Souls College, Oxford. Courtesy of the Library, All Souls College, Oxford.

> When he arrived, (there still remained) in the library an ancient geographical map which the famous cosmographer and Reverend Father Mauro of the same monastery[1] drew in his own hand at least three centuries previously. Since this could be most useful in teaching an account of modern geography, I have decided to place it in the public domain, being persuaded particularly by some of all those who declared unwaveringly that such an undertaking would be welcome to the academic world.
>
> [Cum venit adhuc in Bibliotheca tabula antiqua Geografica quam ante tria ad minimum saecula propria manu delineavit ipsius clarus RP Mauro[ensis] ejusdem cenobii, cosmographus inclytus; quae utilissima esse possit ad historiam Geografie recentioris docendam, ipsam publici juris facere praeposuimus, suasus praesertim universorum[2] aliquot qui tale negotium viris doctis gratum fore constanter asseruerunt].

Although we have no evidence of an eighteenth-century copy of the Fra Mauro map commissioned by Folkes, it would not have been out of character for him to have done so, due to his fascination with cartography. The antiquary Thomas Hearne noted that 'Martin Folkes, Esq., hath been for some time inquisitive about the antiquity of the Geographical maps of our Island'.[3] The surveyor and cartographer, John Rocque (1709–62), and the engraver, John Pine, Bluemantle Pursuivant at Arms and Engraver of Seals, etc. to His Majesty, dedicated a 'Plan of the cities of London and Westminster and borough of Southwark' to Folkes.[4] As a recent article by Heffernan has indicated: 'the origins of John Rocque's great survey and map of London, published in 24 sheets in 1746–7, can be traced directly to ... [the] suggestion that London's spatial pre-eminence would only be conclusively proved by an accurately surveyed map of the English capital, based on techniques [Guillaume] Delisle (1675–1726) used for his 1716 map of Paris'.[5] Just as Folkes went to Venice to demonstrate the primacy of English prisms and instrumentation, he also gave a map to demonstrate England's cartographic prowess and London's pre-eminence.

Lastly, as a Freemason, antiquary and scientist, Folkes's diary demonstrates that he utilized precision instrumentation in his Grand Tour to measure Roman and Renaissance monuments, architecture, and coinage, to understand the role of ancient standard measures and weights in their creation. Folkes, in a larger sense, considered the historical context of the material culture of the past, anticipating in his antiquarianism the nascent development of archaeology's precise empiricism.

For example, Folkes investigated past methods and metallurgy of minting coins at the Venetian *Zecca* (Mint) as part of his proposed numismatic history for the Society of Antiquaries, an homage to his mentor, Newton, who was Warden, and then Master, of the Royal Mint. Hearne wrote of Folkes's numismatic work:

> The book that Martin Folkes Esq., F.R.S., is doing, about our English coins, is to be in 4°. 'Tis to begin with William Conq., and to come to the present time. Many Plates are done. Mr.

[1] RP Mauroensis; the medallion struck shortly after Fra Mauro's death calls him 'Maurus S. Michaelis Moranensis', that is, Maurus of (the monastery of) St Michael on Murano (an island in the lagoon).

[2] There seems to be an omission of 'a/ab' before 'aliquot', which is assumed for sense.

[3] Hearne, entry of 15 March 1730/1, *Remarks and Collections of Thomas Hearne*, X, p. 394.

[4] 'To Martin Folkes esq.r, president of the Royal Society; Plan of the cities of London and Westminster and borough of Southwark', Harvard University, https://curiosity.lib.harvard.edu/scanned-maps/catalog/44-990102051310203941 [Accessed 12 July 2023].

[5] Heffernan, 'The Scale of Two Cities'.

> Vertue is the ingraver. So Mr. Vertue himself told me at my room at Edmund Hall last night. In this work he will give an account of the true value of each coin from time to time, of its Alloy, and of the alterations each hath suffered. He will also distinguish the marks made use of in medals, and of those in coins. He is of opinion that what is called the 5 shilling piece of Hen. VIII was a coin, not a medal, but I take it to be a medal. It seems it was rather a 3S. 6d. piece, if it were really a coin. As for marks of the mint to prove this and some other pieces coins, I take the argument to be of no validity. For formerly even what was a medal, was nothing often but a minted piece, and the distinction of the mint was not then as now observed with respect to coins, so as those are to be reckoned as coins only which have the mint mark. There will be an interruption to this work for some time, during Mr. Folke's [*sic*] absence, who is going into Italy for about a year with a son of his, that is about 13 or 14 years of age.[1]

If we consider Folkes's interests in numismatics more closely, they reflected his concerns for precision weighing, metrology, and instrumentation. One of his mots that he used to sign *alba amicorum* or friendship books was the same as Newton's: *Numero pondere et mensura, Deus omnia fecit* (God creates all things through number, weight, and measure). Folkes's mot reflected his research in mathematics and his interest in the metrology of past architecture and artefacts, for example, determining the exact dimensions of the Roman foot.[2] Regarding numismatics, Folkes recorded in his diary: 'I have employd my self some days in understanding the Venetian money and the changes that have been made in it as also the affairs of their Bank', noting 'they keep all their accounts here in the Roman figures which seems an argument of its antiquity'.[3] He went to the Palazzo della Zecca, a building designed by Sansovino, that housed the Venetian Mint, recording the weights and measures of the bars used to make ducats.[4] He returned on another occasion to witness the coin stamping, observing they usually gave 'about 12 strokes to a half Ducat and 14 to a ducat', but deciding in comparison to English and particularly Newtonian coinage that 'the money is very ill stampt'.[5] Folkes did calculations for fellow virtuosi on the value of Florentine specie, spoke with the Procurator of Venice about the art of engraving coins, and borrowed a gold ducat from the Venetian consul to study.[6]

Folkes was also engaged in the calculation of the exact length of the Roman foot and Venetian palm. Queries about ancient metrology began in 1573 with the publication of *De mensuris et ponderibus Romanis et Graecis* by Lucas Paetus. That publication spawned an active debate about the exact length of the Roman foot or pes. Dissatisfied with inconsistencies in textual evidence, John Greaves, Savilian Professor of Astronomy at Oxford, visited Rome in 1639 to 'examine as many ancient measures, and monuments, in Italy, and other parts, as it was possible', and then compare 'these with as many Standards, and Originals, as I procure the sight of'.[7] He thus measured brass measuring rods in Roman ruins and the foot measures on the tomb of Titus Statilius Aper and on the statue of Cossutius. Greaves also derived the foot that the Jesuit priest and architect,

[1] Hearne, entry of 2 March 1732/3, *Remarks and Collections of Thomas Hearne*, XI, p. 166.

[2] BL, Add. MS 52362, f. 4r.

[3] See below, p. 84. See Appendix B concerning Venetian coinage.

[4] See below, pp. 84–5.

[5] See below, p. 90.

[6] See below, pp. 89, 96.

[7] Greaves, *A Discourse of the Romane Foot*, p. 20.

Juan Bautista Villalpando, calculated from the Congius of Vespasianus or Congius Farnese.[1] The Congius was:

> a possibly counterfeit Roman vase belonging to the collection of Cardinal Farnese in Rome. The Congius had attracted and would attract the attention of many antiquarians ... a key object because according to the inscription on its exterior, it contained exactly ten Roman librae. Therefore it could be used as a standard from which one could derive the equivalence of librae in terms of contemporary units.[2]

The congius could also be used to measure length, as an Amphora quadrantal was equivalent to a cubic Roman foot, and a congius was equivalent to a half cubic Roman foot or ⅛ quadrantal.[3] Greaves then noted with some pleasure that these foot measures exactly corresponded with the distance between the milestones on the Appian Way.[4] He then made comparisons between the Roman foot and the iron standard of the English foot in the London Guildhall, and published his findings in his 'Discourse on the Romane foot and Denarius' in 1647, concluding that the Cossutian foot was the 'true' Roman pes. Greaves also claimed that he thought he could find a relationship between the pes and the sacred cubit, something which Isaac Newton also noted in his own treatise on the cubit, which was historically an ambiguous standard of measure.

Folkes followed the same methodology as Greaves in his own travels. The diary of Sir Andrew Mitchell (1708–71), politician and fellow traveller in Rome, recorded that, on 1 June 1733, he, Celsius and Folkes, went to see the Marquess Caponi, whose palace courtyard had a similar standard stone known as the Capponian foot 'engraved with the tools of a [Masonic] workman', with a bottom rule that was '969 parts of the English foot'.[5] The stone had been previously dug up on the Via Aurelia. In 1736, Folkes published and identified the Root Canonical Greek foot as 1.0057142 times the English measure, and the Roman pes as .966 times the English foot, noting them as being engraved on a Standards stone tablet at the Roman Capitol that might have been the actual record of the *pes monetalis*. His paper about the stone was subsequently published in the *Philosophical Transactions of the Royal Society*; Folkes recorded 'setting the point of my Compasses' in the lines in the stone that represented the measures and noted that 'my chief Attention was given to the Roman foot, as of greater Consequence than the other measures'.[6]

No doubt influenced by Giovanni Poleni's work on Vitruvius and the cult of Palladianism, engendered by consul Joseph Smith in Venice, Folkes also employed precision instrumentation in his Grand Tour to measure both Roman and Renaissance monuments and architecture to understand the role of standard measures in their construction.[7] Folkes's measurements of the various 'feet' were done for this purpose. It was a lifetime obsession. When Folkes returned from his journey and became President

[1] Villalpando depicted the Congius and its precise measures in a copperplate in the apparatus of his treatise, *Ezechielem explanationes*. See, Pastorino, 'Early Modern Study of Ancient Measures', p. 133; de Prado and Villalpando, *Ezechielem explanationes,* III, *Apparatus urbis ac templi Hierosolymitani* (1604), pp. 501–2.

[2] Pastorino, 'Early Modern Study of Ancient Measures', p. 133.

[3] My thanks to Cesare Pastorino for this information.

[4] Greaves, *A Discourse of the Romane Foot*, pp. 20, 23.

[5] British Library, Add. MS 58318, f. 109r. The stone is now in the Capitoline Museum in Rome.

[6] Folkes, 'An Account of the Standard Measures', pp. 262–3.

[7] Poleni, *Exercitationes Vitruvianae; Poleni, M. Virtuvii Pollionis Architectura.*

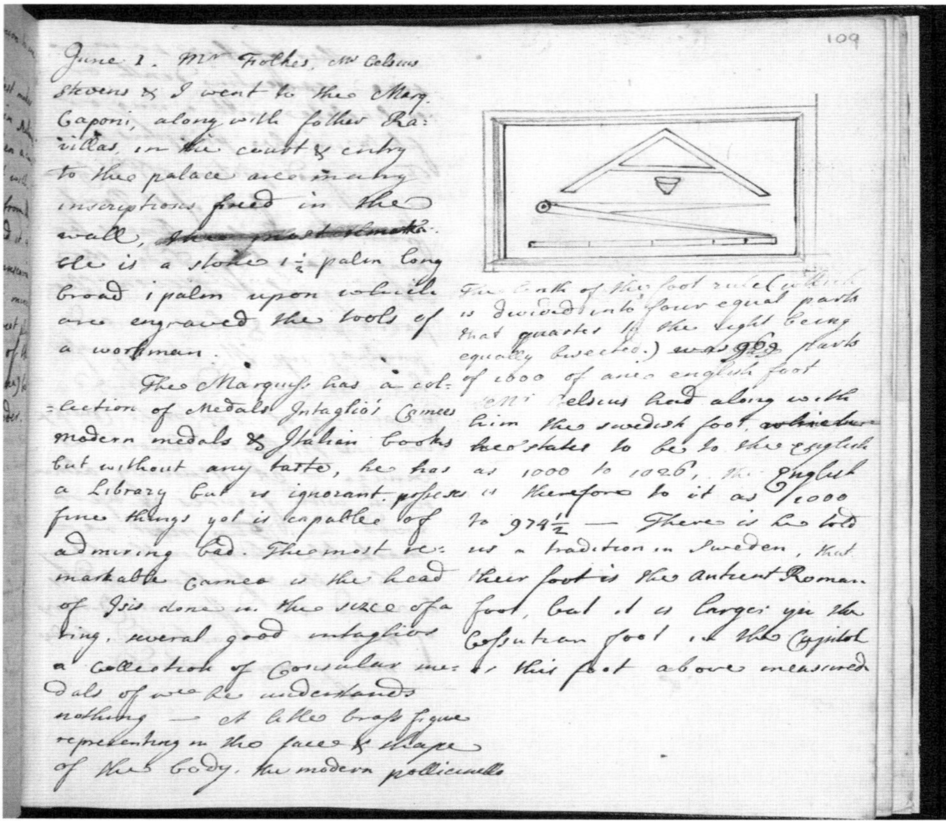
June 1. Mr Folkes, Mr Celsius
Skoens & I went to the Marq.
Caponi, along with father Ra:
villas, in the court & entry
to the palace are many
inscriptions fixed in the
wall,
there is a stone 1½ palm long
broad 1 palm upon which
are engraved the tools of
a workman.

The Marquis has a col:
:lection of Medals Intaglio's Camees
modern medals & Italian books
but without any taste, he has
a Library but is ignorant. possesses
fine things yet is capable of
admiring bad. The most re:
markable Cameo is the head
of Isis done in the size of a
ring, several good intaglios
a collection of Consular me:
dals of which he understands
nothing — A little brass figure
representing in the face & shape
of his body. the modern pollicinello

The length of this foot rule (which
is divided into four equal parts
that quarter of the right being
equally bisected.) was 969 parts
of 1000 of an english foot
Mr Celsius had along with
him the swedish foot, which he
has stated to be to the english
as 1000 to 1026, the English
is therefore to it as 1000
to 974½ — There is an old
tradition in Sweden, that
their foot is the antient Roman
foot, but it is larger yet the
Capitoline foot in the Capitol
is this foot above measured

Figure 7. Measuring standard observed by Folkes in the Villa Caponi, Rome, Diary of Andrew Mitchell in Rome, Add. MS 58318, f. 109r. The British Library Board. Courtesy of the British Library Board.

of the Royal Society, the Académie royale des sciences at the request of the Royal Society sent a measure of half the toise (the Paris foot is one sixth part of the toise), from which George Graham determined the proportion of the Paris foot to that of London.[1] Folkes himself had a London foot measure made by Sisson, according to the calculations of George Graham.[2] One of Folkes's first tasks when he arrived in Venice was to measure the height of the Campanile by its shadow, concluding it was 140 feet high, and then measuring again to confirm his conclusions; to do this, Folkes figured out the conversion factor between Venetian palms and English feet. Persevering despite suffering from a heat rash, Folkes proceeded to measure the Rialto Bridge in Venice to confirm his determination and comparison of past standards of measurement. Folkes used packthread and a plumb bob, carefully comparing his measures of Antonio da Ponte's Bridg to those taken by Francesco Sansovino, the son of Jacopo Sansovino (1486–1570), the proto or chief architect of Venice who was responsible for the rebuilding of the Fabbriche Nuove di Rialto complex (the seat of the magistrates in charge of customs and duty). Disturbed

[1] Raper, 'An Enquiry', p. 778.
[2] Ibid., p. 783.

that his measurements of the bridge were different from Sansovino's, he wrote: 'I would readily think [*sic*] may be partly owing to defect in my own measure, partly perhaps on inaccuracy in the building the bridg itself to the intended model, and perhaps again in the not having to perfect these proportion[s] between the English and Venetian foot.' After making some enquiries after the original standard of the Venetian measures, which he found to be fruitless, he 'took some measures of S[t] Geminianos Church in order to have made something out by them'. However, upon comparing the measures he had taken, Folkes found the church 'was not built by the Venetian foot but by the Roman Architectonick palm whose length comes out very exactly by it'. It was a perceptive comment. St Geminiano was in the Piazza San Marco, part of the complete remodelling of the piazza, the piazzetta, and surrounding buildings in the classical style by Jacopo Sansovino during the Renaissance; as historian Fiona Kisby has noted, the 'effect was to superimpose an evocation of ancient Rome on the existing Byzantine elements'.[1] St Geminiano, directly across the piazza from San Marco, was a focal point of the architectural complex designed by Sansovino in Vitruvian proportions, to evoke the ancient Roman forum. Hence, Folkes suspected and proved that Sansovino used Roman measurements to create his building, particularly as Vitruvius in his third book of *De architectura* gave a thorough description of the Roman foot or pes and palms (one-quarter of the foot). In this manner, principles of measure and geometry would ensure harmony and aesthetic appeal in the building's design. In other words, Folkes was investigating architectural metrology to provide what we would term archaeological information about the architects and builders of Renaissance Venice and Rome.

During his time in Venice, Folkes also met English diplomats, including the Resident, Elizeus Burges. Burges regarded Folkes as a 'very ingenious man, and loves and understands well most of the Things People come into this Country to see'.[2] This reference to Folkes's ingenuity and keen empiricism extended even to his appraisal of popular 'tourist sites' on his journey. For example, on his way to Rome, which he arranged with a vetturino 'to carry me to Rome in my own berlin and find another for my servants and baggage', he visited the Cascata delle Marmore:[3]

> the highest falls in Europe, all of 165m, delivering thousands of cubic meters per second. These fascinating falls are not the work of nature: they are artificial, built in 271 B.C. by the Roman Consul [Manius] Curius Dentatus who wished to put an end to the floods in the Sabine valley and so built what is certainly among the most spectacular public hydraulic works – artificial falls that directed the waters of the River Velino into the River Nera, away from the plains of Rieti ... [these] peculiar 'natural' sights [were] praised by Ancient Romans and celebrated during the seventeenth and eighteenth century by the travelers of the Grand Tour.[4]

Joseph Addison remarked of them, 'I think there is something more astonishing in this cascade than in all the water-works of Versailles'.[5] Whilst Folkes was awed by its experience, he does not have the conceptual framework of the picturesque and its tropes

[1] Kisby, *Music and Musicians*, p. 31.

[2] The National Archives, Kew, PRO/State Papers 99/63, f. 230 (Burges, 17 June 1733).

[3] See below, pp. 160–62. For *vetturino,* see p. 108, n. 5. For 'berlin', see p. 108, n. 6.

[4] 'Cascata delle Marmore and Valnerina', UNESCO, https://whc.unesco.org/en/tentativelists/2031 [Accessed 23 March 2023].

[5] Hurd, ed., 'Remarks on several parts of Italy', *Works of Joseph Addison*, I, p. 412.

that characterized Grand Tourists later in the eighteenth century in their reaction to natural wonders. Rather, his approach was natural philosophical. He records trying to: 'measure the height of the fall in some degree by throwing down large stones, and upon several repeated tryals by my second watch, I found them to be better than 5 seconds and coming down whence the height should not be less than 400 feet '.[1] He also observed the single and double rainbows, his interest in optics ever present.

5. Rome

Folkes's first view of St Peter's in Rome on the Via Flaminia, unsurprisingly, was via his 'glass' or his telescope.[2] He and his family stayed at Giuseppe's on the Corso, and he spent much of his time exchanging visits with other English travellers and residents. His friend, Charles Lennox, 2nd Duke of Richmond, provided Folkes with a number of rather blustery letters of introduction, one to Alessandro Albani, the leading collector of antiquities in Rome, whom Lennox described as: 'a very odd Curr, Ignorant enough, and proud as Hell; butt has the finest library, one of them, in Europe, and without exception one of the very best collection of bustos, in the World. You must flatter him upon his learning.'[3] Lennox also wrote a letter of introduction for Folkes to Teresa Grillo Pamphili (d. 1763), the Genoese poet who led an important literary salon (the Arcadia) in Rome, referring to her rather uncharitably as the: 'Ugliest Bitch, in the World, Damn'd proud also, and stark staring mad, butt a Develish deal of Witt, some knowledge, and altogether Une Maitresse ferme, sans un brin de religion [a resolute woman without a touch of religion]'.[4]

Folkes continued his metrological studies in Rome, measuring Trajan's Column, the Antonine Column, and taking notes on the statue of Marcus Aurelius; all three monuments became the subject of his publications. He also used as his inspiration works such as Antoine Desgodetz's *Les Edifices antiques de Rome* (1682), a treasury of ancient Roman architecture and their measures:[5]

> [Desogdetz] measured the buildings just as they stood without any attempt at restoration, incorporating as many details as possible, the whole undertaking being carried out with a degree of thoroughness and accuracy never before know. Governed by the rationalistic trend of his age, he was well equipped to create a standard work not to be superseded for a long time.[6]

Folkes took detailed measurements of the monumental Tuscan Doric Antonine Pillar, noting: 'The pillar of Antoninus Pius, now commonly called of Monte Citorio is now lying, along; the shaft of one piece, of red oriental granite, is about 48 English feet long, as I grossly measured it; the pedestal of one piece of marble was dug up not far

[1] See below, p. 161.

[2] See below, p. 167.

[3] RS/MS/865/12, Charles Lennox to Martin Folkes, 31 July 1733.

[4] Ibid. The letter to Pamphili is RS/MS/865/13. For Pamphili's role as host of the Arcadia, see Graziosi, 'Women and Academies in Eighteenth-Century Italy', pp. 103–19.

[5] The auction catalogue of Folkes's library shows he owned a copy. See *Catalogue of the entire and valuable library*, p. 126, lot 4255.

[6] Herrmann, 'Antoine Desgodets', p. 24.

from it.[1] He also ascertained the original diameter of the shaft to determine if the pillar had any diminution over its length or tapering, which he concluded it did not.

In a more detailed study, Folkes measured the height of the Trajan Pillar, 'using an exact two-foot rule brought from London, and holding a plum line to every stone for the directing our measure upright'.[2] He accomplished these measurements with 'Mr Bowman', or Walter Bowman (1699–1782), of Logie, Fife, FSA (1735), Dilettanti (1736), FRS (1742), antiquarian, traveller, and friend of Horace Walpole. From 1732 to 1724, Bowman accompanied Simon, 2nd Viscount Harcourt on his travels. As Ingamells has related, on 12 November 1733 from Rome, Bowman acknowledged the receipt of some Latin verses from Samuel Haynes: 'Tho I live a Sort of an old Life among Ruins yet I dare not aspire to answer you in the Language of the Gods. My muse is damnably hoarse by living underground.'[3] Folkes was one of the proposers to his election to the Royal Society in 1742, writing: 'We Accordingly propose him, and recommend him as a Gentleman perfectly well known to us, of great knowledg and Curiosity, Well versed in Mathematical and Philosophical Learning, and every way deserving the honour of being a Fellow of the Same.'[4] Folkes was particularly interested in the column because the height of its shaft from the plinth of the base to the abacus of the column was said to measure exactly one hundred Roman feet, a *columna centenaria*.[5] The column was thus ideal to determine the dimensions of the Roman foot.[6]

Between 29 May and 8 June 1734, Folkes visited various Roman collections, including the Villa Mattei, with Andrew Mitchell; the Villa Mattei (now Villa Celimontana) on the Caelian Hill had another standard measuring stone on a monument of M. Aebutius unsurprisingly known as the Aebutian foot. His resolutely empirical nature at times failed him when attempting to describe the emotional reactions to what he saw, for example, when he visited the temple of Jupiter Tonans or the Sistine Chapel. Folkes clearly had a good eye due to his understanding of optics, he had expert knowledge on what to collect, but his engagement with art seems more analytical then emotional. He thus indicated in the diary he would refer to prints he owned rather than taking detailed notes, trusting the artists' reproductive engravings of works of art to trigger later memories of his journey.

It is not surprising that, given his keen aesthetic sense, Folkes was a connoisseur of etchings, engravings, and mezzotints, noted particularly for his collection of the works of Rembrandt.[7] The posthumous auction of his prints and books collection, which took several months, also indicated a fairly lavish outlay on works to do with Italian art and antiquities. Folkes also owned Bonaventura van Overbeek's *Reliquiae antiquae urbis Romae* (Amsterdam, 1708), a book of the major Grand Tour sites of Rome. He bought several prints as well, for example of Michelangelo's work in the Sistine Chapel, including:

[1] Folkes, 'On the Trajan and Antonine Pillars', p. 121.

[2] Ibid., p. 119.

[3] Ingamells, *Dictionary*, pp. 113–14.

[4] Royal Society Library, EC/1742/16, 'Certificate of Election'.

[5] Boni, 'Trajan's Column', p. 96.

[6] Folkes, 'An Account of the Standard Measures', p. 266.

[7] Many Rembrandt prints from Folkes's collection are in major galleries. *The Raising of Lazarus* by Rembrandt from Folkes's collection is in the British Museum, museum number F,4.152. The National Gallery, Washington DC, has *Canal with an Angler and Two Swans*, number 1943.3.7227, https://www.nga.gov/collection/art-object- page.10023.html.

'four, of M. Angelo's last judgment, and 5 others'; a book of prints of Giovanni Battista Falda's *Le Fontane di Roma* (Fountains in Rome) published by Giovanni Giacomo Rossi; one print of Raphael's *Grotesques* in the Vatican, and another set of Raphael's *Scripture* prints. There were also several prints of works by Federico Barocci in his collection.[1]

The diary ends abruptly, with Folkes making a repeat visit to Trajan's Column and the tomb of C. Publicius Bibulus, who was a plebeian *aedile* circa 100 BC, buried at the base of the Capitoline Hill. The last thing Folkes records was the arrest of his acquaintance, William Dugood. FRS (fl. 1728–67). In 1728, Dugood was nominated for membership of the Royal Society by John Theophilus Desaguliers, Edmond Halley and Folkes, and made magnetic measurements off the coast of Portugal in 1733. In August 1728, Dugood was also present at the Royal Society when optical experiments were made concerning the work of Rizzetti who denied principles of Newtonian optics, claims Folkes refuted when in Venice with optical demonstrations. Ingamells notes of Dugood that he was in Florence, Leghorn,[2] Parma, and Rome from 1731 to 1734, and that Dugood acted as a double agent in Italy while practising as a jeweller and was arrested for his activities in Rome during Folkes's visit.[3] Folkes laconically noted that Dugood was 'put in prison by the Popes order and commanded to leave his dominions in 3 days he sent to us, but we have here no Protector and knowing nothing of his affairs can do nothing for his service'.[4] Perhaps Folkes's diary may have ended abruptly due to sensitive information about espionage, but we also have to consider it was simply because Folkes's notebook was full. However, happily for Dugood, he was eventually released. By 1 May 1734, Brinley Skinner, the British Consul at Leghorn, had procured Dugood a passage to Lisbon, and then to England, where, Skinner suggested, 'he might prove a useful source of information concerning Jacobite activities in Italy'.[5]

6. Afterword: A Wife and a Life – Folkes's Tour and Lucretia Bradshaw

Although the diary ended abruptly, Folkes's later correspondence allows us to continue to follow his journey. Folkes spent a considerable time in Rome and afterwards some months in Florence before returning to England. When he was in Florence, he struck up a close friendship with the natural philosopher and physician, Antonio Cocchi, and on 9 January 1735, he was made a member of the Florentine Academy.[6] Cocchi had visited London in the 1720s, and met Newton there. When he returned to Italy, he became Professor of Medicine at Pisa University and Professor of Philosophy and Anatomy at the University in Florence, and was elected FRS in 1737.[7] A proficient English speaker, Cocchi served as a doctor for several Englishmen on their Grand Tour, had frequent correspondence with fellows of the Royal Society, including its secretary, Cromwell Mortimer, Newton's physician, Richard Mead, and, of course, Martin

[1] *A Catalogue of the Prints and Drawings*, pp. 8, 12, lots 37, 48.

[2] Livorno.

[3] Ingamells, *Dictionary*, p. 317.

[4] See below, p. 199.

[5] Ingamells, *Dictionary*, p. 317.

[6] Ibid., p. 366; Wynne, 'Members', p. 537.

[7] RS EC/1733/14, Election Certificate of Antonio Cocchi.

Folkes.[1] Their correspondence concerned tragic, private details of Folkes's relationship with his wife.

On 18 October 1714, Folkes married Lucretia Bradshaw, a Drury Lane actress, at St Helen's Church in London. It is not known how they met, although before Folkes inherited his father's townhouse on Queen Square he lived on Southampton Street in Covent Garden, where several actors and actresses lived, including Anne Oldfield (resident there from 1709 to 1713), Robert Wilks (b. 1709), and Colley Cibber (resident from 1714 to 1720).[2] Lucretia Bradshaw is a shadowy figure, purportedly the daughter of a Drury Lane box keeper,[3] born sometime between 1680 and 1685.[4] Lucretia became a protégé of the 'incomparable Mrs Barry' who, as a theatrical shareholder in her own right, wielded considerable influence in the troupe. Elizabeth Barry was known for the naturalistic quality of her acting, and Lucretia learned to make 'her self Mistress of her Part' leaving 'the Figure and Action to Nature'.[5] With Barry's backing, Lucretia became the leading lady of Drury Lane; her contemporary, Charles Gildon, remarked rather cuttingly (or jealously): 'Mrs Bradshaw ... if she be not the best Actress the Stage has known, she has hindered Mrs Barry from being the only Actress.'[6] Nonetheless, Thomas Betterton's short *History of the English Stage*, published in 1741, related that 'It was the opinion of a very good Judge of Dramatical Performers' that Lucretia 'was one of the greatest, and most promising Genij of her Time'.[7]

For Lucretia, marriage to Folkes was her way out of Drury Lane and into gentility, a contemporary reporting that she was 'taken off the Stage, for her exemplary and prudent Conduct, by Martin Folkes, Esq; a Gentleman of a very considerable Estate, who married her; and such has been her Behaviour to him, that there is not a more happy Couple'.[8] The nineteenth-century commentator, John Doran, in his work *'Their Majesties' Servants'*, wrote about the marriage:

> At this period (about 1714) the stage lost a lady who was as dear to it as Queen Anne, namely Mrs. Bradshawe. Her departure, however, was caused by marriage, not by death; and the gentleman who carried her off, instead of being a rollicking gallant or a worthless peer, was a staid, solemn antiquary, Martin Folkes, who rather surprised the town by wedding young Mistress Bradshawe.

Despite any public scandal that ensued, the marriage through the 1720s was purportedly quite happy, the antiquary William Cole reporting that Folkes was 'infinitely fond' of

[1] RS/MS/790/27 (Cocchi to Folkes, 26 October 1737); RS/MS/790/25 (24 Dec 1735); RS/MS/790/26 (31 March 1736); RS/EL/C3/37 (Cocchi to Mortimer, 26 Oct 1737). The Letters of Antonio Cocchi are in the Archivio Baldasseroni in Florence, catalogued by Anna Maria Megale Valenti (*Le Carte de Antonio Cocchi*). The catalogue indicates several letters from Folkes to Cocchi in the 1730s and 1740s.

[2] 'Southampton Street and Tavistock Street Area: Southampton Street', in Sheppard, ed., *Survey of London: Volume 36, Covent Garden*, pp. 207–18. British History Online, http://www.british-history.ac.uk/survey-london/vol36/ [Accessed 21 Nov 2019].

[3] An attendant at the boxes in a theatre. See *OED*.

[4] Highfill et al., *A Biographical Dictionary*, II, p. 283.

[5] Gildon, *The Life of Mr Thomas Betterton: The Late Eminent Tragedian*, p. 41.

[6] Ibid.

[7] Betterton, *The History of the English Stage*, p. 62.

[8] Ibid. As for Drury Lane, Hester Santlow, 'ultimately took over nine roles that had belonged to Lucretia ... acquiring most of them during 1714–15'. Goff, *The Incomparable Hester Santlow*, p. 56.

[9] Doran, *'Their Majesties' Servants'*, p. 112.

Lucretia. Lucretia and Martin had two daughters, Dorothy (b. 1718) and Lucretia (1721–58), as well as a son, Martin (1720–40), and there were lively gatherings at their house on Queen Square.

Unfortunately, by the 1730s Lucretia manifested symptoms of mental illness, and during their tour of Italy, had symptoms of erratic, almost manic behaviour. Tom Hill, the 2nd Duke of Richmond's steward, wrote to his lord:

> With much ado I obtain'd leave to transcribe the following account relating to Mrs. Folkes out of a letter that came from abroad, having first sworn not to tel the person that sent it. There is come hither a Lady with her husband, three children, and a monkey, who are no more exempt from obedience to her, one than another, and all seemingly fellow-sufferers alike. I happen'd to be at a visit when she came in. In all my life did I never hear such an insupportable creature, nor so much nonsense in so small a space of time. You will be surpris'd when I tell you the husband is reckon'd as clever a man as any in England. His name is Folkes (Martin Folkes as she cals him) who used to be very much with the Duke of Richmond. The lady he married is very wel known in England. He designs making the tour of Italy and France, by which time I don't doubt but she wil turn out the most accomplish of fine Ladys. She did think indeed of bringing a little dog and a cat to keep poor pug company, but that they could not possibly find more room in the coach. Such characters are no where to be met abroad, whatever they may be in England, and even there I never saw one come up to this. This is al that was read to me out of the letter. I could not help saying, what I fancy you'l join with me in, Poor Martin! In an evil hour didst thou take to thy bosom this Lady Mar-all.[1]

One could attribute this report to spiteful gossip, but Stukeley noted that, while in Rome, Lucretia Folkes 'grew religiously mad'.[2] Stukeley's assessment was confirmed by melancholy letters written between Folkes and his friend Cocchi, which indicated the severity of Lucretia's disorder. In November 1735, at which point Folkes had returned to London with his family, he wrote to Cocchi:

> I must freely tell you the situation of my own affairs is on many account so disagreeable to me I know not what in world to do; my wife is nothing so outrageous as when abroad, which lays me under really an incapacity to make my self easie by confining her, and on the other hand it is not possible for me to live with her or easie if she has it in her power to make me and all my ffriends uneasie.[3]

Perhaps Folkes had thought the travel would do her some good or halt gossip in his social circles in London, but ultimately these hopes were dashed. On his return to England, Folkes even entertained thoughts of moving to Florence permanently and leaving Lucretia in England in an asylum, writing to Cocchi about settling in Florence. Cocchi replied the following month:

> We are glad to hear Shee [Lucretia] is better and I believe if she could be brought to a proper method of cure She might entirely recover. As for living with her, I well conceive it may be very troublesome to you. I have some experience of People who Suffered any disorder in their mind, and I think the Surest way for quiet and ease is to break all commerce with them.[4]

Cocchi then stated: 'after having settled what justice and humanity demands of us, so that I approve mightily of your design of coming abroad which will be a good pretext to

[1] Gordon-Lennox, *A Duke and His Friends*, p. 253.

[2] Stukeley, *The Family Memoirs*, I, pp. 99–100.

[3] BL, MS Facs 589, 584B, Martin Folkes to Antonio Cocchi, 6 Nov 1735.

[4] RS/MS/790/25, f. 1r, Antonio Cocchi to Folkes.

separate your dwelling'.[1] In his account book of his travels, Folkes subsequently recalculated the costs of going to Italy without his wife and children, but he ultimately decided against settling in Florence.[2] In 1735, the Roman Freemasons' lodge ran foul of the Inquisition, and Freemasonry was subsequently prohibited by Pope Clement XII in 1738.[3] Folkes was no martyr for reasons of faith.

With his plans to relocate to Italy defeated, Folkes decided to confine Lucretia to a lunatic asylum in Chelsea. His will, written in 1751, noted:

> And Whereas the unfortunate State of the health of my dear Wife Mrs Lucretia ffolkes render it absolutely necessary several years since according to the joint advice of several Eminent physicians and of all my most particularly ffriends, whom I consulted on that occasion to confine her where she now remains at Chelsea in the County of Middlesex.[4]

But which asylum? After years of searching for Lucretia's ultimate fate, I discovered what happened in an unlikely place, the account book Folkes's brother William kept of the family expenses in the Norfolk Record Office.[5] There were a cluster of private asylums in Great and Little Chelsea run by various operators such as Michael Duffield (d. 1798).[6] Duffield's asylum was a handsome building erected by a Mr Mart, located by Lord Shaftesbury's house and Lord Wharton's park.[7] Starting in the mid-1730s, when Folkes and his family returned from their Grand Tour, there were periodic, six-monthly payments of slightly over £100 to the Duffields, the first one on 10 November 1736, and the last one on 23 May 1754, the year of Folkes's death.[8] The work of Andrew Scull and Jonathan Andrews has shown that in 1743, Michael Duffield charged a patient about £25 a quarter for lodging, clothing, and other expenses, so about £100 a year for basic treatment. Folkes paid double that for Lucretia's care.[9]

In his will, Folkes wrote that he had, 'constantly been informed by those under whose care and direction she is, that there is but small prospect or probability of her recovery from her present state and condition', and thus he set in trust investments that would pay £400 per year for her 'support and maintenance'.[10] Folkes charged his daughters, Lucretia and Dorothy, to administer the payment for his 'said dear Wife their Mother who was ever whilst God gave her the ability a fond Carefull and most Indulgent Parent to them and whom I have ever most dearly and affectionately loved'.[11] It was a tragic end

[1] Ibid.

[2] Norfolk Record Office, NRS 20658, Martin Folkes, 'Memoranda in case he should go abroad again, respecting his Estates, foreighn money & Places'.

[3] Eisler, 'The Construction of the Image, Part I', pp. 20–21.

[4] The National Archives, Kew, PRO/11/809/301, Will of Martin Folkes, 3 July 1754.

[5] Norfolk Record Office, NRS 8739, 21D4 'William Folkes, Account of his brother's affairs while abroad'. William Folkes wrote on the flyleaf: 'My Brother ffolkes of his family having proposed to spend some time beyond Sea, and he having desired me to receive and pay all monies for him \during his Absence/ according to written directions under his Hand hereafter follows the account of his Affairs kept by me with the greatest Exactness and Truth I am Capable of, W. ffolkes. March: 24th. 1732.'

[6] Andrews and Scull, *Customers and Patrons*, p. 9.

[7] Ibid.

[8] Norfolk Record Office, NRS 8739, 21D4, f. 31v. 'Paid Mrs Duffield, £74.2s.5p'.

[9] Smith, *Private Madhouses in England*, p. 175.

[10] The National Archives, Kew, PRO/11/809/301, Will of Martin Folkes, 3 July 1754.

[11] Ibid.

for a brilliant actress, from star of Drury Lane to a figure who, in Folkes's words, had 'in her power to make me and all my ffriends uneasie'. Folkes kept a plaster sculpture of the Roman heroine Lucretia in his study at Queen Square, possibly bought abroad for his wife. It was a sad memorial to Folkes's life and to his Grand Tour.

7. Editorial Conventions

The aim of this edition is to provide transcriptions and translations (where necessary) of the diary of Martin Folkes, reproduced accurately to consistent standards and accessible to the modern reader.

Dates are in the Julian Calendar, unless otherwise specified, with use of the double date when necessary (from 1 January to Lady Day, 25 March). For ease of use, the document reference of the text reproduced has been provided (that is, the reference used to call up the document in the relevant repository), consisting of repository name, collection name and reference number, item number, or inclusive folio numbers. Abbreviations for frequently used repository names are as follows:

BL	British Library, London
RS	Royal Society Library, London
SAL	Society of Antiquaries of London

Original orthography, punctuation, and capitalization have been reproduced, although the thorn (y) is expanded to 'th', 'Sr' has been expanded to 'Sir', and modern practice is used for long 's' where there is no ambiguity. 'i'/ 'j', 'u'/ 'v' have been retained, as well as accents in Latin and other languages. Diphthongs in Latin or other ligatures have been replaced by two vowels or consonants. Titles of works referenced published at the time or later (if the titles can be identified as such) are italicized, and where the original has underlining this has been replaced with italics. '&c' has been expanded to 'etc'. For the sake of consistency and to adhere to the Hakluyt style guide, all incidences of 'B.C.' and 'A.D.' from quoted sources have been amended to 'BC' and 'AD'.

For alterations to the letter, illegibility or loss of text through damage to the document, we are using the following editorial interventions:

- [ ] = editorial insertions; examples might include [*sic*] or letters or words necessary for sense but omitted in the manuscript
- strikethrough = followed by words struck through (crossed out) by Folkes.
- [[ ]] = illegibility or damage: in short instances one may be able to indicate by xxx (how many letters are illegible; in longer instances [[*several words*]] was used
- \ / = insertions above the line by the writer. A footnote records insertions made by someone other than the writer

Translations have been provided for Latin and French text. Original illustrations with an intended purpose in the diary have been scanned and placed appropriately in the text; non-substantive doodles have been omitted. Footnotes have been used also to explain points in the letters, identifying such things as individuals, places, events, subjects, and works cited, as well as providing a short biography for people who appear frequently. When, to date, a person has proved to be unidentifiable, this has been indicated in the footnote.

THE TRAVEL DIARY OF MARTIN FOLKES

[Flyleaf]
[*Annotation in pencil:*] Journey from Venice to Rome
[*Bookstamp:*] Sir T. P | Middle Hill[1]
[*Archivist Annotation in pencil:*] 2930
[*At page bottom (not shown):*] P[urchased]. 1966.

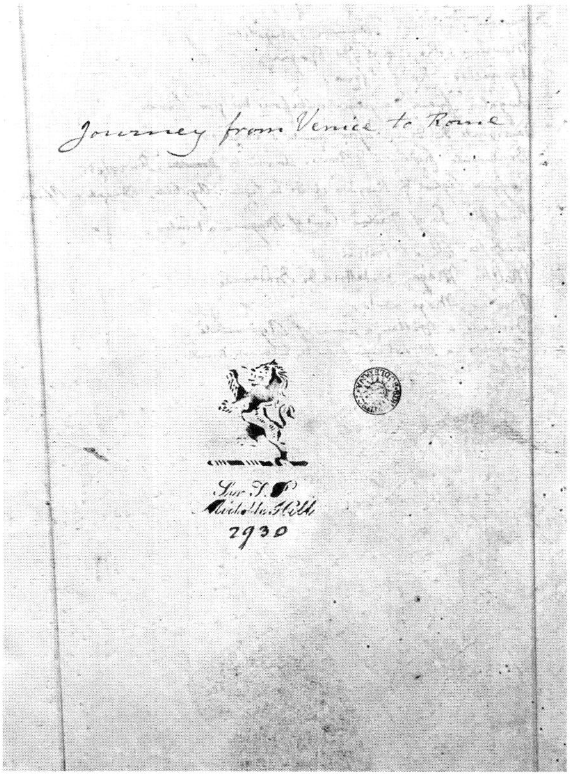

Figure 8. Book Stamp for Sir Thomas Phillipps (1792–1872) of his library at Middle Hill, and the number a reference to his published book catalogue: *Catalogus librorum manuscriptorum in bibliotheca d. Thomae Phillipps, Bart,* Middle Hill, Worcs, 1837, p. 33, no. 2930, identified as 'Journal of an Englishman at Venice', Bodleian Library, Oxford. Courtesy of the Bodleian Library, Oxford.

[1] Sir Thomas Phillipps (1792–1872), was one of the most pre-eminent English collectors of books and manuscripts, acquiring over 60,000 manuscripts which he maintained on his estate of Middle Hill, at Broadway, Worcestershire. For more on Phillipps and the diary's provenance, see the Introduction, pp. 2–5 above.

[f. 1v]
Orlando Conte d Angoulesme inamorato di Angelica[1]
Angelica Reina di Cathai[2]
Rinaldo Conte de Montauban figliolo d Amone \Conte di Chiaramonte and Duch of Dordona/ inamorato d Angelica[3]
Ferrau. Saracino. Amoros. d Angelica.[4]
Marsilio Re Sarac[in]. di Spagna[5]
Agramante. Re d Africa[6]
Argalia frere d' angelique autrefois tuè par ferrau[7]
Sacripante Rè di Circassia. amante d' Angelica[8]
Bradamante figliola d' Amone. Amorosa di ~~Brunello~~ Ruggiero[9]
Ruggiero figliol[o] de Reggiero et de la figlola d Agolante, educato in Africa[10]
Pinabello. Son of Anselmo Count of Maganza a traitor[11]

[1] This is a list of characters in the Italian poet Ludovico Ariosto's epic *Orlando Furioso* (1516), a work consisting of 46 cantos, written in Italian using octaves and hendecasyllables. The work takes inspiration from the French medieval *Chanson de Roland*. In the tale, 'Saracens' and Christians, in the days of Charlemagne, are at war for the possession of Europe, and the tale relates the adventures of one of Charlemagne's knights, Roland (Orlando, Count of Angoulême), and his fellow warriors. The main plot line concerns Orlando being driven mad by his unrequited love for the lady, Angelica. Angelica ultimately marries the Muslim soldier, Medoro, with whom she flees to India. A parallel love story concerns the Christian warrior maiden, Bradamante, who falls in love with the Muslim hero, Ruggiero. It was one of Folkes's favourite books, his library boasting eight 16th-century Italian editions, and a 1634 edition in English. See Kolbe, 'Godfather to all Monkeys', p. 40. For the annotation, I have used throughout the Harvard University Press edition of 2009, edited by David R. Slavitt and Charles S. Ross.

[2] Angelica, Queen of Cathay, who lures Orlando away from his military duty defending Europe against the African 'Saracens' and ultimately drives him to madness with her beauty.

[3] Rinaldo, the brother of Bradamante and son of Amone, the Duke of Dordone (Dordogne in France). Rinaldo and his cousin Orlando both fall in love with Angelica, causing rivalry.

[4] Ferraù, a 'Saracen' paladin, invulnerable except for his navel. Passionately in love with Angelica, he was eventually killed by Orlando.

[5] Marsilio, Muslim king of the 'Saracens' and Sargossa, who dies of his wounds whilst fighting Charlemagne's armies.

[6] Agramante, 'Saracen' king of Africa.

[7] Argalia, brother of Angelica and son of Galafrone, king of Cathay. Possessor of a ring which makes the bearer invisible when carried in the mouth. In the tale, Argalia is slain by Ferraù.

[8] Sacripante, King of Circassia and a prominent Saracen knight. Another knight in love with Angelica who offers to be her protector, but was rebuffed.

[9] Bradamante, a female Christian knight whose adventurous romance with Ruggiero is a major plot device in *Orlando Furioso*. She is the daughter of Amone, the Duke of Dordone.

[10] Ruggiero, a noble Muslim knight, said to be a descendent of Alexander the Great and Hector, who falls in love with Bradamante. He is the son of a Christian knight (Ruggiero II or Reggio Calabria) and a 'Saracen' mother (Galaciella, daughter of Agolant, king of Africa). It was prophesied that Ruggiero and Bradamante would wed and found the d'Este line, which became one of the major families of medieval and Renaissance Italy. Opposed to this prophecy is Atlantes, an African wizard, who imprisons Ruggiero in Atlantes's castle. Bradamante, however, pursues Ruggiero regardless, and the Bradamante-Ruggiero story eventually takes centre stage in Ariosto's work.

[11] Pinabello, one of Charlemagne's vassals, and nephew and friend of the knight Ganelon. Pinabello tricks Bradamante into stepping off a cliff. Bradamante narrowly avoids death, and later kills Pinabello for his treacherous act. Anselmo was actually King of Braganza, so there is some slight misidentification on Folkes's part.

Gradasso. K. of Sericana[1]
Melissa Maga protettrice di Bradamante[2]
Merlino Mago morto
Brunello a Villain, a servant of Agramantè[3]
Prasildo. a Kt that came from the East with Rinaldo
Iroldo his friend.[4]

[f. 2v] By the waters of Babylon, we sat down and wept, when we remembered thee o Sion

If I forget thee o Jersualem let my right hand forget her cunning

If I do not remember thee, let me tongue cleave to the roof of my mouth, yea if I prefer not Jerusalem in my mirth.[5]

[f. 3r] Venice, June 15 Monday. The feast of St Vito[6] being one of the 4 feasts the Doge makes every year, was a day of Masking,[7] which is here accounted the top diversion of the place, but very little to strangers who know hardly any body, the habits are almost all alike being nothing but a cloak bahout[8] white mask and hat but both for men and women who no otherwise distinguist than by the petycoat, there is no sort of humour in talking or any seeming affair but sauntring all day about the town especially about the place St Mark.[9] before dinner we were like others about the Doges palace to see his tables set out etc. he went in a large barge about 11 a clock to the Church of St Vito,[10] and after his return to St Marks, the senators at last a good number of them went in barge with him or another like it, and then followed him up to dinner, I had a full view to day of himself a very reverend well looking old gentleman of about four score,[11] his name is \Carlo/

[1] Gradasso, Muslim King of Sericana, described by Ariosto as the 'bravest of Pagan knights', who fights Charlemagne and drives him back to Paris with 100,000 vassals, all 'discrowned kings'.

[2] Melissa, a good sorceress, who engineers the love affair of Ruggiero and Bradamante. She was said to have been an apprentice of the dead magician Merlin, to whom Folkes next referred.

[3] Brunello, a dwarf and thief, who works for the 'Saracen' King Agramante and meets a sticky end.

[4] From *Orlando Furioso*, canto 4.40: 'and there is Prasildo, who came with Rinaldo from the East | And also Iroldo, another friend'.

[5] Psalm 137:1, 5–7, *Book of Common Prayer*. It is noteworthy that Folkes inscribed this in his travel notebook, as he was a known deist and freethinker. One wonders if this show of piety was a cover for some of his more subversive thoughts in his diary in case it was intercepted abroad.

[6] Folkes is referring to the annual feast of St Vitus on 15 June. Vitus was the son of a pagan senator of Lucania, and a child martyr during the reign of Roman Emperor Diocletian. The saint is appealed to for cures of epilepsy, referred to as St Vitus's Dance, and he is one of the 14 Martyrs who give aid in times of trouble. 'Sts. Vitus, Modestus, and Crescentia', *Catholic Encyclopedia*, https://www.newadvent.org/cathen/15490b.htm [Accessed 5 Dec 2022], *Venice,* pp. 115–16; see also Crouzet-Pavan, *Venice Triumphant*, pp. 192–3.

[7] The men's mask was called the *larva*, a white half-mask of waxed carton that extended below the nose, 'wedged against the forehead by a three-cornered hat'. If a woman went hatless, she wore a *morèta*, a black velvet or lace mask, held in place by clenching a small button between the teeth. The mask could be bejewelled. Usually, masks were worn with the *tabàro*, a black cloak, and *bauta*, a 'close-fitting hood that encircled the face and hid the neck'. See Johnson, *Venice Incognito,* p. 8.

[8] A *bahut* or dress for masquerading, a *domino*. See *OED*.

[9] When Folkes refers to the 'place' throughout the diary, he is referring to San Marco and its piazza.

[10] San Vio, built in 912 and originally dedicated to St Vitus and his partner in martyrdom, St Modestus. See Cotton, 'San Vio', http://churchesofvenice.com/dorsoduro2.htm#sanvio [Accessed 11 Dec 2023]. The church was closed in 1808 and demolished in July 1813, its fragments used to build the oratory. The oratory is now a private residence.

[11] Four times 20, or 80 years of age. See *OED*.

Ruzzini.[1] he was in his robes with the Corno on his head and a sort of Coif of white linnen under it;[2] when I saw him before he carried the corno in his hand and had only the coif on his head. 2 of the Senators Sig^{r} Marcello[3] and Sig^{r} ~~Cappello~~ \Pesaro/[4] who speak ffrench and the ~~better~~ \former/ some English were pleasd to come up to us \to/ take very kind notice of us. After dinner I receivd a visit from Mr Ducane[5] and his 2 sons who are lately come hither from Rome but go away to morrow morning. this Gentleman was my schoolfellow at Mr Cappels[6] above 36 years since and I never saw him before from that time. he expressed a great deal of pleasure at meeting so old an acquaintance and gave me some account of 2 more at the same School, whom I shall hope to see if it please God I live to see my home again. he thinks we 4 are all the survivours of that company, he seems an exceeding friendly good man and is now a very eminent Merchant in London. In the evening we were again at the Piazza[7] where we saw the masks walking backwards and forwards, and their were chairs set by people who let them for about 5 sols[8] a piece to those who wanted to set down, towards night, the Resident,[9]

[1] Carlo Ruzzini (1653–1735), a Venetian diplomat to Madrid, Vienna, The Hague, and the Ottoman empire. He was elected Doge on 6 June 1732, serving until his death in 1735.

[2] Folkes is describing the *corno ducale*, the unique hat of the doge. It was a stiff bonnet that was horn-like, made of brocade or cloth of gold and worn over a *camauro*, or cap of linen, made by the nuns of the convent of St Zechariah's church.

[3] Probably Benedetto Marcello (1686–1739), brother of Alessandro Ignazio (for whom, see below, p. 56, n. 1). Marcello's 'formal occupation was as a lawyer, and from 1706 he served on the Great Council of the Republic of Venice'; and as a provincial Governor of Pula 1730–33. See Berger, ed., 'Introduction,' *Cassandra by Benedetto Marcello*, p. ix. Marcello finished his career as a chamberlain of Brescia, where he died of consumption in 1739. He was also a gifted amateur musician, prodigious composer (several hundred cantatas are attributed to him), and poet, who could sing, and play harpsichord and composed for violin. See Selfridge-Field, *The Music of Benedetto and Alessandro Marcello*, pp. 5, 9.

[4] A member of the Venetian patrician family best known for the Palazzo Pesaro, built in 1679 by Baldassare Longhena, architect of the Salute, and for Titian's *Pesaro Altarpiece, S. Maria Gloriosa dei Frari* (1518), which 'made politically and socially significant statements about a prominent Venetian and his family'. The family was one of the most illustrious in Venetian history, coming to Venice in 1225, descended from Jacopo Palmieri of Pesaro. Their palace is now a modern art museum, the International Gallery of Modern Art. See Howard, 'Venetian Architecture', p. 789.

[5] Richard Du Cane (1681–1744) of London and Coggeshall, Essex. He was an MP 1715–22, and Director of the Bank of England 1720–30. Du Cane was in Venice with his sons, Peter (1713–1803) and Richard (1715–44). See 'Du Cane, Richard', in Ingamells, *Dictionary*, p. 316.

[6] Du Cane, like Folkes, was educated by the Huguenot, James Cappel (1639–1722), who had held the Professorship of Hebrew in the University of Saumur before coming to England as a religious refugee. James Cappel taught Folkes Greek, Latin, and Hebrew in the humanist tradition, as well as conversational French; this instruction would serve Folkes well when he assisted in editing Newton's work on biblical chronology. Cappel wrote to M. Le Clerc from Hillington Hall in February 1707 that his pupil Folkes was 'a choice youth, of penetrating genius, and master of the beauties of the best Roman and Greek writers'. Agnew, *Protestant Exiles from France*, II, p. 229; see also Roos, *Folkes*, pp. 26–7.

[7] Piazza San Marco. See Plate 2.

[8] *Soldi*, plural of soldo. See Appendix B for a table of Venetian coinage.

[9] In 1727–8, George II renewed the appointments of Elizeus Burges as 'Resident to the Republick of Venice', and of Neil Brown as Consul. See *The London Gazette* no. 6621, Saturday 4–Tuesday 7 November 1727, p. 1; no. 6646, Tuesday 30 January–Saturday 3 February 172[8], p. 1. Colonel Elizeus Burges was Secretary Resident twice: from 1719 to 1722, and from 1727 until his death on 14 Nov 1736. A resident was a diplomatic agent whose rank was below that of an ambassador, and 'a key part of this role as resident was to keep the secretary of state in England informed of notable visitors to Venice', and was of 'particular interest to the British government at this time as James Francis Edward Stuart, the Old Pretender, was attracting Jacobites to his cause in Italy'. See National Archives, *In Their Own Words*, II, p. 260.

Consul,[1] Mr Mace[2] and my self joynd at a fresco[3] shop till it was time to go home. this is the hottest day we have yet had.

Tuesday 16. I visited the Consul and the Resident. I was also at some glass grinders one named Selva[4] has some reputation here for optick glasses, he has made some small reflecting telescopes,[5] into which one looks behind but not throu[g]h a hole in the principal reflector but just above it the other directing the reflection obliquely. I saw some writing pretty distinctly against the side of a house, but design to try more per-[f. 4r] fectly by looking at Jupiters Satellites[6] which the Instrument if good should be capable of seeing. Mr Baillie[7] was to see me whilst I was out, I passd the evening first with Mr Mace at the Piazza, and then at Sigr: Giustinianis[8] where were besides ffather

[1] Neil Brown (*c.*1660–1740), Scottish merchant and British Consul in Venice from 1716 to 1738. From the *Register Cinque Savij alla Mercanzia* (15 April 1723), as likewise from Burke's, *Extinct Baronetage*, it may be inferred that this Neil Brown had also the Christian name of Robert, and that, having acted as Secretary Resident, he assumed that title in the patent, which created him a baronet, 1731–2, although the official post which he filled at Venice limited itself apparently to the consulate, so far as the evidence of Venetian documents is concerned. His wife, Lady Margaret Brown, became a leading patroness of Italian opera when they returned to London. See 'Attested Diplomatic Agents in Venice', in *Calendar of State Papers, Venice, Vol. 1*, pp. cli–clii; *British History Online*, http://www.british-history.ac.uk/cal-state-papers/venice/vol1/cli-clii [Accessed 22 Oct 2020]. For Brown's musical patronage, see Chrissochoidis, 'True Merit always Envy rais'd', pp. 79–83.

[2] Possibly William Mace (d. 1767), who would become Professor of Civil Law in Gresham College, and who was successfully nominated by Martin Folkes in 1745 as a Fellow of the Royal Society.

[3] The serial, *The Present State of the Republick of Letters*, indicated that a 'fresco shop' was one 'where they sell Limonade, and other cooling Liquors'. See *Present State*, VI, p. 253, in footnote. My thanks to Cesare Pastorino for alerting me to this source.

[4] The Selva family was established in Venice by 1700, one of 'several dynasties of Italian opticians who made simple telescopes and microscope lenses, in response to the demands of naturalists and the medical profession'. Domenico Selva's (d. 1758) optical workshop was near Piazza San Marco. His son Lorenzo documented his father's work in a book, the *Esposizione delle comuni*, among other works on his productions. These instruments included surveying levels, refracting and reflecting telescopes. Lorenzo became the *ottico publico*, public optician of the Republic of Venice, after his publication of a book on the manufacture of achromatic lenses using flint glass, which enabled him to build superb telescopes. See 'Domenico Selva', Museo Galileo, https://catalogue.museogalileo.it/biography/DomenicoSelva.html; see also McConnell, 'Instruments', pp. 347–8; Lualdi, 'Venetian Makers', pp. 10–13.

[5] The reflecting telescope, of course, was an invention of Newton's, presented to the Royal Society in 1671, followed the next year by his paper published in *Philosophical Transactions* about light and colour. 'In his catadioptrical telescope, as it was called, Newton placed a concave spherical mirror at the base of the tube, and near the other end he placed a very small, plane secondary mirror inclined to the axis so that it would reflect the rays through a hole in the side of the tube, which contained a plano-convex lens for viewing the image.' Shapiro, 'Newton's Optics', p. 167.

[6] Otherwise known as the Galilean satellites, Io, Europa, Ganymede and Callisto, first seen with a telescope by Galileo Galilei in December 1609 or January 1610, recognized by him as satellites of Jupiter in March 1610. As they were the first objects found to orbit a planet other than the Earth, their existence was a challenge to geocentrism. Galileo published his discoveries in the *Sidereus Nuncius* (1610), where he called the satellites the 'Medician Stars' in honour of his patron Grand Duke Cosimo II of Tuscany, a member of the Medici family. See Galileo, *Sidereus Nuncius*, ed. Van Helden, pp. 14–16.

[7] Possibly George Baillie (1664–1738), husband of Lady Grisell Baillie, who travelled to Italy from 1731 to 1733 with his numerous children and grandchildren. The Baillies arrived in Venice on 11 June 1733, and were in Padua by 22 June, returning to Venice for the month of July. See 'Baillie, George', in Ingamells, *Dictionary*, p. 40.

[8] Girolamo Ascanio Giustiniani (1697–1749), member of the first family of Venice, Capitano di Verona 1728, Vice-Podestà di Verona 1729, Capitano di Padova 1733. Verona belonged to the Republic of Venice, and the Capitano, usually an army officer appointed from Venice, was the commander of the local garrison responsible for public order and military preparedness. The podestà was the ruler of the city state, governing in

Crivelli,[1] Prince Gonzaga,[2] Sig[r] Loredano,[3] and another Noble Venetian whose name I forgot.[4] Sig[r] Giustiniani is as well bred a good natrd gentleman as I have seen, and is well versd in most parts of Polite learning. Father Crivelli brought me a compliment from Count Schuylenberg[5] with an invitation to see his pictures, that are a celebrated collection here.

Wednesday ~~16~~ 17. I returned visits to Dr Michaelotti[6] and Sr Rob. Brown,[7] as also to

---

Venice's name, an outsider, 'disinterested and temporary', to diminish governmental corruption; the vice-podestà was his right-hand man. O'Connell, *Men of Empire*, pp. 42, 57. In 1716, Giustiniani married Adriana, daughter of Alvise Barbarigo, a patrician of Venice. He was a man of vast culture, friends with Antonio Conti, a violinist and lyricist for the compositions of Benedetto Marcello, and known for his salons for intellectuals and scholars from different social backgrounds. The Palazzo Giustinian overlooks the Grand Canal next to Ca' Foscari. See Litta, 'Tavola III', in *Famiglie celebri di Italia*. When Folkes returned from Italy, Giustiniani wrote to him on 7 July 1737 to ask him for a silver watch made by the London horologist George Graham, as well as to affirm his deep friendship with Folkes and the esteem he had for him. RSL, RS MS/790/63.

[1] Venetian Somascan cleric, Father Giovanni Crivelli (1691–1743). The Somacans or Somaschi was the name of a 'charitable religious congregation of regular clerics, in the sixteenth century founded by St Jerome [Gerolami] Emiliani with the motherhouse at Somasca (Venice), whence the name.' 'Somaschi', *Catholic Encyclopedia* https://www.newadvent.org/cathen/14140a.htm [Accessed 11 Dec 2023]. Crivelli wrote *Elementi di fisica* (1731), a work, 'characterized by a relatively elementary mathematical approach; however, his precise and concise textbook provided a historical perspective that effectively framed the development of Newtonianism' in Italy. Folkes would come to a dispute with Crivelli, however, about Newtonian optics, as recorded later in his journal. See Macuglia, 'Newtonianism and Information Control', p. 121. Crivelli was recommended for election to the Royal Society by Folkes on 25 Oct 1733. RSL, RS/EC/1733/08. The certificate was signed by John Machin, Sir Hans Sloane, and Cromwell Mortimer. Crivelli was elected to the fellowship on 24 Jan 1733/4.

[2] Possibly Giuseppe Maria Gonzaga (1690–1746), the last reigning Duke of Guastalla and member of the House of Gonzaga.

[3] A member of the Loredan family, possibly Francesco Loredan (1685–1762), who served as the 116th Doge from 1753 until his death. He was head of the Santo Stefano branch of the family. The Loredan family was amongst the first families of Venice, a political dynasty that produced many doges and statesmen.

[4] Folkes later wrote in the right-hand margin in pencil 'Sig[r] Vendramin', a member of a rich merchant family of Venice, known for their status as art collectors, patrons and theatre owners. Gabriele Vendramin was one of the first private collectors of Italian drawings in Northern Italy in the early 16th century. Karet, *Antonio II Badile Album*, p. 158. The Teatro Vendramin, built in the 1720s, is now the Teatro Godoni.

[5] Johann Matthias Reichsgraf von der Schulenburg (1661–1747), a German aristocrat and field marshal who served in the Venetian army at the beginning of the 18th century. In 1716, Schulenburg defended the Venetian island of Corfu, which was threatened by the Ottoman empire. In this terrible siege, the Turks were said to have lost 17,000 men, 56 pieces of artillery, and most of their war materials. At the close of the war, Schulenburg introduced improvements to the Venetian armies. He retired as a patron of the arts in Venice. For more biographical information and his relationship with Folkes, see Introduction, pp. 17–18 above.

[6] Physician, Pietro Michelotti (1673–1740), FRS, member of the Institute of Bologna, Professor of Philosophy and Medicine at Pisa, and the premier doctor of the Venetian Republic. Michelotti was a disciple of Bernoulli and, during the 1720s, had kept a correspondence with Royal Society secretary, James Jurin, about the muscular movement of animals, the separation of fluids in the body, and foot and mouth disease in cattle, sending him several of his books. See RSL, RS/EL/M2/61. When he was in Italy, Folkes added a note of introduction to the Royal Society, appending it to Michelotti's paper concerning a cure for madness. The paper was read to the Royal Society on 12 December 1734. See RSL, RS/EL/M3/32. One wonders if the paper was of interest to Folkes as his wife, Lucretia, was increasingly evincing signs of mental illness.

[7] Sir Robert Brown, 1st Baronet (d. 1760), of Wiggenhall St Mary, Norfolk, was a politician and self-made merchant who was in Venice *c.*1698–1734. At Easter 1725, he married Lady Margaret Brown, née Cecil, sister of Dr Charles Cecil, Bishop of Bangor. On 11 March 1731, he was made a baronet. The Browns owned a country house at Paluello, where Folkes would see them in October 1733. Brown's brother was Venetian Consul, Neil Brown. Brown and his family left Venice in September 1734. Ingamells, *Dictionary*, p. 140.

Mr Smith[1] just come to town, whilst I was out I was honourd with a visit from Sig[r] Giustiniani which is lookt as a very great favour he seldom stirring abroad. S[r] Ro. Brown and his Lady were likewise at our house. I endeavourd at a workmans to get the true dimension of a Venetian foot[2] but have not succeeded, as[3] [*sic*] I could do was that a watchmaker set me out on my ruler an inch from his own scale, I make it 1 27/100 of an English Inch. at his rate the foot is 13 English Inches 620/1000 pretty near the measure told me in round numbers by the Consul. which was 13 inches 5/8 or 625/1000.

I passed the evening at the place S[t] Mark, with The Resident Mr Mace, Mr Haynes,[4] etc, and at my return found the Consul at my house who supt with me he communicated a paper of the observations he had made on the money of his country.[5]

Thursday 18. I have been very ill all night of a pain in my side however I went this morning, with Mr Feake[6] and Mr Mace, to see the Palace[7] of Sig[r] Sagredo,[8] it is a fine

[1] Joseph Smith (*c.*1674–1770), the merchant banker, art collector, virtuoso, publisher and dealer. He lived in Venice for 70 years, working as a merchant banker and for some years was British Consul (1744–60). Smith, however, is best known as an art patron and dealer. The most prominent part of Smith's collection was contemporary paintings and drawings, particularly those by Rosalba Carriera and Canaletto. Most of his collection of paintings, gems, coins and medals was sold for £2,000 to King George III in 1760, negotiated by James Stuart Mackenzie (who had been British Envoy in Turin). Smith's collection remains in the Royal Collection today, except for books and manuscripts which were transferred to the British Museum in 1823. See Ingamells, *Dictionary*, pp. 869–71. See also Vivian, *The Consul Smith Collection*, pp. 11–41; Vivian, *Il Console Smith.*

[2] The Venetian foot or *piede* (plural *piedi*) had a length close to 347.76 mm, and this was divided into 12, giving an inch or *oncia* of 28.98 mm. Folkes published works on standard measures in different countries, including those used in coinage. See Cotty, *Aide-mémoire*, II p. 899; See also O'Brian, *Historical Italian Metrology,* http://www.claviantica.com/Historical_Italian_Metrology.htm [Accessed 27 Oct 2022]; Folkes, 'An Account of the Standard Measures'; Folkes, *Tables of English Silver and Gold Coins.*

[3] Folkes may have meant 'all I could do'.

[4] Possibly Samuel Haynes (d. 1752), eldest son of Hopton Haynes. He went to King's College, Cambridge, in 1719, receiving a Doctorate of Divinity in 1748. Haynes was rector of Hatfield, Herts from 1737 until his death, as well as serving as a canon of Windsor from 1743 until his death. Ingamells, *Dictionary*, p. 478.

[5] Folkes had in his own collection 13 ounces of Italian money for his studies of numismatics, including, 'money of several popes from the scudo to the half paul 47, weight'. See the auction catalogue of his collection by Langford: *A catalogue of the collection ... of coins*, p. 17. See also Appendix B on Venetian coinage.

[6] William Mace travelled with Samuel Feake, senior administrator of the East India Company and member of the Dilettanti, arriving in Venice by 22 May 1733. (The Society of Dilettanti was formed in 1732. Initially a dining club, the Society became a significant cultural force, organizing key archaeological expeditions, as well as having a role in the formation of the British Museum and the Royal Academy. See Kelly, *Society of Dilettanti*, for the best recent overview.) They passed through Padua on the way to Rome, where Folkes renewed his acquaintance with them in November (31 Oct; 1 Nov 1733). Before coming to Venice, Feake spent much of his career in India. He was Deputy Chairman of the East India Company in 1739, 1743, and 1746; and was twice Chairman – in 1740, and again in 1744. Cotton, 'Editor's Note Book', p. 64. See also Ingamells, *Dictionary*, p. 350.

[7] The Palazzo Morosini Sagredo. This 13th-century Byzantine-Gothic palace was purchased in the early 18th century by Gherardo Sagredo (d. 1676), the grand-nephew of Doge Nicolò Sagredo (d. 1676). See Brown, *The Venetian Bride,* p. 218. Gherardo Sagredo modernized the interior, a process continued by Zaccaria Sagredo (1659–1729), who commissioned stuccoes by Abbondio Stazio and Carpoforo Mazzetti-Tencalla and ceiling decorations by Gasparo Diziani. Zaccaria was nephew of Niccolò Sagredo, Doge of Venice 1675–6, and Alvise Sagredo, Patriarch of Venice 1678–88 (the ordinary bishop of the Archdiocese of Venice, and one of four patriarchs in the Latin Church of the Catholic Church). Zaccaria's family wealth allowed him to assemble Venice's largest private holding of paintings and drawings, adding to paintings of the 16th and early 17th centuries that may have come to him by inheritance. It is speculated that he was one of the early patrons of Tiepolo, Piazzetta, and Canaletto. Peck, *Period Rooms,* p. 69.

[8] Gherardo Sagredo the Younger? (d. 1738), nephew and heir of Zaccaria Sagredo. Folkes was fortunate to see his house and its collection intact. After Gherardo's death, the collection was left to his aunt, Cecilia Griani

house with several noble rooms some of which are newly fitted or fitting up. the Staircase is painted with the war of the Giants, in a bold but somewhat extravagant manner;[1] Several rooms are furnisht with paintings some very good, there are a great many of Bassan,[2] some but I think not very capital of Titian and Paulo,[3] with 2 or 3 called correggio,[4] 2 Salvator Rosas viz the Democritus and Plato,[5] but black and unpleasant. there is a small Library neatly fitted up, and another room furnisht with drawings and frames and glasses,[6] some very good, one with a pen very fine of the last supper of Rafael.[7]

Sagredo, who began to dispose of it. We do not know exactly when the dispersal began, but 'it must have been subsequent to [Francesco] Algarotti's presumed efforts to act as a middleman for its acquisition by the Elector of Saxony, a project destined to fail'. We do know that in 1743, when Algarotti made his inventory of the collection's drawings, it was still intact; the inventory listed 4,115 drawings bound in 50 albums, including many by Veronese, as well as works on paper by Rembrandt and Guercino. Consul Joseph Smith and his successor, Consul John Udny, acquired some of the choicer parts of the collection of drawings, some of which were sold to George III and are now part of the Royal Collection in London. See Rearick, 'Drawings from the Sagredo Collection', p. 133. See also Binion, 'Algarotti's Sagredo Inventory', for a transcription of the inventory of drawings; Borean, 'Paolo Veronese', p. 95.

[1] Folkes is describing *La Caduta dei Giganti*, or *The Fall of the Giants* (1734), a fresco covering the ceiling and walls of the grand staircase in Palazzo Sagredo, which was completed by Pietro Longhi (1701–85), under the tutelage of Antonio Balestra. It depicts Jupiter's defeat of the rebellion of the Giants against the Gods. Although a bold and grandiose piece, it was a relative critical and artistic failure. The failure of this fresco is one reason Longhi moved from history painting to genre scenes, becoming known as the 'Venetian Hogarth'. He was the main painter of intimate, everyday-life scenes in 18th-century Venice. He also painted *capricci,* 'views in which the factual was suggestively fused with the fanciful'. Favila et al., 'Venetian Art', pp. 849, 854–5. See also Plate 3.

[2] Probably Jacopo Bassano (*c.*1510–92), also known as Jacopo dal Ponte, an Italian painter. He was one of the first genre and pastoral landscape painters in Italy, some seeing him as a pioneer pointing the way to the Dutch landscape paintings of the 17th century. See Rearick, 'Drawings from the Sagredo Collection', p. 132; Aikema, *Jacopo Bassano,* p. 4; Gottardo, 'Il gusto collezionistico', p. 246.

[3] Probably Paolo Veronese (1528–88), the Italian Renaissance master based in Venice. Several of his drawings were also in the Sagredo Collection, as described by Rearick, 'Drawings from the Sagredo Collection', *passim.* In 1750/1, Consul Smith bought a volume with 22 drawings by Paolo Veronese from the Sagredo, which was sold to the Royal Collection. See Gottardo, 'Il gusto collezionistico', p. 241.

[4] Antonio da Correggio (1489–1534), the leading painter of the Parma school of the High Italian Renaissance. Algarotti noted his drawings in his inventory of the Sagredo Collection. Zaccaria Sagredo bought the drawings from the Bonfigilioli Collection in Bologna. Binion, 'Algarotti's Sagredo Inventory', p. 395.

[5] These paintings were part of a series of pendant portraits of ancient philosophers that Salvator Rosa (1615–73) painted in his mid-thirties. See da la Fuente Pedersen, 'Salvator Rosa's *Democritus*', p. 2. Their tenebrous quality was due to Rosa's exploration of the melancholic temperament and genius. Nicolò Sagredo was known to collect such pendants by Rosa for the Sagredo Collection; as Golahny has noted: 'in 1652, while serving as Venetian ambassador to Rome [Nicolò] Sagredo purchased two large pendant canvases from Rosa, *Democritus in Meditation* and *Diogenes Throwing Away his Bowl* (1651)', now in the Statens Museum for Kunst, Copenhagen. Golahny, *Rembrandt*, pp. 208, 210. These are probably the paintings to which Folkes referred. See Plate 4.

[6] This could refer to ornaments made of glass, or to looking glasses, or mirrors.

[7] This is probably the *Last Supper* (*c.*1514), attributed to Raphael, now in the Royal Collection in London, accession number RCIN 912745. The sheet's poor condition indicates it was framed and displayed for several years, unglazed. It is possible that it 'might therefore be one of two drawings listed in the 1696 inventory of Palazzo Bonfiglioli in Bologna, the source of a number of old master drawings now in the Royal Collection'. If this is true, and the drawing was in the Bonfiglioli family collection in Bologna, this collection was bought by Zaccaria Sagredo in 1728, and from Folkes's description, subsequently displayed in the manner presumed. Consul Joseph Smith then acquired the drawing, from whom it was bought by George III. See *The Last Supper* (*c.*1514), The Royal Collection Trust, https://www.rct.uk/collection/912745/the-last-supper [Accessed 2 Nov 2022]. See also Plate 5.

a pretty many of Julio,[1] Parmegiano, and others the 4 Scenes of Canals, pretty large.[2] in the same room are about 360 large Volumes bound in Turkey, and said to contain Prints and drawings, but being in Glass cases lockd I could not see any of them.[3] there are about his palace some very pretty modern statues, one particularly struck me it is a sort of vestal by Corradini, there is a veil over the face through which the features appear in a very easie agreable manner.[4]

Father Crevelli did visit me to day, and after dinner we waited on Marshal Schuylenberg, who received me in an exceeding civil kind manner, he is a very agreable nobleman, and talks very freely on all subjects, particularly religion which I was surprizd at in [f. 5r] this place. he is a protestant, and has done very great services to the Republic particularly in defending Corfu against the Turk. he shewd me 2 horse tails he had taken from the infidels. he is now very old but takes great delight in shewing civilitys to strangers particularly to the English, he talkt to me with great kindness and offered me letters to any part of Italy.[5] he walkd all over the house to shew me his collection of pictures which is extreamly large, there are some very good but I do not think upon the whole he has collected with the greatest judgment. he has particularly been very curious in Plans of all sorts, of which he has a vast collection and has been at an immense expense to procure some of them, he shewd me two very large solid ones[6] like those the K of France has in the Galery of fortifications they are of Corfu[7] and [*blank*][8] in Dalmatia

[1] Probably Guilio Romano, original name Giulio Pippi (1499–1546), a late Renaissance Mannerist painter, and the principal heir of Raphael.

[2] The work of Giovanni Antonio Canal (1697–1768), known as Canaletto, renowned for his *vedute*, or views, of the canals of Venice. Zaccaria Sagredo was one of Canaletto's earliest patrons, first seeking and acquiring his work in the 1720s.

[3] The Biblioteca Sagredo, which overlooks the Grand Canal, still possesses the 18th-century curtained bookcases in white and light green wood with gilded carvings, and the letters of the alphabetical directory painted on the cases still stand out clearly. 'La Biblioteca Sagredo', Ca' Sagredo, https://www.casagredohotel.com/moments-awards/back-to-the-past/museum-arts/ [Accessed 28 Oct 2022]. The Sagredo drawings were kept in albums of fine Venetian paper to which the artwork, 'sometimes several to a folio, were tipped down with four heavy pulp tabs', with a detailed inventory on the margin of the mount next to the drawing likely by Zaccaria Sagredo. Rearick, 'Drawings for the Sagredo Collection', p. 132.

[4] Antonio Corradini (1688–1752), a Venetian sculptor noted for his sculptures of veiled women, the *donne velate*, the drapery inspired by that of classical statuary. Klemenčič indicates that, following Zaccaria's death in 1729, his collection was scattered and the sculpture unidentified. However, 'in 1755 the inventory evaluators Giambattista Tiepolo and Antonio Gai recognised two reclining statues of *Venus and Adonis*, as well as a veiled statue of *Religion* as the work of Corradini ... there is significant – albeit circumstantial – evidence that the Sagredo *Religion* is identical with the one now in the Louvre.' The Sagredo *Religion* thus appears to be the sculpture that Folkes describes in his diary. See Klemenčič, 'Antonio Corradini', pp. 103, 113–14. For the Louvre statue, see Cogo, *Antonio Corradini*, pp. 170–77, cat. 10; for its provenance, see Klemenčič and Clemente, 'Caritas Romana'; Bacchi, 'Antonio Corradini', pp. 132–3.

[5] Letters of introduction.

[6] Early modern technology resulted in the production of large-scale reliefs of plaster and papier mâché, including a series of extremely detailed models of fortresses in the Stato da Mar dating to 1692. The models are now at the Naval Museum in Venice. Cosmescu, *Venetian Renaissance Fortifications*, p. 7.

[7] Fortezza Vecchia or the Old Fort of Corfu, which covers the promontory that had originally contained the old town of Corfu. The fort successfully repulsed all the major Ottoman sieges of 1537, 1571, and 1716. After 1718, Schulenburg supervised the building of further defences around Corfu, having two large forts built on the Salvatore and Abraham hills. 'Due to these Venetian defensive efforts, Corfu was the largest and most important territory in the classical Greek world never under Ottoman rule (along with the majority of the Ionian islands).' Cosmescu, *Venetian Renaissance Fortifications*, p. 140.

[8] Folkes had forgotten the name of the fort, and he left a blank space here.

both of which have been fortifyd by himself; the former is both by nature and art one of the stongest places in Europe, and the Venetians have spared no expence as it is the great bulwark against the turk, his plan is above 20 foot one way and very well done so as to be like seeing the place it self. When I took my leave of the Marshal I went with Father Crivelli to the Padri Sommaschi of the Madonna della Salute,[1] which congregation he was of before he was in his triennial office of Rector of the Mendicanti,[2] here he showd me the Convent which is very neat and their Library very handsomely fitted up a long room with a galery, the shelves and pillars of Cypress, and out of the Galery here is an other ~~g~~ long room which was their old library they are both almost full and of pretty good books, the Library keeper an old reverend man of fourscore received me with great civility and showd me what they had most considerable, particularly their Mathematical books. and a very large and good collection of Drawings which were collected some time since by a Noble Venetian who lodgd them in this Library, they are now putting them in order in Volumes, they have also a great number of Prints of the same collections but these I had not time to look at. not being very well I returnd early to get to bed.[3]

ffryday. 19. having had a good night and being much better of my indisposition I went out about 11 a clock to the Church of S[t] Catherine[4] when I saw the fine picture of her marriage by Paul Veronese it is a noble piece, and in high preservation, they have built a dome over the Altar which lights it very well,[5] the dome is painted with her Apotheosis by Brusaferro;[6] the whole Church is full of pretty good paintings, several of her history are pretty pictures. the consultation of S[t] Catherines mother about marrying of her is by

[1] Santa Maria della Salute, a Baroque votive church located at Punta della Dogana in the Dorsoduro *sestiere,* or district, of Venice. Designed by Baldassare Longhena (1598–1682), its central plan and Palladianism has made it a recognizable symbol of Venice. See Hopkins, *Santa Maria della Salute.*

[2] Crivelli had entered the Somascan Order as a regular cleric at the Santa Maria della Salute, and then went to the seminary of Murano to teach eloquence and rhetoric, where he became rector. See 'Crivelli, Giovanni Francesco', *Dizionario Biografico degli Italiani*, https://www.treccani.it/enciclopedia/giovanni-francesco-crivelli_%28Dizionario-Biografico%29/ [Accessed 4 Nov 2022].

[3] The convent library of Santa Maria della Salute, famous for its size and valuable holdings, disappeared following the French invasion of Italy in 1796 and the suppression of the convents. Folkes referred to seeing volumes of prints and drawings. One of the volumes of drawings in the collection was by Giambattista Tiepolo (1696–1770). Tiepolo gathered his drawings into an album and presented them to his son, Guiseppe Maria Tiepolo, a priest, for the convent library. See Lehman and Szabó, *Eighteenth-Century Italian Drawings*, no. 78.

[4] Santa Caterina dei Sacchi, an 11th-century church in the Cannaregio district of Venice, founded by the order of the Friars of the Penitence of Jesus Christ, called the Sacchini friars due to their sackcloth robes. It was rebuilt in the mid-15th century, and in 1807, the church and monastery were suppressed by Napoleon. See Cotton, http://churchesofvenice.com/cannaregio2.htm#santacat [Accessed 5 Nov 2022].

[5] This is the *Mystic Marriage of Saint Catherine* (*c.*1565–75), by Paolo Veronese (1528–88), which was painted for the high altar as a means for the nuns to reflect on their own chastity. The painting was taken from Santa Caterina and given to the Accademia in 1918. Folkes's comment that the dome was well illuminated was corroborated by his contemporary Zanetti, who, in his catalogue of publicly accessible pictures in Venice, noted the Veronese painting was, 'perfettamente illuminata, e veduta [perfectly illuminated and seen]' by the dome. Zanetti, *Descrizione*, p. 386. See Plate 6.

[6] Girolamo Brusaferro (1679–1745), a Venetian artist who painted the saint in all her glory, as well as the four virtues in the corners of the nave. He was a pupil of Niccolò Bambini and Sebastiano Ricci. In 1733, he was described by Zanetti in his catalogue as an attentive and disciplined painter whose painting combines beauty and intelligence, and who is worthy of much praise. Zanetti, *Descrizione*, p. 60.

Palma.[1] and here is also a good picture of Tobit curd of his blindnes by Pietro Vecchia.[2] the Nave of the Church is painted with stories of the Scripture by Andreas Vicentinius.[3] A nunnery [f. 6r] joyns to this church: we saw several of the Nuns at the Grate by the chancel. from hence I went to visit Count Bilgre, and whilst I was out Dr Michaelotti was to see me. after dinner I was with my wife at Sig^r Cananis garden in the Guidaica,[4] it is pretty enough and has a good many exotick plants, the house has been pretty well painted in ffresco on the outside but it is now worn out. I passd the evening with the Resident and Mr Mace at the Piazza.

Saturday. ~~19~~ 20 I was visited in the morning by Dr Michaelotti[5] who presented me with his book,[6] after he was gone I went to see Sr Rob Brown. and in the afternoon I went with the Resident and Mr Mace to the Island of Lido.[7] This is a long Island, about 14

[1] Painter and engraver Jacopo Palma, called Palma Giovane (1548–1628). Palma was from a family of artists, the son of painter, Antonio Palma, and the great-nephew of the more well-known painter, Jacopo Palma called Palma Vecchio. Rylands, 'Palma family', *Grove Art Online*, https://doi.org/10.1093/gao/9781884446054.article.T064911 [Accessed 5 Nov 2022]. See also Langé and Scola, *La dimora dei patriarchi*; Zeri and Gardner, *Italian Paintings*, pp. 45–7. Folkes was referring to Palma's *Saint Catherine's Mother Consulting Wise Men about her Daughter's Marriage* (now in the Palazzo Patriarcale, home of the patriarch of Venice). Zanetti confirms Folkes's observations, but also indicates Palma did a cycle of the story of St Catherine's life and her burial. Zanetti, *Descrizione*, p. 387.

[2] Pietro della Vecchia (1603–78), painter, director of an art academy, connoisseur of ancient drawings and paintings, and member of the libertine Accademia degli Incogniti (Academy of the Unknowns). Muttoni, 'Pietro della Vecchia', National Gallery of Art, Washington D.C., https://www.nga.gov/collection/artist-info.6893.html [Accessed 5 Nov 2022]; See also Aikema, *Pietro della Vecchia*. Folkes is referring to the biblical story where the archangel Raphael tells Tobias that the burnt heart and liver (of a fish, which had attacked him), can drive out demons and the gall can cure blindness; Tobias uses this information to cure his father Tobit's eyesight.

[3] Andrea Vicentino (1542–1617). Vicentino moved to Venice in the mid–1570s and registered in the 'Fraglia', or guild, of Venetian painters in 1583. The cycle of stories painted on the ship nave of the Santa Catarina included scenes from the life of Moses and other Old Testament stories. As Zanetti indicated, in the middle nave, there are Old Testament stories in various sections divided with ornaments of colonnades and various chiaroscuric statues; all the works are by Andrea Vicentino. Zanetti, *Descrizione*, p. 388.

[4] Probably Count Carl Gustaf Bielke (1683–1754), Governor of Västernorrland in Sweden from 1727 to 1739, Ambassador to France and one of the ten children of Nils Tureson Bielke (1644–1716) and Eva Horn af Björneborg. He was also a renowned map collector and bibliophile with a collection of over 8,000 books, still housed in Skokloster Castle in Uppland. He also seemed to have had Masonic proclivities, so he and Folkes had many common interests. See 'Map Collecting in Sweden-Finland', in *The History of Cartography*, p. 778; Lindberg, 'Bibliotecket på Skokloster', pp. 133–72. The Guidecca refers to an island to the south of the city proper, which in the 18th century had many gardens and country houses. As the 18th century progressed, the Venetian nobility abandoned such island villas for country houses on the Brenta for the *villeggiatura* (holidays). See Goy, *Venetian Vernacular Architecture*, pp. 216–17.

[5] A cross was put at the end of this line in the right-hand margin.

[6] The auction catalogue of Folkes's library showed that he owned, *Jacobi Jurini regiæ scientiarum Londinensi societati à secretis dissertationes De motu aquarum fluentium contra nonnullas Petri Antonii Michelotti animadversiones defensio* (1724). James Jurin (1684–1750), a physician and mathematician, who discovered the principles of capillary rise, was an ardent Newtonian, defending Newton on the issue of *vis viva*. (For a discussion of the *vis viva* dispute, see Introduction, p. 12 above.) Jurin also debated Michelotti about fluid dynamics, and this book by Michelotti was a response. It seems likely this was the book that Michelotti gave Folkes. See *Catalogue of the entire and valuable library*, p. 61, lot number 1893. For Jurin, see Rusnock, ed., *Correspondence of James Jurin*, pp. 8–22.

[7] One of two barrier islands to Venice, the other is Pellestrina. They both form the central part of the coastline of the lagoon on the Adriatic. The Lido served mainly as a sea wall. As Edward Wright commented, 'the Word in the general Acceptation signifies no more than Shoar; and this Lido ... keep[s] the main Force of the Sea from much affecting the *Lagune*'. See Wright, *Some Observations Made in Travelling through France, Italy, &c*, I, p. 44.

miles they say,[1] but on the end twas not above 3 or 4 hundred yards over it thats in the Lagune from the Adriatick and all ships that come in pass a narrow opening at one end of this Island,[2] there is a fort on another Island at the end next venice,[3] and a town called Malamoco[4] at the further end. near here the ships lie that are making their quarantine.[5] upon the Island is at the end twas a Convent[6] and also the Barracks for the Venetian soldiers none being allowd to come into the city of Venice.[7] there is a platform with watch towers, round that end of the Island and beyond it a rampart to divide this part from the rest. here the English and German of the reformd religion are allowd to bury,[8] there are some gravestones somewhat like those of Nuremberg[9] over 5 or 6 German Gentlemen of distinction but none that I saw any of the English. a Brother of Ld Bolingbroke[10] lies here. at my return I went to Sigr Giustinianis where I passd the remaining part of the evening.[11]

[1] The Lido is, in fact, 11 km or 7 miles long.

[2] The Lido inlet is at the north end of the island, which separates it from Cavallino/Punta Sabbioni. At the southern end there is the Malamocco inlet, which separates it from Pellestrina. See the Atlas of the Venetian Lagoon, http://cigno.atlantedellalaguna.it [Accessed 6 Nov 2022].

[3] The defence of Venice via fortifications took into account geography as well as the depth and shape of the lagoon's seabeds. Towards the end of the Venetian Republic, in 1797, there were five fortifications positioned to 'defend the harbour entrances (from north to south: the forts Sant'Andrea, and San Nicolò at the northern end of the Lido, the Malamocco fort, also called Alberoni at the south end of Lido, the fort of San Pietro in Volta just north of the coastal strip of Pellestrina, and, initially the fort of San Felice in Chioggia) in addition to that series of mini-octagonal fortifications.' Marzo, 'A Theme', pp. 46, 57–8.

[4] The village of Malamocco on the lagoon side coast, in the central southern part of the island, was the only significant settlement on the Lido until the 19th century.

[5] The 40-day isolation (*quaranta*) from infectious disease imposed by the Venetian health office (*Provveditori alla sanità*). The time period of 40 days was taken from Hippocrates, who thought it would take 40 days for acute illnesses to reveal themselves. In the 15th century, the Lazzaretto Nuovo was established, where passengers from suspected locales were kept in quarantine before being allowed to enter the city. The Venetians also instituted *bollettini* or *fedi di sanità*, health passports, to control the ingress of travellers. Salzberg, 'Renaissance Lockdown', *History Workshop Digital Magazine*, https://www.historyworkshop.org.uk/venice-plague/ [Accessed 11 Nov 2022]; See also Bamji, 'Health passes', pp. 441–64.

[6] Monastery of San Nicolò, a Benedictine church, dedicated to Saint Nicholas of Myra which was founded in 1044 and rebuilt in 1626. See Cotton, 'San Nicolò', http://churchesofvenice.com/islands.htm#sannic [Accessed 8 Nov 2022].

[7] The 'Guglielmo Pepe' Barracks, formerly Palazzo dei Soldati in the Venetian Republic, constructed in 1591. The barracks enclosed a large fortified square in the centre of which there is a well, and housed the 'Fanti da Mar', a militia to carry out naval combat and landing operations, some of the first amphibious troops.

[8] The Protestant cemetery of Venice was originally established on the Lido in 1684. This was where the English who died in Venice were buried, including the Resident, Elizeus Burges, in 1736. See 'Venice In Peril', https://www.veniceinperil.org/religious-art/tombstones-of-sir-joseph-smith-and-other-british-residents-in-the-lido-cemetery/ [Accessed 11 Dec 2022]. (The tombstone has since been moved to the Catholic cemetery on the Lido, due to the construction of the Lido Airport.) In 1719 the German community was granted a burial site on the Island of San Cristoforo, which was enlarged and used until 1807. At this point, Napoleon ordered all churchyard cemeteries in Venice closed, as well as ordering a central burial ground on San Cristoforo. See *Venetian Protestant Cemetery*, https://www.veniceprotestantcemetery.com/en/home [Accessed 8 Nov 2022].

[9] Folkes spent three days in Nuremberg before he continued his travels to Italy. His travel account book details the stopping places of his entire journey and days spent in each locale. See Appendix A, a transcription of this portion of MS NRS 20658, Norfolk Record Office.

[10] Henry St John, 1st Viscount Bolingbroke (1678–1751), leader of the Tories. His brother was George St John (1693–1716), who died in Venice when his parents were still living.

[11] A cross was put at the end of this line in the margin.

Figure 9. Isabella Piccini, *The Mausoleum of the Barbarigo Family in Venice* (1692), engraving, 43.2 × 59.7 cm. Biblioteca Nazionale Marciana, Venice. The engraving is from the printing house of Vincenzo Coronelli. The medallions of the two doges in their distinctive corno ducale are at the bottom of the print. Showing its original position in the Santa Maria della Carità where Folkes saw it. Courtesy of the Biblioteca Nazionale Marciana, Venice.

Figure 10. Master of the Barbarigo Reliefs, *Sculpture of the Kneeling Doge Agostino Barbarigo* (1485–1515), marble, 120 cm. Sacristy Museum, Santa Maria della Salute, Venice. This is the last remnant of the Barbarigo family monument that Folkes saw in Venice in Santa Maria della Carità. Author's photograph.

Sunday 21. I went in the morning no further than to take leave of Mr Baillie and his family who go away to morrow. After dinner I called at the Church of la Carita,[1] where on the left hand going in is a magnificent tomb of the Brothers Marcus and Augustine Barbadico [*sic*] 72ᵈ and 73ᵈ Dukes of Venice[2] the Monument takes up one side of the Church and consists of 3 noble Arches in the midmost is an altar and in the 2 others the tombs of the 2 brothers with their effigies; lying along in their ducal robes;[3] on the other side the Church is another fine Monument for Nicol da Pontè 86ᵗʰ Duke,[4] it is also a sort of triumphal Arch in the middle part is his sarcophagus adornd with his bust, and 2 statues in niches \on either side/ between the Cornithian Columns that support the Pediment.[5] The pictures in this Church are not very considerable, the best is a raising of Lazarus by one of the Bassano,[6] but hard by adjoyning is the Scuola de la Carita in which are several good modern paintings, and one old celebrated one, the presenting of the [f. 7r] Virgin by Titian,[7] she is on the steps of the temple the figures are hardly 3 feet high there are several

[1] Santa Maria della Carità, consecrated in 1177 by Pope Alexander III, and the site of an annual procession on 3 April by the doge and Signoria, which enhanced the church's revenue and status. Goy, 'The Glory of God', pp. 1, 20. In the 15th century, the church was rebuilt, with the sculptor-mason, Bartolomeo Bon, playing a vital part. Adjacent to the church stood the Scuola (later Scuola Grande) della Carità, the first of the six original *scuole grandi* to be founded in the city (founded in 1594 and granted official recognition by the Council of Ten in Venice in 1597). The term, *scuola*, originally meant, 'rest and free use of one's own spiritual forces', or an institution with a purpose of education and instruction. The Scuola were wealthy lay confraternities, associations of non-clerical citizens or *confratelli*, usually from the middling classes, who came together for devotions and to carry out charitable works. The Scuola were controlled by an assembly of brothers in a Chapter, headed by the *Guardian Grande* (or president). The second most important member was the *Vicario*, who could act for the president when circumstances permitted. See Franzoi and Lugato, *Scuola Grande dei Carmini*, pp. 7, 14–15; Goy, 'The Glory of God', pp. 1, 20.

[2] Antonio Rizzo's tomb of Marcus and Agostino Barbarigo, both of whom had served as Doge in the 15th century, was known for its enormous size (10 × 15 metres), as well as being the first 'double monumental tomb in Venice'. See Munman, 'The Last Work of Antonio Rizzo', pp. 89–90. The tomb was demolished in 1807 after the church was deconsecrated and incorporated into the complex of the Accademia delle Belle Arti. A copper engraving by Isabella Piccini, and a preserved relief showing Christ's resurrection, now set above the altar in the Scuola de San Giovanni Evangelista, give a good impression of the tomb's appearance at the time of Folkes's visit. See Figure 9. Undoubtedly, the most important surviving sculpture from the Barbarigo tomb is the depiction of the kneeling doge, who – as Sister Isabella Piccini's 1692 engraving shows – remained in prayer in front of the altar. The sculpture is now in the Sacristy Museum of Santa Maria della Salute. See Roeck, 'Zu Kunstaufträgen', pp. 4–6; McAndrew, *Venetian Architecture*, pp. 77–8. See Figure 10.

[3] Folkes's description of the tomb is reasonably accurate. See Munman, 'The Last Work of Antonio Rizzo', p. 90.

[4] Doge Nicolò de Ponte (1491–1585; r. 1578–85). 'His imposing monument, overseen by Marc'Antonio Barbaro (1518–95), was erected in Santa Maria della Carità in only two years (1582–4).' Jones, 'The Business of Sculpture', p. 21.

[5] Again, Folkes is quite accurate here. The monument combined a masterly portrait bust of the Doge by Alessandro Vittoria with a number of allegorical statues by Girolamo Campagna, the leading sculptor of Venice. Jones, 'The Business of Sculpture', p. 55. See also Simane, *Grabmonumente*, pp. 82–95.

[6] Zannetti noted in his *Descrizione* this 'very well-regarded work' of Leandro Bassano. Zanetti, *Descrizione*, p. 342.

[7] Titian's, *The Presentation of the Virgin at the Temple* (1534–8). The oil on canvas was commissioned by the fraternity based in the Scuola Grande di Santa Maria della Carità for the *Sala dell'Albergo* (boardroom). The painting is now in the Galleria dell'Accademia. The painting portrays the 3-year-old Mary entering the Temple of Solomon in Jerusalem, while the left-hand side of the composition features portraits of the governing board of the confraternity. 'Hood, 'The Narrative Mode', p. 130. The definitive article on this painting is Rosand, 'Titian's *Presentation of the Virgin*', pp. 55–84.

ritrattos[1] as of Titian himself and his friend in the crowd, but it is in a bad light in a lesser room going out of the chief Scuola, which is a fine room the cieling richly gilt but too low.[2] I went from hence to hear the musick of the girls at the Incurabili,[3] here are said to be some of the best performers and voices in Italy, one they call Isabella[4] was much admired when I was there, they sit in a gallery guarded from sight by Wires,[5] and all who please come into the church to hear, paying 2 sols[6] to a man who brings a chair, there is also a purse carryed about for Charity to which any one gives or not as he pleases, few of the Italians did, foreigners generally put in some small matter. the ceiling here is pretty well painting [*sic*] particularly one or 2 of the compartments by Padovanino,[7] there are also some paintings over the Altars but very much spoild by time and the smoke of the ~~watch~~ wax lights. I went from hence to the Piatza where I sat with the Resident till suppertime. It lightened to night I think the most I ever saw and there was some thunder and rain towards morning.

Monday. 22 I went to visit Mr Stroud[8] and Mr Smith[9] two English Gentlemen just

[1] 'Ritrattos' are portraits.

[2] Giambattista Tiepolo was commissioned on 21 Dec 1739 to decorate the ceilings of the *Sala Capitolare* in the Scuola Grande di Santa Maria della Carità, well after Folkes's visit to Venice. See Franzoi and Lugato, *Scuola Grande dei Carmini*, p. 76.

[3] In 18th-century Venice, a large repertory of sacred music was performed at San Salvatore, the church annexed to the hospital called the Incurabili (Ospedale degl'Incurabili), designed by Sansovino. The Incurabili took in patients with incurable diseases, such as bubonic plague or syphilis. The Ospedali Grandi, charitable hospices, were also recognized for educating talented female orphans to professional standards of musicianship, predecessors to music conservatories. The Incurabili church was known for its excellent acoustic characteristics. See Howard and Moretti, *Sound and Space*, p. 165. In the 1730s, the female chorus and orchestra of orphans was led by such eminent maestri as Nicola Porpora, followed by Johann Adolph Hasse. See Hansell, 'Sacred Music', p. 284.

[4] Maestro Nicola Porpora dedicated an *Ave Maria*, dated 2 March 1733, for an alto named Isabella, who has not been identified, but may have been the singer that Folkes heard perform. The music is now in the Selden-Goth Collection at the Music Division of the Library of Congress. See Hansell, 'Sacred Music', p. 286, n. 19; Gisella Selden-Goth Collection, Library of Congress, https://www.loc.gov/collections/gisella-selden-goth-collection/about-this-collection [Accessed 15 Nov 2022].

[5] The musicians, called the *Figlie del coro,* meaning 'daughters of the choir', performed in raised galleries, which had thin crepe and gratings to hide the musicians from the audience. They could be heard, but not seen. Many of the *figlie* were masters of several instruments, part of the teaching method at the Incurabili. Many were disabled or disfigured from diseases such as smallpox. Vivaldi wrote many of his concertos for the *figlie.*

[6] *Soldi.* See Appendix B for Venetian coinage.

[7] Alessandro Varotari, known as il Padovanino (1588–1648), who painted *The Parable of the Five Wise and the Five Foolish Virgins* (see Matthew 25: 1–13). The painting is now in the Gallerie dell'Accademia. See Pagotto, 'Varotari, Allessandro, known as Padovanino', *Dizionario Biografico degli Italiani*, https://www.treccani.it/enciclopedia/varotari-alessandro-detto-il-padovanino_%28Dizionario-Biografico%29/ [Accessed 15 Nov 2022].

[8] William Strode (*c.*1712–55), of Ponsbourne, Hertfordshire, and Mapledurham, Oxfordshire, eldest son of Samuel Strode; MP 1740–47, 1754–5. In 1736, he married Lady Anne Cecil, daughter of the 5th Earl of Salisbury. In 1730–32, Strode made the Grand Tour in France, Flanders, and Holland, but was also in Italy. Strode was in Florence between August 1732 and April 1733, Rome (by 22 April 1733), and Bologna and Venice (22 June–mid-July), travelling with his friend Arthur Smyth on their homeward journey. As we will see later in the diary, on 22 June 1733, Folkes went to call on 'Mr Stroud and Mr Smith two English Gentlemen just arrived here from Rome and last Bologna'. Both Strode and Smyth became members of the Society of Dilettanti in 1736. Ingamells, *Dictionary*, p. 909.

[9] Arthur Smyth (1706–71), the eighth son of Thomas Smyth, Bishop of Limerick. He received his MA in 1727 and DD in 1741 from St John's College, Oxford. Subsequently Smyth became Dean of Raphoe (1743–1744), of Derry (1744), Bishop of Clonfert (1752), of Down (1753), of Meath (1765), and Archbishop of London in 1766. Smyth went abroad for his health in 1732, arriving in Florence in October 1732 to see his friend, Joseph Spence. See Ingamells, *Dictionary*, pp. 874–5.

arrived here from Rome and last Bologna; I also visited the Baron [*blank*] a Swedish Gentleman who has been here sometime, and others. I also met Count Bielgne.[1] after dinner I made some visits with my wife and spent the evening at Sr Giustinianis.

Tuesday 23. I was at the Church of the Frari,[2] where is a very magnificent tomb of Giovanni Pisarro 102^{d} Doge of Venice,[3] it is against the north wall of the Church like a triumphal Arch, the door of the church is on the middle of the lower order which is supported by 4 Caryatides, of stones the head and feet etc black marble with white draperys of a Gigantick size, between the pillars on either side is a brass figure of death holding a monumental table, with his Epitaph. The upper order is supported by Corinthian pillars, and in the middle sits the doge in his chair under a rich canopy, with his inscription, revixit 1669, viz 10 years after his death when his monument was set up by his heirs, on either side are allegorical figures etc in honour of him, I have seen few so good monuments; there are other good monuments also in his church particularly another Doge.[4] their are also some good pictures two particularly of Titian.[5] a fine picture of the presentation by Salviatti but much damaged.[6] 2 antient pictures of Bellini,[7] and Vivarini[8] curious as having their names, to give an idea of the old manner of painting, there is a great mixture of stiffness yet something well turned in the area of the heads especially.[9] after dinner I was visited by the English Gen-[f. 8r] tlemen just arrived, as also by the Resident. Fat[h]er Crevelli calld upon me to go and hear a disputation[10] in Philosophy of a young Jesuit in the church of St. Jeremy,[11] the respondent was placd in a

[1] Probably Count Carl Gustaf Bielke.

[2] Basilica di Santa Maria Gloriosa dei Frari, the largest church in Venice, in the San Polo district. The original church was built in the 13th century after Franciscan friars came to Venice, and the current church was begun in 1340, finished in 1442, and consecrated in 1492. It has a brick Gothic facade.

[3] The colossal Baroque monument in the third bay of the left aisle for Doge Giovanni Pesaro (Doge 1658–9), designed by Baldassare Longhena (1598–1682). The monument is of polychrome marble, and pedestals of black-and-white marble support statues of Africans holding up a Doric entablature. Between them bronze skeletons appear, holding draperies with the funerary eulogy. Colombo, 'Baldassare Longhena's Funerary Monument', p. 49.

[4] This could be the monument to the Doge Francesco Foscari (1457) on the right wall of the Presbytery, or the impressive grand sepulchral monument (1476–80) of Doge Nicolò Tron, by Antonio Rizzo on the left wall of the Presbytery. The latter is considered one of the greatest Renaissance sculptures in Venice.

[5] Titian, *Madonna di Ca' Pesaro* (1519–26). The other painting is Titian, *Assumption of the Virgin Mary* (1516–18), known as the *Assunta*.

[6] Giuseppe Porta Salviati's, *Purification of the Virgin* at the Basilica dei Frari (*c.*1548).

[7] For instance, the *Madonna with Child and Saints* (1488), by Giovanni Bellini in the Sacristy of Pietro Pesaro's Chapel.

[8] Folkes is probably referring to *The Apotheosis of St Ambrose* (1503) by Alvise Vivarini, in the Cappella Milimesi in the Frari, with its inscription stating that it was completed posthumously by one of his pupils, Marco Basaiti Bartolomeo Vivarini (*c.*1432–*c.*1499), whose *St Mark Enthroned* is in the Capella Corner, and whose *Madonna and Child with Saints*, forms the altarpiece in the third south choir chapel, also painted mostly in tempera. Bartolomeo executed the first oil painting in Venice, so Folkes is witnessing a definite transition in style and technique.

[9] The Bellini altarpiece is much more symmetrical and formal in its composition than works by later artists.

[10] Disputation is a mode of debate that arose in the universities. Typically, a 'convert's eagerness for scholarly disputation arose from a fusion of religious with intellectual assurance: certainty in faith, and in the academic process'. Rodda, 'Condition and State of a Scholar', p. 391.

[11] The 12th-century church San Geremia in the Cannaregio *sestiere* of Venice, famous for housing the relics of St Lucy, which was demolished and rebuilt in the Baroque style in 1754. There was another Jesuit church in Venice, I Gesuiti, known as Santa Maria Assunta, consecrated in 1728.

sort of tribune towards the middle of the Church, and the master of Philosophy whose scolar he was had a chair on his ~~right~~ \left/ hand, there were chairs round a long square coverd with a carpet before him where sat several religious of different orders, his theses was printed consisting of several positions in different parts of Philosophy. then 3 fathers from their chairs one after another each relected one of his positions and disputed with him sitting, the master who sat by him moderated and sometimes took up the argument. the first was a Theatin[1] who opposed the Generation of insects only ab ovo.[2] The second was a Jesuit who undertook a point about the nature of Gods ~~deg~~ decrees. and the 3d was of another order, and defended against the respondent the cartesian definition of matter, as having partes extra partes with 3 dimensions, which was immediately run into the body of Christ in the Sacrament, which the respondent distinguist to be really matter, and to have partes extra partes, quoad se sed non quod locum, and existing in the bread realiter sed non localiter.[3] The other questions were managed in the same scholastic manner with metaphysical distinctions and argument rather drawn from the general conclusions of the old writers than from the nature of the subject. there was much action of the hand in disputing and an affectation of great vehemence, but it appeared to me upon the whole but a very poor exercise. after it was over most of the religious from other houses complimented the Master on the good performance of his pupil, and particularly on his zeal for the Catholic faith in the third question. after it was over Mr Mace who was with us went with the ffather and my self to the Gardens of St George where we walkt all the evening.[4] the Father talkt to me with great freedom on the state of learning in the several religious around him. he said that formerly the Benedictines had great reputation having the sole education, and the preferments of not only Divinity but also law and physicks among them, since learning having spread somewhat they have lost much of their reputation, neither are the Jesuits here in great credit; the other orders he says know nothing and being mostly kept so many hours of a day to sing in the choir, think it no ways incumbent on them to learn any thing farther. upon the whole he agrees with me there are infinitely too many religious and those under no regulation that is productive of any advantage to learning or society.

Wednesday 24. Count Bielgne and the baron were to see me in the morning. I was afterwards again at St Giovanni and Paolo;[5] finely hung out with damask and gold fringe

[1] The Congregation of Clerics Regular, a religious brotherhood founded in the Counter-Reformation by Archbishop Gian Pietro Carafa, on 14 September 1524, to promote clerical reform via asceticism and apostolic work.

[2] The ovist doctrine of insect generation. Under the influence of naturalist, Antonio Vallisneri's, *History of Generation of Men and Animals* (1721), ovist preformationism was adopted by leading Italian naturalists from the early decades of the 18th century in opposition to epigenesis which was the doctrine held by more conservative clergy. However, preformationism was difficult to reconcile with the empirical tradition of Italian natural history. Hence, the debates that arose in Italian academic circles. See Dal Prete, 'Cultures and Politics'.

[3] Descartes's changes in the definition of substance, and his identification of bodily substance with extension, reinvigorated the theological debates about transubstantiation that had occurred during the Protestant Reformation. The Catholic doctrine of transubstantiation in the Eucharist (the transformation of the whole substance of the bread in the body of Christ) was a 'stumbling block for Cartesian physics'. Armogathe, 'Cartesianism and Eucharistic Physics', p. 547. A cross was written in the right-hand margin at this point.

[4] Probably the grounds of the Palladian Benedictine monastery, San Giorgio Maggiore, on the island of the same name.

[5] San Zanipolo, a church in the Castello *sestiere* of Venice, completed in the 15th century. It was the site of the funeral services of the doges. See below, p. 74, n. 9.

Figure 11. Agostino Carracci, *The Crucifixion*, after Paolo Veronese (1582), engraving, 30.9 × 22 cm. Metropolitan Museum of Art, New York, The Elisha Whittelsey Collection, The Elisha Whittelsey Fund 1959. Folkes recorded seeing the original painting in Venice and noted Carracci's engravings of it. Courtesy of the Metropolitan Museum of Art, New York.

for the feast of St John.[1] I calld as I came home at the church of St Maria fformosa,[2] but found nothing very remarkable here, tho there are pretty good pictures some in every Church they are generally so spoild as to give very little pleasure [f. 9r] I did not go out in the afternoon till past dark, when I sat an hour with the Resident Consul and Mr Mace at the Piazza.

Thursday 25. Mr Mace calld upon me and we went together to the Church of St Sebastian, where Paulo Veronese is buried,[3] and which is filled with his works. over the

[1] Feast of St John the Baptist.

[2] The Church of the Purification of Mary, erected in 1492 under the design of Mauro Codussi. Folkes may have been a bit presumptuous in his judgement of the paintings inside the church, as they include the *St Barbara* polyptych by Jacopo Negretti, called Palma il Vecchio (*c.*1480–1528), considered his greatest work.

[3] The Venetian church of San Sebastiano is known for its artwork by Veronese, Titian and Tintoretto. Constructed between 1506 and 1548, the church was designed by Antonio Abbondi, the architect known as Scarpagnino. See 'San Sebastiano', Save Venice, https://www.savevenice.org/project/church-of-san-sebastiano [Accessed 16 Nov 2022].

Great Altar[1] is The ~~Martyrdom of the~~ Saint with other saints and the Virgin and Christ above of which there is a print,[2] on either side are 2 large paintings the one is the Death of St Mark and St Marcellinus[3] an admirable picture, the other represents St Sebastian bound and going to be put to death[4] but I prefer infinitely the former.[5] there are prints of both these.[6] a great deal of this church has been painted in ffresco by Paulo but that is in a manner ruind,[7] the cieling is his in compartments and very fine the story of Esther.[8] there is over \one of/ the side altars the crucifix[9] which Agostin Curati [*sic*] has gravd;[10] but here said to be by Carlino the son of Paolo [*sic*][11] the figures underneath his ffathers. The Organ case[12] is painted all over by Paulo,[13] the Shutters on the outside, the Presentation which Villamena has gravd,[14] the inside of the same the healing the Paralytick gravd by

[1] Folkes is referring to Paolo Veronese's, *Virgin and Child in Glory with Saints Sebastian, Francis, Peter, John the Baptist, Catherine, and Elizabeth* (1562), painted for the high altar of San Sebastiano.

[2] Folkes possessed in his collection prints of the *Twelve Histories* by Veronese, as well as various prints. See the auction catalogue of his collection, *A catalogue of the prints and drawings*, p. 3, lot 30; p. 14, lot 49; p. 16, lot 54.

[3] Veronese, *Saints Mark and Marcellinus Being Led to Martyrdom* (*c.*1565), San Sebastiano, Venice. The painting is on the wall of the presbytery; the iconography is based on the *Golden Legend* of Jacopo da Voragine. See 'High Altar,' Save Venice, https://www.savevenice.org/project/san-sebastiano-presbytery [Accessed 16 Nov 2022].

[4] Veronese, *Martyrdom of St Sebastian* (1565).

[5] Folkes was correct that the *Saints Mark and Marcellinus Being Led to Martyrdom* has a superior composition, with fewer figures detracting from the pathos of the scene. Seeing both pieces in situ, there is an inherent dynamism in this work leading the eye from left to right as the brothers process to their martyrdom with onlookers witnessing from balconies above.

[6] Giuseppe Maria Mitelli etched a pair of prints from Veronese's depictions of the martyrdom of St Sebastian, and, in 1666, *St Sebastian exhorting the martyrs Mark and Marcellian not to be swayed by the prayers of their families*. See Bertarelli, *Le incisione di Giuseppe Maria Mitelli,* nos. 40 and 41. See also *The Martyrdom of S. Sebastian*, after Veronese, 1666, British Museum, https://www.britishmuseum.org/collection/object/P_W-9-84 and https://www.britishmuseum.org/collection/object/P_W-9-85 [Accessed 18 Nov 2022].

[7] For instance, Paolo Veronese depicted St Sebastian's final martyrdom in two frescoes towards the west end of the church.

[8] The ceiling of San Sebastiano has a wooden framework to hold canvases of several Old Testament scenes, including those dedicated to the story of Old Testament heroine Esther. The Coronation of Esther was presented in medieval arts as a parallel to the Coronation of the Virgin. See Kahr, 'The Meaning of Veronese's Paintings', p. 240.

[9] *The Crucifixion*, attributed to Veronese, is in the chapel of the Garzoni family and was created between 1560 and 1565.

[10] In his first visit to Venice in 1582, Agostino Caracci (1557–1602) engraved six of Veronese's paintings, including the Garzoni *Crucifixion*. See Cocke, *Paolo Veronese*, p. 9. For Carraci's engraving of the *Crucifixion*, see Figure 11.

[11] Carlo Cannovaro Carliari (1570–96), the youngest son of Veronese. The most talented member of his father's workshop, he undoubtedly painted many works attributed to his father.

[12] In 1558, a contract was made with one Maestro Domenico from Treviso to build a wooden organ loft. It was agreed that the loft would be made according to drawings by Veronese and that the gilding and carvings were to be created by Francesco Fiorentino and Bartolomeo Bolognese. Save Venice, https://www.savevenice.org/project/church-of-san-sebastiano-organ-loft [Accessed 18 Nov 2022].

[13] Veronese painted the organ shutters with two images: the *Presentation of Jesus in the Temple* when the shutters were closed, and when opened, the *Miracle at the Pool of Bethesda*. 'Church of San Sebastiano Organ Loft', Save Venice, https://www.savevenice.org/project/church-of-san-sebastiano-organ-loft [Accessed 18 Nov 2022].

[14] The engraving is by Francesco Villamena, *Presentation of Christ at the Temple* (1597). See Plate 7.

le Febra,[1] both these are in a fresher colouring then any of the rest. I was pleasd in this church with a Virgin Bambino and St John in marble by Lombard,[2] as pretty a thing as most I have seen; there is also a Christ at Emaus[3] by Andrea Schiavone,[4] and a St Nicholas by Titian[5] but this is in great measure spoild. there are abundance more good things in this little Church and Sacristy especially of Paulo, and it as well deserves returning to as any I have seen in Venice. I went then to St Nicolo de Frari[6] a very small church but rich in painting the cieling is paulos; the Magi offering in the middle,[7] and a history of St Nicholas in the division next the altar are admirable, I do not so much like the St Antonio in the division next the door. The great Altarpiece is said to be one of the fine works of Titian[8] but spoilt entirely (there is a print of it of which more is to be made than of the painting). The Annuntiation is finely painted in fresco on both sides the Organ by [[~~xxx~~]] Marcus Titiani,[9] but it has certainly been divided since the painting the organ taking away the middle part of the picture which has a vile effect. there are many other very good pictures in this church. I then went as I have done several times before among the book sellers but find nothing there; it is incredible how little opportunitys there are of having the most common usefull books, and as they understand nothing of those they have, whatever a stranger asks the price of they know not what to ask for it as they suppose it must be a curiosity. After dinner I was at Mr Smith where I met Sigr Martinelli of fflorence,[10] and went with him to take the air, in the evening was almost a continued lightning till I went to bed, but [f. 10r] without thunder at least any that I heard, there

[1] Folkes refers to the Flemish artist Valentin Lefèvre (1637–77), who produced the 'Opera Selectiora' of etchings after Titian and Veronese. The series was left unfinished at his early death, and it was published in 1682 by Jacques van Campen. In 1682, Lefèvre also created an etching of the *Presentation of the Infant Christ at the Temple*, originally painted on the organ shutters in San Sebastiano by Veronese. See Wellcome Collection, https://wellcomecollection.org/works/w657mf48 [Accessed 17 Nov 2022]. See also Ruggeri, *Lefèvre*, for the catalogue raisonné.

[2] Tommaso Lombardo (in Venice 1536–7), *Madonna and Child and Saint John* (mid-16th century), Melio da Cortona altar, San Sebastiano, Venice. See Plate 8.

[3] Emmaus.

[4] Andrea Meldolla, known as Schiavone (*c.*1510–63). His work is in the Grimani Chapel in San Sebastiano.

[5] Titian's, *Saint Nicholas of Bari at San Sebastiano* (1560s). The painting is located on the right side of the atrium, before entering the nave of San Sebastiano.

[6] The lost church of San Niccolò ai Frari (San Nicoletto), one of the smallest of the 34 Venetian conventual churches that were closed by Napoleon in 1806. It was richly decorated, including not only a celebrated high altarpiece by Titian, but canvases by Veronese and his workshop, Palma Giovane and Paolo Fiammingo. The Titian was sold in 1770, and at the closing of the church, the other works of art were dispersed and the building demolished. Humfrey and Sherman, 'The Lost Church', p. 247.

[7] Paolo Veronese painted for the ceiling's centre, the *Adoration of the Magi*, the *Consecration of St Nicholas* for the chancel end, and for the entrance end, the *Stigmatisation of St Francis*. The Four Evangelists were depicted at the four corners. Humfrey and Sherman, 'The Lost Church', p. 258.

[8] Titian, *Virgin and Child in Glory with Six Saints* (*c.*1520–23), now in Rome, Mvsei Vaticani, Pinacoteca.

[9] There was a pair of paintings (now lost) on either side of the choir by Marco di Tiziano (Marco Vecellio) of the *Virgin Annuniciate* and *A Prophet*. See Humfrey and Sherman, 'The Lost Church', p. 251. There was an illegible word that Folkes crossed out before he wrote 'Marcus Titiani'.

[10] Possibly Cristino Martinelli (1653–1732), a Venetian patrician, member of the Sarotti Academy, and sometimes mathematician. He was also a keen botanist and collected books of natural history, corresponding with Marsigli about the nature of coral. See McConnell, 'The Flowers of Coral', p. 63. This person also could be Vincenzo Martinelli (1702–85), a Tuscan from Montecatini with a degree in civil and canon law from Pisa University, who met English noblemen in Florence, and came to England in 1748 to teach Italian. Thorne, 'Vincenzo Martinelli in England', p. 92.

was some appearance I was not acquainted with, in almost every flash that seemed to come from behind a cloud like a sheet of light there appeared as it were a shot of fire not as running along but as all contemporary at the same instant. and this appeared several times in the same place but ever in a flash of the other light tho it did not look as if the other light proceeded from it, but as if the light discoverd it. however odd and imperfect this account is being what seemed the same to the Consul and Mr Mace who went in a boat with me into St George's channel \about 10 a clock from the Piazza/ to take a view of it, I think it worth making this memorandum of.[1]

Fryday. 26. I went in the morning with Mr Mace to Murano an Island or rather a host of Islands about a mile and ½ or somewhat better from the City it resembles Venice exactly and here are the glass works for which formerly venice was so famous;[2] there are a great number of houses; in one we saw the making colourd bugles,[3] what I took notice of the colour does not go through but a lump of christal or clear glass when just blown to be somewhat hollow is dipt into a furnace of coloured glass, which when spread all over it by blowing and rowling on a marble stone, is again dipt into a furnace of clear glass that covers it with a glazing of glass above the colour. this done it is again heated to a perfecte degree of fusion and an Iron tube being also fixed to the other end two men draw it by running different ways into a tube of 60 or 70 yards length which is hollow all the way and that which is afterwards cut into bugles, or little beads which taller are no other than short bugles again put into a state of fusion. in another glass house they were making a sort of glass teacups of coloured glass, and we were at a third where the work the looking glass plates, but had not the fortune to see them work, we saw the metal in fusion which seemd in a very perfect state of fluidity. but they were not ready to make the plates till evening, they told us the largest they make are of about 8 palms[4] the longest way. before we left Murano we want to the Cornaro Palace[5] here pretty much celebrated, the body of the house is small but with a very handsome Portico, but what is most remarkable, are continued galerys from it round several courts and gardens that in the whole they say make the length of an Italian Mile but

[1] Folkes would become fascinated by electricity after his return to London, even having his own electrical apparatus created by royal optician and instrument maker, Francis Watkins, for gentleman natural philosophers, advertised as 'a most compleat Electrical Machine, which, with its whole Apparatus, is contained in a box exceeding the Size of an ordinary Tea-chest'. *The Remembrancer,* 14 May 1748.

[2] Venetian glass was made from the 13th century, with a glassblowers' guild extant since 1225. In 1291, the glasshouses moved to Murano, a mile north of Venice. Coloured glass was perfected by the 16th century, as well as *millefiori*, where canes of coloured glass were bonded together and sectioned to make multicoloured beads resembling flowers. From the 15th century, Venetian mirror making attained distinction as an applied art. There were also those that had skill with optical instruments and spectacles. By the time Folkes visited Murano, competition from Bohemian glass had diminished the prestige of Venetian glassware. See 'Venetian glass', *Encyclopaedia Britannica*, https://www.britannica.com/art/Venetian-glass [Accessed 22 Oct 2020].

[3] Tube-shaped beads made of glass which were globally traded. 'Early in the eighteenth century four or five hundred chests of Venetian beads, of ten hundred-weight each, were annually shipped to Cairo.' 'Murano Glass Work', p. 563.

[4] This is the palm, the Venetian measurement for length. It was a little over a foot, at 13.68 inches (34.74 cm). Folkes recorded many such instances in his diary, reflective of his interests in Freemasonry and metrology. See Carrington, *Foreign Measures,* p. 99.

[5] The Cornaro were the most notable patricians on Murano, owning four large houses, two of which were luxurious villas. Folkes probably saw what was known as the house of the Queen of Cyprus. As Goy has indicated, the villa was Palladian, three storeys high, and in the centre of the symmetrical and rusticated facade was 'a classical portico two storeys high, and with hexastyle columns probably of the Corinthian order, surmounted by a triangular pediment'. Goy, *Venetian Vernacular Architecture,* p. 206. Zanetti, *Guida di Murano*, p. 282.

is I really believe none than half an English one, these galerys have no beauty but their length, they are all adornd with pictures or busts, but I saw nothing excellent in the whole, they mostly look into the garden consisting of tall trees and the walks bordered with some 100[s] of statues, at the end of the Galerys is a Pavilion of a few rooms, from whence we were let out a back way where our boat waited for us on an other canal.

As we went home we calld at the Convent of S[t] Michael[1] of the orde[r] Carmeldunenius[2] \on the Island of the same name./ (white Benedictines)[3] their Church is pretty [f. 11r] good, and adornd with several paintings some very good of modern Masters. Particularly the story of the Brazen serpent by Ant[onio] Zanchi[4] and the golden Calf by Lazarini;[5] there is also a Madonna and Christ, an old picture of Bellini one of the best I have seen of him;[6] one of the ffathers showed me the Library a good room tollerably furnisht with books, here was a good copy of the first printed bible of 1462 on Vellam.[7] I also took notice of an old map which took my curiosity; It is drawn in one Hemisphere of about 6 foot diameter and as they say by the hand of a brother of this convent whose effigies they showed me with his inscription.[8] *Frater Mauras, S[t] Michaelis Muranensis, Venetus, Ordinis Camaludunensis* [*sic*] *Cosmographus insignis.*[9]

[1] The original Convent of San Michele (now S. Michele in Isola) was founded in the 10th century, and it is on an islet between Murano and Venice. A monastery was also built and housed monks of the Benedictine Camaldolite order from 1212 to 1818. Cotton, 'San Michele in Insola', *Churches of Venice*, http://churchesofvenice.com/sminisola.htm [Accessed 22 Oct 2020]; see also Meneghin, *San Michele in Isola di Venezia*; Cornaro, *Notizie storiche delle chiese e monasteri di Venezia*, pp. 637–45. Since the early 1800s, the islet has served as Venice's only official Christian burial ground; in 1813, after the blockade of Venice by the Austrians, the public cemetery was created, and the vineyards of the abbey were destroyed.

[2] The Benedictine Camaldolese Order was founded in 1012 in Italy by St Romuald, and it promotes a very ascetic form of common life. The order combines the community life of Western monasticism with the hermetic life of Eastern monks. Its reputation for strictness and learning attracted several natural philosophers to its fold, such as Guido Grandi (1670–1742), historian of the order and famous mathematician, friend and correspondent of Newton. See 'Camaldolese: The Congregation of Murano', *Catholic Encyclopedia*, https://www.newadvent.org/cathen/03204d.htm [Accessed 29 October 2020]. The order also had an elaborate and sophisticated liturgical music programme, the nave of St Michele divided by the barco or raised monk's choir.

[3] The monks wore white robes as a symbol of spiritual purity.

[4] *Moses raises the Bronze Serpent* (1712), by Baroque master, Antonio Zanchi (1631–1722).

[5] *Adoration of the Golden Calf* (*c.*1700–7) by Gregorio Lazzarini (1657–1730), teacher of Giambattista Tiepolo. See Da Canal, *Vita di Gregorio Lazzarini.*

[6] This is the *Madonna and Child with Four Saints and Pietro Priuli* (1511–13), by Giovanni Bellini and assistants. The altarpiece was originally in the Priuli funerary chapel, the Capella del Croce. It is now in the Staatliche Kunstakademie Düsseldorf, on permanent loan to the Museum Kunst Palast, Düsseldorf. Fletcher and Mueller, 'Bellini and the Bankers', p. 14.

[7] The magnificent library at San Michele was confiscated after the suppression of religious orders in Italy in 1810. After the more notable works were relocated to public libraries, the remaining 18,000 volumes were sold at public auction. See 'Camaldolese: The Congregation of Murano', *Catholic Encyclopedia*, https://www.newadvent.org/cathen/03204d.htm [Accessed 29 Oct 2020].

[8] Among the works of greatest worth and interest is the extraordinary, *Mappamondo* (*c.*1400–1464), of Camaldolese monk Fra Mauro, a geographic monument made in San Michele in the mid-15th century and now conserved in the Biblioteca Nazionale Marciana, Venice. The map is dated 26 August 1460. See Fra Mauro's World Map, Museo Galileo, Florence, https://mostre.museogalileo.it/framauro/en/ [Accessed 28 Oct 2020]. For a further discussion of the map, see Introduction, pp. 18–21 above, also Plate 1.

[9] Frater Maurus S. Michaelis Moranensis de Venetiis ordinis Camaldulensis chosmographus incomparabilis, or 'Fra Mauro of the Venetian monastery of St Michael in Murano of the Camaldolese Order, an incomparable cartographer'. Folkes muddled some of the Latin.

they thought he livd 400 years ago but the following inscription fixed to the map seems to settle its date.

Haec Tabula Geographica, cum per centum nonaginta quatuor[1] *annorum curricula, partim in ecclesia partim [vero] in aula quae suo nomini dicata erat, et dicebatur il Mappamondo fuisset appensa. tandem jussu Rev.*[ssi] *Patris Doni Francisci Gherardi dicti Errici Abbatis huius monasterii, in hac Bibliotheca ab ipso instaurata ditata et exornata, translata et collocata fuit Anno 1655. Die 20 Dec[embris].*[2]

This inscription seems to fix the date of the map to the year 1460, and that is an age sufficient for it to be somewhat of a curiosity. The North is at the bottom,[3] and the countrys in the Mediterranean viz Italy and Islands Greece Asia minor and Spain, are very nearly in the figure Ortellius's[4] and other maps give them, the taste somewhat like that of the map of Utopia in the first Edition of that book.[5] The figure of England is more imperfect, and that of Sweden and the baltick sea yet more so, Afric is throughly represented, but the filling in beyond the coast of Guinea looks somewhat like the Red sea on the other side.[6] Madagascar seems divided by little more than a river. the Aut[h]or notes upon Babylon which he places on the Tygris, but he did it from the testimony of eyewitnesses, and therefore is not to be blamed in departing from the antients who placd it on the Eufrates.[7] he observes about Mesopotamia etc he had left 30 maps that he had drawn with great care of the countrys hereabouts. the Fathers told me they believed there were some imperfect scetches of his among their archives but nothing compleat but this map. which I will see again if I can, as I look upon it somewhat considerable as to the history of the revival of Geography among the moderns.[8] Whilst I was out Sigr Marcello

[1] 'Quatuor' should be 'quinque'.

[2] After August 1460, the date on which the work was finally completed, the map continued to be kept at the monastery of San Michele, albeit in different positions, as evidenced by the parchment inscription Folkes recorded affixed to the right side of the map in 1655. This reads: 'This map hung for one hundred and ninety-four years, at times in the church and at times in the hall which was dedicated to its name, and was called the Mappamondo. Finally on the instructions of the most Reverend Father Don Francesco Gherardo, the abbot of this monastery, who was known as Erricus, it was moved and placed in this library, which he had personally restored, enriched and decorated, in the year 1655, on the 20th day of December, 1655.' The map was first placed in the church, then – following the rebuilding of the church itself in the period 1469–77 – in an adjoining room, called 'the Globe', and finally, from December 1655, in the premises of the complex used as a library. The planisphere remained there until May 1811, when it was finally transported to the Biblioteca Nazionale Marciana, Venice. See Falchetta, *Biblioteca Nazionale Marciana*, p. 59. There is a cross inked in the right-hand margin by Folkes's transcription from the map.

[3] The planisphere's orientation is with the south at the top, mostly likely because Venetian compasses were south pointing. See 'Fra Mauro World Map', British Library, https://www.bl.uk/maps/videos/fra–mauro–world–map [Accessed 2 Dec 2020].

[4] Abraham Ortelius (1527–98), Flemish cartographer, who published in 1570 what is described as the first modern atlas, the *Theatrum orbis terrarum.*

[5] Sir Thomas More's *Utopia,* first published in 1516; the work featured a map of the imaginary island of Utopia (nowhere).

[6] Fra Mauro was the first cartographer to speculate that Africa was its own continent and not terrestrially connected with South Asia.

[7] Folkes would have been sympathetic to Fra Mauro's methodology, similar to the Royal Society's *nullius in verba* [on the word of no one, or take nobody's word for it].

[8] Folkes was correct. Fra Mauro's work was considered the first 'modern' world map for its questioning of ancient sources. In a letter tipped into his diary, Folkes also proposed to publish a new edition of the Fra Mauro Map. See Lippincott, 'A Chapter', p. 286, n. 87. See also, Introduction, pp. 19–21above.

a Noble Venetian[1] was to see me, he saw my Wife[2] and promisd his company again in the afternoon.

[f. 12r] In the afternoon we visited Miss Edwin her uncle and brother (so they speake her;) she has been travelling these 3 years in Italy and is now going to Vienna [*series of nine dashes*][3] We then went with Signr Marcello who calld upon us to the Convent of S[t] Maria Celeste[4] where he has a daughter a Nun, a young Lady of beauty and merit, a relation of his is also Abbess they received us with great civility at the grate, the young Ladys name is Sister Alessandra Maria Marcella, we saw also several others of the Nuns, who are well drest and all persons of quality, after a pretty long long [*sic*] conversation at the grate, the Abbess and young Lady came to the door of the convent where they talkd with us again on the threshold.[5] we were in a handsome Church belonging to this monastery.[6] he then carried us to another S[ta] Maria \della/ Vergine,[7] where we were

[1] Possibly one of the brothers, Alessandro Ignazio Marcello (1673–1747), or Benedetto Marcello (1686–1739) (for whom, see above, p. 36, n. 3). Both were polymaths, poets and composers. The Marcello were one of the oldest Venetian noble families. Folkes's 'Sigr Marcello' seems unlikely to be Benedetto as he died without issue, and the diary later refers to a daughter placed in a convent, but Folkes does refer in the diary later on 26 July to an 'Alex Marcello'. Alessandro, on the other hand, married Lodovica Bettoni; the couple had six children. As a Venetian patrician, Alessandro held various judicial and administrative positions. As a composer, he is still known today, thanks to his oboe concerto in D minor of which Johann Sebastian Bach edited a transcription for a keyboard instrument (BWV 974). Although admired for his gifts by his contemporaries, Charles-Louis de Secondat, Baron de Montesquieu, was not so complimentary. Montesquieu met Alessandro during a stay in Venice in 1729, and learned of his transalpine travels; however, the French philosopher, far from admiring the eclectic genius of the Venetian patrician, defined him in no uncertain terms as, 'une espèce de fou' [a kind of madman], and 'un omnis homo pour les demi-talents' [a common man of half talents]. This may account for Folkes's later disparaging comments about him in his journal. See Marco Bizzarini, 'Alessandro Ignazio Marcello', *Dizionario Biografico degli Italiani,* https://www.treccani.it/enciclopedia/alessandro-ignazio-marcello_(Dizionario-Biografico)/ [Accessed 4 Dec 2020].

[2] Lucretia Folkes née Bradshaw (d. *c.*1755), a former Drury Lane actress. During this journey, she began evincing symptoms of mental illness (see Introduction, pp. 30–32 above). For Bradshaw's biography, see also Roos, *Martin Folkes*, pp. 49–72.

[3] Catherine Edwin (d. 1771), daughter of Samuel Edwin of Llanmihangel Plas, Glamorgan. Catherine Edwin went to Italy for the sake of her health, accompanied by her only brother, Charles, and her uncle (unidentified, possibly Humphrey Edwin, 1673–1747). See 'Catherine Edwin', *The Grand Tour*, Adam Matthew Digital, http://www.grandtour.amdigital.co.uk [Accessed 3 Dec 2020].

[4] Santa Maria della Celestia, built 1119–1239, by the Celsi family. The Cistercian convent of the Celestia (Saint Mary of Heaven) was closed in 1810, the monastery building incorporated within the Arsenal and the church demolished. It is now home to the municipal historical archive and University offices. See Schulz, *Venetian Painted Ceilings of the Renaissance*, p. 69.

[5] As dowries greatly increased from the 15th century onwards in Venice, and since usually only the eldest son in a Venetian aristocratic household would marry, superfluous daughters were routinely put in convents. Convents were also boarding schools for girls. Laven suggests that approximately 50 per cent of Venice's nuns were forced into a convent, often to accommodate familial pressures, namely the marriage strategies of aristocratic families and the costs of dowries. Laven, *Virgins of Venice*, pp. 39, 43. As Sperling has indicated, convent parlours, legally reserved for nuns' relatives only, became centres of entertainment, 'salonlike gathering places', where 'purity faced the perils of corruption'. Sperling, *Convents and the Body Politic,* p. 158. Nuns also indulged in elaborate dress, curled their hair, and wore heels. For commentary on the luxurious fashions of the nuns, see also Campagnol, *Forbidden Fashions;* Evangelisti, *Nuns: A History of Convent Life 1450–1700*; Laven, *Virgins of Venice.*

[6] The church was rebuilt by Vincenzo Scamozzi and consecrated in 1611; its simple Latin-cross shape may have appealed to Folkes. See Cotton, 'Santa Maria della Celestia: La Celestia', *The Churches of Venice*, http://churchesofvenice.com/demolished.htm#santamariadealcel [Accessed 3 Dec 2020].

[7] Santa Maria delle Vergini, Venice was under the *jus patronatus* of the doge, but belonged to the order of San Marco Evangelista di Mantova and followed the Augustinian rule. Lowe, *Nun's Chronicles and Convent Culture*,

again entertained by 2 young Ladys of great beauty and distinction, these nuns are yet better drest than the former, they are in white with workt stays, and hoops,[1] Sister [Beatrice?][2] one of the young Ladys that met with us had a gold watch by her side in both these convents they wear a sort of yellow gause[3] cap on their heads, which just covers the crown and lyes under the chin, as they are all of the first quality of Venice they have a pretty deal of liberty, and there was a great number of people to see them at the grates, there are abundance in both the convents so that almost each sister may have a grate to her own company.[4] but the beggars[5] here are as troublesome as else where coming freely into the parlour and importuning people whilst they are in conversation with the Nuns, which the Sigr tho a Senator could not prevent that being a part of the Venetian Liberty. The Sig[r] came home and dranketh tea with us as he speaks English and ffrench passably he is very proud of showing himself polite to strangers, and is indeed most exceedingly civil, tho he puts me not a little in mind of my Lord Anthonie in Otway:[6] for he takes all opportunity of telling us of the weight of state affairs on his shoulders; and the pains he takes to get time from his [august?][7] occupations, and learned studies to wait on strangers: he has been known I perceive to all the English that have been here. ~~been here.~~

Saturday 27. I returnd Sig[r] Marcellos visit, after which I waited on the Resident with whom and Sigr Martinelli who came in hence I passd the morning. This afternoon I was

---

p. 184. Although the resident Augustinian canons had spiritual authority over the convent, the canonesses there were noblewomen and were not required to follow the enclosure rule. As Glixon has noted, these arrangements caused serious problems with discipline, 'either because the monks and friars were unable to control the nuns, or because they participated in the nun's inappropriate activities'. Glixon, *Mirrors of Heaven*, p. 3.

[1] According to several 18th-century sources, the dress was white, tight-fitting with large flowing sleeves secured by bows at the shoulders. The face was framed by a wimple. See Campagnol, *Forbidden Fashions*, p 75.

[2] This name is unclear.

[3] 'Gauze'. In his *Historical Dictionary of All Religions* (1742), Broughton indicates that Benedictine nuns had similar attire: 'It is true, they wear the gown and black scapulary, like the other Benedictines; but their hair is curled, and is not hid by a kind of veil of yellow gauze, which they tie under their chin'. Broughton, *Historical Dictionary of All Religions*, I, p. 442.

[4] As Campagnol states: 'Venetian convents had a sort of osmotic relationship with the city, and the parlors constituted a key space, a borderland where the boundaries of the lives of the encloistered women and those of the people of the outside world could be, at the same time, highlighted and blurred'. Although conversations at the parlour windows with their curtained grilles were supposed to be short and devout, the discussions were often profane in nature. Campagnol, *Forbidden Fashions*, p. 59.

[5] Beggars were ubiquitous in Venice as there were a 'continual influx of peasants driven by poverty to seek work ... between 1760 and 1787, licensed beggars made up 3–5 percent of the population and unlicensed ones another 13–15 percent'. Anderson, *Europe in the Eighteenth Century*, p. 56. Beggars were at the bottom of a vastly unequal society: 'the number of poor and very poor families in seventeenth- and eighteenth-century Venice [was] at slightly below two-thirds of the total population'. Davis, *A Venetian Family*, p. 48. Peter Burke noted a trend towards the 'increasing visibility of the poor' from the 17th to the 18th century in the Venetian census. Burke, *Historical Anthropology*, p. 35.

[6] Lord Antonio, a corrupt and lascivious Venetian senator, a character in Thomas Otway's play, *Venice Preserv'd* (1682). The character was meant to ridicule Antony Ashley Cooper, 1st Earl of Shaftesbury (1621–83). Antonio's scenes with the courtesan Aquiliana, referring to 'Nicky-Nacky', whilst comic, were famously scurrilous, as 'Nicky Nacky' was slang for female genitalia. The play was extremely popular, performed frequently until the 19th century. See Ward and Waller, eds, *Cambridge History of English Literature, vol. 8: The Age of Dryden*, pp. 183–4.

[7] The word is unclear as the pen's ink was running out.

not out, being visited at home by Mr Feake[1] with Mr Mace, Father Crivelli, Sir Thomas Twisden,[2] Mr Stroud, Sr Rob Brown, and his Lady, [f. 13r] the Company just arrived at next door Miss Edwin, etc.

Sunday. 28. I was to visit Mr Feake and Mr Mace, whence the Consul came in just from my house. I understand also Sig^{r} Marcello had been to see me. In the afternoon I was at the Churches of the Jesuits,[3] and St Caterine,[4] in the latter we talkt with the nuns white Augustines, I was also at the little church of the Citelle now on the Guidecca,[5] and went in the Evening to Sr Giustiniani.

Monday. 29. I was in the Morning at the Church of St Pietro di castello.[6] the ffront is very handsome having been built by Palladio, the Church is not large tho the Patriarchal seat, the Throne of the Prelate is in the cappella maggione which we should call the chancel on the right hand the altar.[7] The altar of this Chapel is dedicated to St Lorenzo Giustiniani the first Patriarch. The shrine is marble with his figure on the top. the sarcophagus is supported by angels, and attended by St Peter St John Baptist. St Mark and St Paul the whole handsome of Marble. on either side the Chappel are two large pictures

[1] Probably Samuel Peake (see above, p. 39, n. 1), but just possibly John Freke (1688–1756), elected FRS in 1729, and a surgeon and natural philosopher. In the *Philosophical Transactions* of 1736, he described a case of abnormal bony growth in an adolescent boy, and in 1743, he read to the Royal Society a paper about his invention to reduce shoulder dislocations. See Moore, 'Freke, John (1688–1756)', Rev. Michael Bevan, *ODNB*.

[2] Sir Thomas Twisden, 4th Baronet of Bradbourne, Kent (1704–37). The eldest son of Sir Thomas Twisden, 3rd Baronet, and heiress Anna Musters, was a spendthrift and collector, who after 1730 spent most of his time travelling abroad until his death in Granada on 30 July 1737, where he 'came to a violent end ... eloping with a nun'. See Hatton and Hatton, 'Notes on the family of Twysen and Twisden', p. 47. His younger brother, Sir Roger Twisden, 5th Baronet, continued the family line. Twisden spent nearly four years in Italy accompanied by [the Rev. Charles?] Brown, and was in Venice from 22 May to 16 July 1733, where he saw Martin Folkes. Twisden sat for Rosalba Carriera, had 'open dealings' with Jacobites, and was an attendant at the Jacobite Lodge at Rome. Ingamells, *Dictionary*, pp. 958–9.

[3] The church of Santa Maria Assunta, known as I Gesuiti, built *c.*1155. The church was rebuilt by the Jesuits to mark their return to Venice and completed in 1728. The interior is high Baroque and overdecorated, which may be why Folkes says so little about it, his preference being for Palladianism. See Cotton, 'Gesuiti', *Churches of Venice*, http://www.churchesofvenice.com/cannaregio.htm#gesuiti [Accessed 4 Dec 2020].

[4] Founded in the 11th century by the order of The Friars of the Penitence of Jesus Christ, the complex passed to Augustinian nuns around 1289. The church and monastery were rebuilt in the mid-15th century and subsequently suppressed by Napoleon in 1807. Cotton, 'San Caterina', Churches of Venice, http://churchesofvenice.com/cannaregio2.htm#santacat [Accessed 4 Dec 2020].

[5] Santa Maria della Presentazione known as Le Zitelle, or 'The Spinsters', a 'hospice for "beautiful girls" from poor families whose beauty was thought to put them in danger of falling into prostitution'. See Cotton, 'Le Zitelle', Churches of Venice, http://churchesofvenice.com/giudecca.htm#zit [Accessed 4 Dec 2020]. See Puppi and Ellero, eds, *Le Zitelle*.

[6] San Pietro di Castello is on the island of Olivolo (now known as Castello) in the eastern end of the Venetian archipelago, just beyond the complex of the Arsenale. Palladio was commissioned to rebuild the church in 1556. Although now overshadowed by San Marco, it was the seat of the patriarch of Venice for 1,000 years (until 1807). Despite its seeming marginal placement in Venice, it was considered an important site to visit on the Grand Tour, and the studio of Canaletto painted it *c.*1734–42. See 'Basilica of San Pietro di Castello', *Chorus Venezia*, https://chorusvenezia.org/visita/basilica-di-san-pietro-di-castello/ [Accessed 14 Dec 2022]; Marina, 'From the Myth to the Margins', pp. 353–429. For comments about the bell tower, see Giustiniani, *Historia di m. Bernardo Giustiniano,* p. xlv; Studio of Giovanni Antonio Canal (called Canaletto), *San Pietro di Castello* (*c.*1734–42), National Gallery, London, https://www.nationalgallery.org.uk/paintings/canaletto-venice-s-pietro-in-castello [Accessed 24 Dec 2023].

[7] The high altar in polychromed marble was designed by Baldassare Longhena in 1649. It features an urn, supported by angels above the high altar, with the remains of San Lorenzo Giustiniani (1381–1456), Bishop and the first Patriarch of Venice. His festival was set by Pope Innocent XII for 5 September.

over the Patriarchs Throne, is the S[t] giving alms by Lazarini a fine picture. In another chappel is a ~~fine~~ \good/ picture of Luca Giordan of the souls coming out of Purgatory but somewhat extravagant.[1] these are what best pleased me in this church. I observed the Patriarchs arms are crowned with a hat of a Cardinals form but green instead of red, that being I apprehend peculiar to him.[2] I then returnd S[r] Th. Twisdens visit, but not finding him out till dinner with Mr ffeake and Mr Mace who are leaving this place. they are both very good people and I shall have a great loss of the latter especially being the most conversable man I have met abroad.

being early in the afternoon at the place I went up the Campanile of S[t] Mark, one has from hence a very good view of the City and Neighbouring Islands, the Laguna, Terra firma, and Adriatrick over the Island of Lido. it is remarkable from hence the town seems to stand very compact and on one large Island for the Canals even the great one are entirely hid by the height of the houses, here being so many churches convents and large Palaces gives the city a very grand appearance; tho the size is not large and even less than any other town equally inhabited, the generality of the canals being narrower than streets and the houses all in a manner without any sort of opening behind them.[3] I passd the evening in the place with the usual company.

[f. 14r] Tuesday 30. I went before noon to the ~~town~~ Piaza and at 12 a clock markd as near as I could the shadow of the top of the angel on the Campanile[4] in order to try by measuring the shadow whereabouts the height is. I found it difficult to be exact by reason of the large penumbra from so great a height, in the evening I measured the distance of this mark as near as I could to the center of the tower and found it about 140 English feet. from when I compute the height [*blank*][5] of the same feet. I dind this day with Mr ffeake and Mr Mace who leave Venice early to morrow, I was afterward at the place where I passd the even[in]g as usual. Sig[r] Marcello was at our house, and gave a ample caution to my wife that if I designed the noble venetians should keep company with me, I must not be familiar with the resident of our nation, as if it was not of farr greater consequence to me to be under the Kings protection than to be well with such people as by me.[6]

[1] *Virgin Mary and Child with Souls in Purgatory* (1650), by Luca Giordano (1634–1705), a Neapolitan artist who was a painter of large mythological and allegorical scenes. See Ferrari, 'Luca Giordano', pp. 168–9. The *Virgin Mary and Child with Souls in Purgatory*, an impressive piece almost 2 metres tall and 1.5 metres wide, was a recommended artwork for gentlemen on the Grand Tour to visit. The painting had a dramatic and memorable subject of the Virgin Mary holding Christ whilst seated over a bed of clouds, the pair floating over pitiful souls in purgatory. This made the work memorable, but clearly Folkes thought it de trop. The painting is in the catalogue raisonné of Ferrari and Scavizzi, *Luca Giordano*, no. A179. It is now in the Museum of Fine Arts, Houston. https://emuseum.mfah.org/objects/64360/madonna-and-child-with-souls-in-purgatory [Accessed 1 Jan 2021]. For an 18th-century appraisal of the work, see Martyn, *An Appendix*, p. 144.

[2] Folkes was incorrect. A bishop's *galero* is usually green with six tassels on each side, so this iconography was not unique to San Lorenzo Giustiniani.

[3] Rosemary Sweet has noted the perception that Folkes and 'others had formed that Venice occupied a uniquely densely inhabited space'. Sweet, *Cities and the Grand Tour*, p. 210; see also the Introduction, p. 7 above.

[4] A gilded wood statue of the Archangel Gabriel, which also served as a weathervane, turning into the wind.

[5] Folkes left a blank space here to be filled in later, and he failed to do so. The author also found it difficult herself to measure the height of the campanile in this manner. The Campanile di San Marco is 98.6 metres or 323 feet tall.

[6] Venetian nobles were, 'prohibited by law to speak with a representative of another government', though they, 'donned masks to do just this in the eighteenth century'. Johnson, 'Debt and Sincerity', p. 406. As to the British residency of Venice, it was not just the king's protection that was valuable. The residency was not merely a sinecure, but also offered a real opportunity to buy and sell art to English Grand Tourists such as Folkes. Joseph Smith served as agent for Canaletto from 1729 to 1735, and he would expand his activities as an art dealer when he became

however he deigns me the honour of eating a dinner with me \after/ to morrow, being more fond of English Gentlemen than I perceive are usually are of noble venetians who excepting a very few are but miserable sort of animals. The Lord Marcello is I fear one of the latter, mais il faut passer par la.[1]

Wednesd. 1st July. it being very hot I did not go abroad till evening nor see any body. at night I just took a turn on the water and the Piazza and went from hence to Sig[r] Giustinianis where I staid till supper time.

Thursday. 2. I visited in the morning the Resident, S[r] R. Brown and the Count, and coming from the latter, saw the decoration of S[t] Marcilians[2] bridg it being his feast, a frame of wood representing a sort of open temple was set over it and, adornd with statues made of pasteboard linnen and paper, with festoons of the same the virgin and bambino above, the whole pretty enough. Sig[r] Marcello dind with me according to his invitation he came before my wife had breakfasted so staid all day. I passed the evening as usual at S[t] Marks place.

Fryday. 3. I did not go out till evening, when I rowd partly round the city, and was set down at Sig[r] Giustinianis, where I passd the evening.

Saturday 4. I visited Mr Smith, who was at my house at the same time. I then was at the Place and went up to see the Antient horses[3] which pleased me better neer than from below, there is a very great spirit in the work tho a want of connection in the design, the feet have been most broken off and are supplyd with some of indifferent [f. 15r][4] work enough, there have also been several other breaches that have been solderd up, but by one not knowingly repaird one seest by one hollow tho of a pretty good thickness. In the evening D[r] Michaelotti carried us to a garden near S[t] Roch[5] for a walk it belongs to a

---

Resident of Venice. Smith also was a patron of the pastellist, Rosalba Carriera. Folkes will later record visiting Carriera in his diary (see below, p. 68). Folkes's relationship with the Resident and his visits with Smith during his stay may have been useful in informing his choices in buying and cultivating his connoisseurship of Italian art. See 'Dictionary and Archive of Travellers', The Grand Tour, Adam Matthews Digital, http://www.grandtour.amdigital.co.uk/ Dictionary/TravellerDetails.aspx?documentid=64931 [Accessed 14 Jan 2021].

[1] 'But that is the way it has to be / but that pain will be required.' In other words, Folkes needed to swallow hard and go through meeting with Marcello, like it or not.

[2] Saint Martial or San Marziale, the first Bishop of Limoges, who reigned in the 3rd century. In 1549, Tintoretto painted, *Saint Martial in Glory with Saints Peter and Paul*, for the parish church of San Marziale in Venice, founded in 1133, but rebuilt 1693–1721. Ilchman, 'The Triumph of Tintoretto', p. 28. The Festa of San Marziale on 1 July was always celebrated by the Venetian Republic, as the day was also the anniversary of three Venetian military victories. See Hare, *Venice*, pp. 180–81. In addition to being decorated for the feast day, the Ponte San Marziale was one of the many sites of battle for the stave (*bastoni*), and later fist (*pugni*) fights, or mock battles (*battagliole*) between the Castellani and Nicolotti factions in the Renaissance, the practice outlawed by the Council of Ten in 1705. Gori, 'Sporting Events', p. 586; see also Davis, *The War of the Fists*, p. 28.

[3] Presumably the Horses of San Marco's Basilica, or *Triumphal Quadriga*, which were on the facade of the loggia above the porch until they were taken as war booty by Napoleon in 1797. Taken from Constantinople by the Venetians in the 13th century, 'the collars on the four horses were added in 1204 to obscure where the animals' heads had been severed to allow them to be transported from Constantinople to Venice'. Houpt, *Museum of the Missing*, p. 32.

[4] Folios 16r–19r comprise a set of Folkes's drawings tipped into the diary (Figures 12–15) so there is a textual gap between 15r and 20r.

[5] This could be a reference to the Scuola Grande di San Rocco, owned by the confraternity of San Rocco and named after the 14th-century saint, who is the protector from plague. The Scuola is best known for its works by Tintoretto, particularly its ceiling – Tintoretto's Sistine Chapel – and his spectacular, *Flight into Egypt* (1587). Or Folkes could have been referring to the adjacent Renaissance church, San Rocco, rebuilt in 1725, which

gardiner and is very good for Venice, there are great numbers of standing figtrees, and good mulberrys, I went from hence to the Place where I met the Resident and sat with him till supper.

Sunday 5. The Count was to see me in the morning and did me the favour to dine with me so that I did not go out till I went with him to walk in the garden at St Georges in the evening,[1] from hence we went together to the Piazza where meeting soon after the Resident[2] and Mr Swimmer[3] we sat together till eleven a clock. there talking of several subjects that of apparitions the second sight etc was talkd of and the Resident who is a man that entirely laughs at all stories of that sort yet told me the following very remarkable one in which his self was concernd, and being a man of undoubted veracity there can be \no/ doubt of so much of the affair as fell under his own observation, and as such I thought it worth preserving a memorandum of. When he was a young man, he had the misfortune to fight a duel with another Gentleman whom he killd, and was tryd with one Captn D'Accord who was his Second; he was himself found guilty of Manslaughter and the Captain was acquitted,[4] the day they were dischargd they dind together the Captn and he, they dind together at a tavern with several other friends among which was Brigadeer Tyrral whom I well know in England.[5] after dinner all the Company went away but the Resident and Cap^n^ D'Accord who staid a little after, and as they were both talking very pleasantly on subjects common with young men, which is mentioned that the discourse had nothing to do with what followd, Captn D Accord of a sudden fell back in his chair and changing countenance, lost his speech for some moments and being askd by the other what ~~wh~~ was the matter could only point towards a corner of the room where some fagots were piled up which doing with great earnestness, the ~~consul~~ \resident/ lookt behind him but could see nothing, the other continuing his action with the same earnestness recovered his surprise and voice by degrees to tell him

---

houses the remains of the saint. The Feast Day procession of St Roch took place on 16 August each year, and it was portrayed by Canaletto in 1735 (now in The National Gallery, London), giving a sense of what Folkes saw during his walk in the adjacent garden. See 'Canaletto: The Feast Day of St Roch, Venice', National Gallery, London, https://www.nationalgallery.org.uk/paintings/canaletto-venice-the-feast-day-of-saint-roch [Accessed 12 Jan 2021].

[1] Probably San Giorgio Maggiore, a 16th-century Benedictine church designed by Andrea Palladio, who also designed the inner garden. The church of gleaming white marble sits on an island opposite the lagoon from the San Marco Piazzetta, which connects the south side of the Piazza to the waterway.

[2] The Resident was Elizeus Burges. See above, p. 36, n. 9.

[3] This was probably Mr Swymmer, English merchant (possibly from Bristol) in Venice, also met by Richard Pococke in August 1737. See 'Dictionary and Archive of Travellers', *The Grand Tour*, Adam Matthews Digital http://www.grandtour.amdigital.co.uk/Dictionary/TravellerDetails.aspx?documentid=65315 [Accessed 13 January 2021].

[4] As Folkes's diary indicates, Burges indeed killed two men, one in a duel and one in a bar brawl, the latter victim a well-known comic actor named Hildebrand Horden. Burges was subsequently arrested for manslaughter and put into the Gatehouse Gaol in Westminster, where he escaped a week later in an organized gaol-break. By the end of Nov 1697, Burges had obtained the King's pardon. See Ford, 'Col. Elizeus Burges', *Royal Berkshire History*, http://www.berkshirehistory.com/bios/eburges.html [Accessed 14 Jan 2021]. For the trial deposition, see Old Bailey Online, Elisias Burges, 23 February 1698, case t16980223-17, https://www.oldbaileyonline.org/record/t16980223-17&div=t16980223-17 [Accessed 19 March 2025]. The incident was long remembered, recorded by the actor Colley Cibber in 1739. Cibber, *An Apology*, I, pp. 302–3.

[5] Lieutenant General and Whig MP, James Tyrell (*c.*1674–1742), who was made a brigadier general in 1727. Folkes may have known him via his friendship with John, 2nd Duke of Montagu, who was made a major general in 1735 and Master-General of the Ordnance in 1740.

what he saw was a bloody hand which came forward from that corner of the room towards the table, and now says he it is on the table grasping at the piece of bread that lyes before you, and imitated with his hand the action he said he saw the hand continue to make, astonsht Mr Burges saw nothing which he never did. after some little time the captn said he saw the hand go back from the table to the place whence it came, he pointing to it all the way: he was some time [f. 20r] before he recovered his surprise; and the Resident is well assured that was real and that the Captain entirely believed and was frightd with what he said he had seen. now what is most remarkable is that very soon after Mr Burges had the misfortune to fall again into the same accident, to fight and kill this man.[1] I confess here circumstances seemed to me so odd that I was willing to write them down immediately, as one of the most particular things of the sort I've met with.

Monday 6. I waited on the Resident who signd a certificate for the Exchequer.[2] in the evening I met at the place Sigr Riva[3] Professor of Astronomy at Padua, with whom and the Consul and Sigr. Martinelly[4] I sat the evening.

Tuesday 7. it being very hot I did not go out in the forenoon, I was visited by the Consul, and old Marcello who dind with me, and after dinner by the Resident, with whom I went to see Miss Edwin and afterwards to take the air towards Lido from whence I cam as usual to the Piazza.

Wednesday. 8. The Arch of the Rialto Brig making a good deal of noise and writers not agreeing perfectly with one another about its dimensions, I went this morning to take its dimensions my self;[5] and in order to be sure what I did I made the scetches on the other side to show the lines I really took the measures of. I then stretcht a fine packthread[6] that would not very sensibly sway down the in the middle curves

[1] Folkes seemed to think the ghost was a premonition of the second murder.

[2] Folkes may have been cashing in a government annuity.

[3] Ludovico Riva (1698–1746), Professor of Astronomy in Padua. Folkes nominated him by letter on 1 Oct 1733 to be a Fellow of the Royal Society, as a: 'Person of great Modesty and knowledge in his way and Author of Several Works, which he hath given me to be presented to the Society, one of which is a dissertation on certain fiery Meteors, that have lately appeared and done a pretty deal of Mischief. He is likewise strongly recommended by the Marquis Poleni and Dr Michaelotty.' Riva was elected on 24 Jan 1733/4. See EC/1733/07, The Royal Society, London.

[4] Probably Cristino Martinelli. See above, p. 52, n. 10.

[5] The Rialto, 'was the most ancient settlement of Venice, known for the presence of a market, and takes its name from the Latin *rivus praealtus* (then *rivus altus*), which means "deep channel", and which can be referred to the present-day Grand Canal, of which the Rialto islands form its shores'. See Bernabei, et al., 'The wooden foundations of Rialto Bridge', pp. 85–6. The bridge was originally wooden, built in the 12th century by Nicolò Barattiero, a Lombardy carpenter. The bridge was the only way to cross the Grand Canal by foot, so it had heavy usage, but also had to allow boats to pass underneath. At the end of the 16th century, the Venetian government launched a competition to rebuild the Rialto in stone, with Vincenzo Scamozzi and Antonio da Ponte proposing a bridge with a single arch; the proposal was selected on 9 June 1588 and the work completed in 1591 with the assistance of architects Antonio and Tomasso Contin da Besso. Folkes was correct that there was frequent discussion, vague description and some disagreement on the measures of the arch. As Maximilien Misson stated, the 'Compass of the Arch makes exactly the third Part of a Circle, and there are Ninety Foot [27.4 metres] from one End, or one Extremeity of the Arch to the other, on the level of the Canal: From which we may gather, that the Arch is near Twenty four Foot [7.31 metres] high'. Misson, *A New Voyage to Italy*, I, p. 262. Thomas Salmon's subsequent publication, *Modern History: or, the present state of all nations*, agreed with Misson. Salmon, *Modern*, X, p. 7. Chamber's *Cyclopædia* noted, 'the span of the arch is ninety-eight feet one half [30 metres], and its height above the water only twenty-three feet' [8 metres]. Chambers, *Cyclopædia*, I, s.v. 'Bridge'.

[6] Packthread is 'strong cord or twine used for sewing or tying up packs or bundles'. *OED*.

what I call the Chord of the Arch. this I did twice, and did not differ more than a quarter of an inch from my self. I then measurd this packthread 3 times with what care I was able indeavouring to stretch it with the same force I had done when I drew it cross the bridg, and found it 99, English feet and 8 inches. I then observed The ornament that runs round the Arch exactly touches the bottom of the Modiglions[1] of the cornish that runs every where under the balustrade, and from this cornish to the water I let fall a plummet and found the height 27 English feet and t[w]o inches, from this height substracting from the depth of the ornament round the arch 2:f. 3 inches and ½ and the depth of the cornish and Modiglions 2.^{f} 7.^{i} ½. and lastly the depth of the water line at that time below the chord of the Arch 3 foot together 7 foot 11 there remains for the height of the Arch sgf:ii English,[2] which measure I dare be confident differ very little from the truth. Now upon these measures I observe that if the Venetian foot be to the English as I have hitherto found it as 109 to 96. this chord of 99^{f}.8 inch or 1196 inches gives Venetian Inches 1053, 3, or feet 87, 8 nearly now were this number 88 feet it would make just 110 Venetian Palms for which I should believe the line was intended, which measure differing little more than 2 inches from mine. I would readily think may be partly owing to defect in my own measure, partly perhaps on inaccuracy in the building the bridg itself to the intended model, and perhaps again in the not having to perfect these proportion[s] between the English and Venetian foot. for the height my measure equals 23 palms and [f. 21r] a half, I should suppose this was intended for 24 Palms, and there may be more reason here for the difference because it is very probably there has been some set[t]ling in the Arch, and besides my own measures of the cornish and work round the arch may be somewhat defective as taken from the small numbers added together, and which being now pretty much flawd as difficult to take exactly. My measures are very different from what it is in sansovino's book[3] I cant help it these are the true ones that ocurred in a tryal made with what exactness I am able. I am the more convincd the palms was the integer made use of by the builder from the frequent occurence of the measure of 11 inches English among the small numbers that quantity being very near equal to the venetian Palm. I was visited the afternoon by Count Bielgne, and I spent the evening at the Giustiniani Pallace.

Thursday. 9. I dind with S^{r} ~~Thom~~ Rob. Brown and afterwards was with him to the Residents where I passd the remainder of the evening. the Count was with us.

Fryd. 10. I went after Dinner with S^{r} Th Twisden and Sig^{r} Gislansoni[4] to see a

[1] Any of a series of projecting brackets placed on the underside of the corona of the cornice in the Corinthian, Composite, and (less commonly) the Roman Ionic orders; also in extended use. See *OED.*

[2] These measurements refer to Folkes's diagrams. See Figures 12–15. An English foot was 12 inches, though other countries had different measures for the foot, a preoccupation of Folkes's metrological research. For the Venetian foot, see above, p. 39, n. 2.

[3] Folkes referred to Sansovino, *Venetia, città nobilissima*, p. 365. He possessed a copy in his personal library. See *A catalogue of the entire and valuable library*, p. 6, lot 197. Folkes carefully compared his own measures of the Rialto to those taken by Francesco Sansovino, the son of Jacopo Sansovino (1486–1570). Jacopo was the *proto* or chief architect of Venice, who was responsible for the rebuilding of the Fabbriche Nuove di Rialto complex (seat of the magistrates in charge of customs and duty).

[4] Patrician of the Venetian noble family Ghizlanzoni, whose Villa Ghizlanzoni del Barco was at Bertesina, a short distance from Vicenza and Venice.

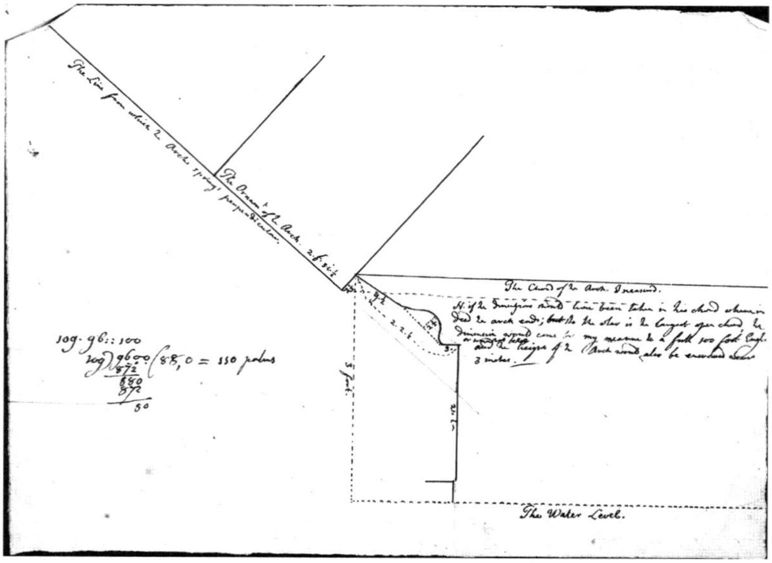

Figure 12. Folkes's measure of the Arch of the Rialto in Venice, f. 16v. He notes 'The Line from which the Arches springs perpendicular', and 'The Ornament of the Arch. 2 f. 31½'. He also notes: 'The Chord of the Arch I measurd'. 'N. if the dimensions should have been taken in this chord where indeed the arch ends; but tho the other is the largest open chord, the dimension would come by my measure to a full 500 foot Engl. \or somewhat better/ and the height of the Arch would also be encased near 3 inches.' Note the mathematical comparison of English feet to Venetian palms on the left. Courtesy of the Bodleian Library, University of Oxford.

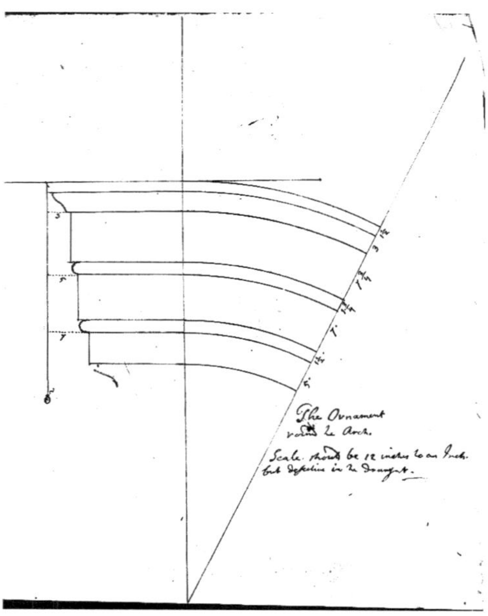

Figure 13. Folkes's drawing of the ornaments above the arches of the Rialto, f. 17v. He wrote: 'The Ornament round the Arch. Scale should be 12 inches to an Inch. but defective in the draught'. Courtesy of the Bodleian Library, University of Oxford.

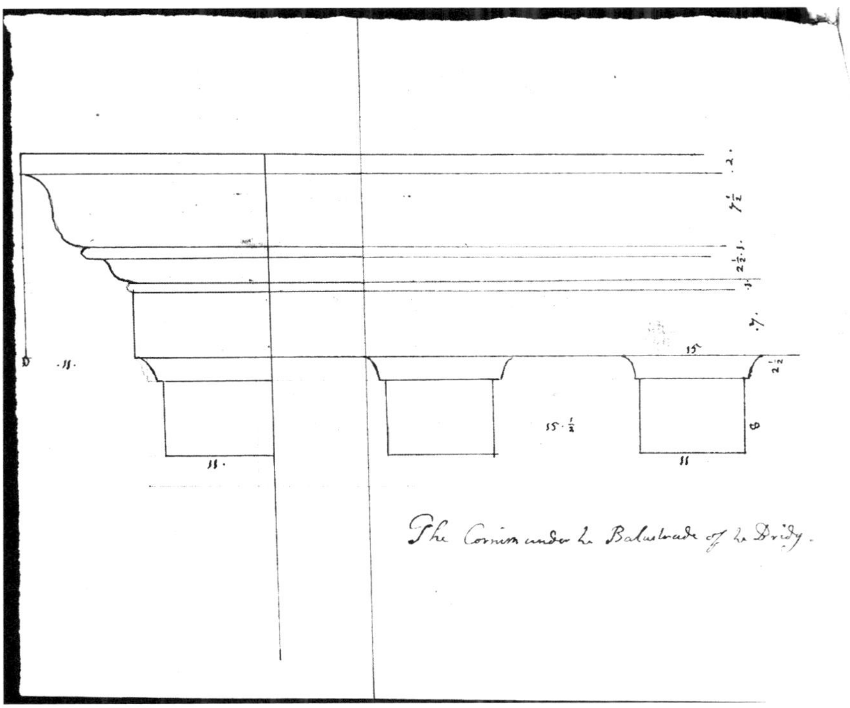

Figure 14. Folkes's drawing of the 'Cornish under the Balustrade of the Bridg', the Rialto in Venice, f. 18v. Courtesy of the Bodleian Library, University of Oxford.

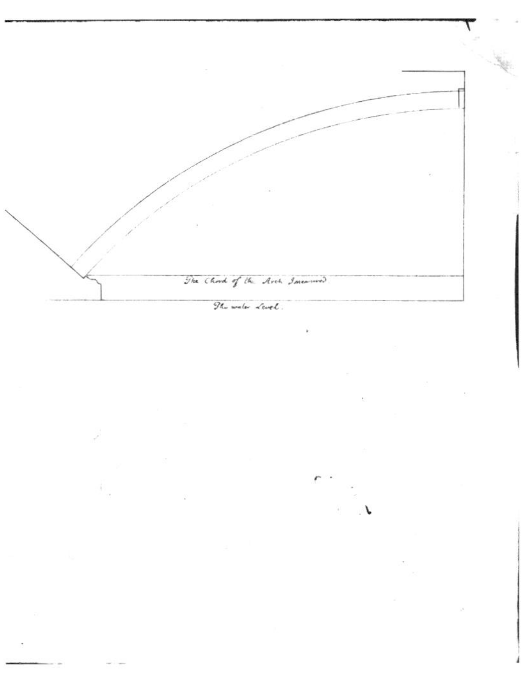

Figure 15. Folkes's drawing of 'The Chord of the Arch I measured. The water level', the Rialto in Venice, f. 19v. Courtesy of the Bodleian Library, University of Oxford.

collection of Antique Busts belonging to the family of Trevisani,[1] which they now want to sell, there is also some few statues bass reliefs altars etc they were collected by the ffamily and cheifly brought from Greece to Morea,[2] and Islands of the Archipelago, they are undoubted antiques but very many have been broke and repaird and there are several but indifferent tho truly antient. they talk of sending them to England.[3] I went after to the place and from hence to Sig^{r} Giustinianis.

Saturday 11. I was for the first time at the Publick Library.[4] for the most part given by Cardinal Bessarion, it takes up a good part of the building over against the Palace St Mark, and is a fine room well adornd in painting and ornaments, but a great part of the Cornish which that was Stucco work is fallen down lately and the rest must be new done[5] they have they say a pretty many MSS. but none very antient,[6] ~~here~~ \the/ paintings in the

[1] The Trevisani family, including Francesco (1656–1746) and Angelo (1669–*c.*1753), and some of the most fashionable portrait painters then active in Venice and Rome. Thomas Coke, 1st Earl of Leicester, sat for Francesco Trevisani during his Grand Tour (1717). Coke was a Freemason; his seat was at Holkham Hall, Norfolk. Since Folkes was Coke's neighbour at Hillington Hall, Norfolk, he would likely have been familiar with the portrait, and thus Trevisani, so it may be why he attended the sale. As Joan Coutu noted: 'Coke transformed the northern plains of Norfolk into a rural classical idyll, a reflection of his profound appreciation and knowledge of classical literature. A committed Freemason – he became a grand master in 1731 – he could well have been aided by his fellow Mason connoisseurs, including his Norfolk neighbours Andrew Fountain and Martin Folkes ...'. Coutu, *Then and Now,* p. 25. See also Difederico, *Francesco Trevisani*; Sutton, 'Augustan Virtuosi', p. 321; Bulman, 'Moral Education on the Grand Tour', pp. 27–34.

[2] Morea, the name of Peloponnese peninsula in southern Greece in the medieval and early modern period.

[3] There was a brisk market in antique, repaired antique and fake antique busts to English Grand Tourists. See Craddock, *Scientific Investigation,* p. 242.

[4] Biblioteca Nazionale Marciana, otherwise known as the Libreria Sansovino. The original collection consisting of 482 Greek manuscripts and 264 Latin manuscripts was donated in 1468 by the humanist, neoplatonist and scholar, Cardinal Basilios Bessarion (1403–72). Bessarion specifically asked 'to let public access for all those who would like to "study and read" his manuscripts'. See Raines, 'Book Museum or Scholarly Library?', p. 34. The Venetian State began constructing the library in 1537. Jacopo Sansovino as architect harmonized classical Renaissance style with the Venetian aesthetic. See 'Biblioteca Nazionale Marciana', https://www.venice-museum.com/biblioteca-marciana.php [Accessed 22 Jan 2021]. For an overall history of the Biblioteca and its founder Bessarion, see the following exhibition catalogue and collection of essays: Fiaccadori, et al., *Bessarione e l'Umanesimo*. On Bessarion see Märtl, et al., *Inter graecos latinissimus*, pp. 1–36. See also Mohler, *Kardinal Bessarion als Theologe*; Mioni, 'Bessarione bibliofilo e filologo', pp. 61–83.

[5] The stucco work on the staircase to the library was accomplished in the 16th century by Alessandro Vittoria. Folkes was not alone in commenting upon the glories of the library's architecture, as well as its decay. As Perry has indicated: 'Foreign travellers of the later seventeeth and eighteenth centuries regularly reported the extraordinary beauty of the architecture, the impressive number of antique statues in the Antisala, the brilliance of the paintings decorating the building, and the disturbing fact the Library itself, despite its rich holdings, was neglected, disordered, rarely frequented.' Perry, 'The Statuario Publico', p. 87. In his 1714 travel memoir, Misson also noted: 'To return to the Library; if it is none of the most numerous, rarest, and best condition'd in the world; yet, to make amends, it is adorn'd with Paintings by Titian, and some other famous Masters, which are much esteem'd. There are also many Greek Statues of admirable Beauty ... It must be acknowledg'd, that this Library is neither kept in good order, nor much frequented.' Misson, *A New Voyage to Italy*, I, pp. 250–51. It would not be until 1736, when Lorenzo Tiepolo (1673–1742) was elected Bibliotecario [Librarian] of the Marciana, that the library was revived and newly catalogued.

[6] Folkes is incorrect. Many ancient and rare texts, including those of Ptolemy, Cyril of Alexandria, Euclid and Stabo, were in the original collection. See Lebowsky, *Bessarion's Library*, pp. 6–7. An inclusive printed manuscript catalogue of the Greek codicies was not published until 1740, and the Latin in 1741, a few years after Folkes's visit, so that might account for his misapprehension. See [Antonio Maria] Zanetti, *Græca D. Marci Bibliotheca; Latina et italica D. Marci Bibliotheca.*

compartments, ~~of~~ the cieling and over the books are of Titian, Paulo[1] and Tintoret etc[2] the particulars are set down in the book sold here. a sort of Antichamber[3] to the Library is filld with antique statues and busts given by 2 of the Grimanis,[4] there are among here some of the best things of that sort I have seen, the Ganimeda and Eagle hanging from the cieling has a very uncommon elegance.[5] A Leda and Swan is also very fine and I was much pleased with 2 bass reliefs representing antient Sacrifices over the door. other things here I may personally take notice of another time. After dinner we were visi[ted by] [f. 22r] M[rs] Edwin her brother etc and M[r] Stroud who are going immediately for Vienna, and after they were gone we removed into their lodgings which we think somewhat more convenient for our ffamily than our own; tho I am sorry to leave Mr Beau Cousin[6] who has treated us extreamly well. I shall not find so good a table.

Sunday. 12. I was to wait on Count Bielgne and the Consul, as also after dinner on S[r] Rob Brown, the Even[in]g I spent at Sig[r] Giustianinis where was a pretty deal of company, particularly Pr Gonzaga,[7] Sig[r] Grimani[8] who has been

[1] Presumably Paolo Veronese.

[2] In 1556–57, 21 *tondi* of the ceiling were painted in for an art competition judged by Sansovino and Titian. Veronese was the winner with his three entries – *Il Canto*, *La Musica*, and *L'Onore*. Perry, 'The Statuario Publico of the Venetian Republic', p. 78, n. 12.

[3] The *Antisala*, devised by Vincenzo Scamozzi. The *Antisala* had previously been used as a classical academy for sons of the aristocracy, but Scamozzi constructed wall niches to accommodate statuary and bas reliefs, so the walls of the *Antisala* could display smaller pieces, and the larger statues, altars, and urns were placed on the floor. See Perry, 'The Statuario Publico of the Venetian Republic', p. 81.

[4] Donated by Domenico Grimani and Giovanni Grimani. Cardinal Domenico Grimani (1461–1523), had an important collection of Roman antiquities, including statues and inscriptions, originally housed at his residence in the Palazzo San Marco. On his death, the Cardinal bequeathed a part of his collection to the Republic of Venice. These marbles (eleven busts, four small statues, and a torso) were subsequently displayed in the Palazzo Ducale, in the room, *Sala delle Teste*, from 1525 to 1586. Giovanni Grimani (1500–93), nephew of Domenico, was also a collector, filling the courtyard and rooms of his palace near Santa Maria Formosa. Grimani also offered to donate his collection to the Republic of Venice, but only with the proviso that it would be a public collection. After Giovanni's death, the collection moved to the antechamber or the *Antisala* of the Marciana Library, where it was displayed from 1593 until 1812; the room was known as the Statuario Pubblico until it was dispersed. These works form the central part of the Museo Archeologico of Venice, displayed in the Procuratie Nuove. See Perry, 'The Statuario Publico of the Venetian Republic', p. 81; See also de Grummond, *Encyclopedia*, pp. 541–2. For a reconstruction of the placement of the statuary of the collection in the Palazzo Grimini which Folkes would have seen, see also Kaplan, Case Studies, https://digitalvenice.wordpress.com/2014/10/01/second-digital-humanities-fall-school-in-venice-case-studies/ [Accessed 23 Jan 2021].

[5] A 1736 manuscript inventory of the antique marbles was compiled by Antonio Maria Zanetti, the 'Custode' of the Marciana, and it confirms Folkes's observation that Ganymede was airborne above the head of Pomona, as well as the placement of the statue of Leda. See Perry, 'The Statuario Publico of the Venetian Republic', p. 137.

[6] Silvestro or Lodovico *Boncousin* (also Beaucousin), proprietor of a hostelry in Venice. It was also the inn where Giacomo Casanova stayed when he returned to Venice. See Casanova, *History of My Life*, II, pp. 217, 230; IX–X, p. 385.

[7] Possibly Luigi II Gonzaga (1680–1746). He was the titular prince and pretender to the throne of Castiglione delle Stiviere, and was a Venetian patrician, serving the Republic from 1723. His son Don Leopoldo (1716–60), also a patrician and Venetian army general, is also a possibility. See Marocchi, *Castiglione delle Stiviere*.

[8] Pietro Grimani, who served as Venetian Ambassador to England from 31 Oct 1710 to 15 June 1715. 'Venetian Diplomatic Agents in England', in *Calendar of State Papers Venetian*, I, pp. cxxii–cxxix. *British History Online*, http://www.british-history.ac.uk/cal-state-papers/venice/vol1/cxxii-cxxix [Accessed 24 Jan 2021].

Embass in England, Young Sig^{r} Barbarigo,[1] young Sig^{r} Tron,[2] and several others.
Monday. 13.

I was visited in the morning by The Consul, Sig^{r} Riva the Professor of Padua, and also by The Resident. Old Sigr Marcello was also at my Lodging when the Resident came in, but immediately walk'd out of the room as frighted, I am told he must go instantly to the Inquisition of State to inform of the accident by not being on any account permitted to be in company with a foreign Minister. I passd the evening at the place with Sig^{r} Martinelli and Sig^{r} Gislanzoni the former is to dine with us tomorrow.

Tuesday. 14. I went to Sign^{a} Rosalbas[3] whose pictures in crayons have been with Justice esteemd the most excellent pieces of art of that sort. She is now ~~between~~ better than 50, and was formerly as deservedly famous for her works in water colours;[4] I was here extremely well entertaind with a great number of fine portraits some of my acquaintance very like. There are 3 sisters which were all together when I was here. Mad[ame] Rosalba herself and one of the others is single,[5] the third is wife to Sig^{r} Pelegrini[6] and was with him in england, for which she expresses a great regard. I saw among others a fine idea representing musick the face was taken I heard from a person now kept by an English Nobleman, but the whole attitude was as fine and well disposed as almost any th[in]g I have seen.[7] Mr Martinelli dind with me and in the evening I saw at the Church of the Apostoli[8] the ceremony of a cristening. it was my Landlords child where I lodg. besides

[1] Gregorio Barbarigo (1711–1801), scion of an aristocratic family of Istrian origin. He would become a significant figure in the Venetian Republic, 'one of the Republic's most rigid conservatives, he was famous for his profound religiosity and intransigent habits as well as for his hardfought victories in the Venetian senate over the innovative proposals of his former friend and ally Andrea Tron regarding the regulation of ecclesiastical institutions and religious affairs.' See Barcham, 'Secular Commissions', p. 153.

[2] Andrea Tron (1712–85), was from an eminent patrician family in Venice, and served as a senator (1778), a Venetian Ambassador to Paris and Vienna, and Procurator of San Marco (1773). Tron used his marriage to the poet Caterina Dolfin, member of a prominent patrician family, to make his political ascent. See Pitteri, *Per un confinazione*; Queruel, *Andrea Tron*; Tabacco, *Andrea Tron (1712–1785)*.

[3] Rosalba Carriera (1673–1757), the famous Venetian portrait pastellist. As Neil Jeffares has indicated, she 'was admitted – on merit, rather than as the *accademica d'onore* title normally given to women artists ... to the Accademia di San Luca in Rome on 27 November 1705, as a "pittrice e miniatrice veneziana"; her reception piece was a miniature on ivory'. In 1709, she painted a renowned pastel self-portrait, which entered the collection of Cosimo III de'Medici, now in the Galleria degli Uffizi, Florence. In England, her followers included William Hoare of Bath, Francis Cotes and John Russell. Jeffares, 'Carriera, Rosalba', *Dictionary of Pastellists before 1800*, http://www.pastellists.com [Accessed 25 Jan 2021].

[4] Before turning to pastels, Carriera had made a career for herself as a portrait miniaturist, painting the lids of snuff boxes. Biographies of Carriera include those of Sensier, who translated her diary. Sensier, *Journal de Rosalba Carriera*; See also Hoerschelmann, *Rosalba Carriera, die Meisterin der Pastellmalerei*; Malamani, *Rosalba Carriera*. The catalogue raisonné is by Sani, *Rosalba Carriera*, and the most recent study is by Oberer, *The Life and Work of Rosalba Carriera*.

[5] Giovanna Carriera (1675–1737), sister and assistant to Rosalba.

[6] Angela Carriera, sister to Rosalba, and wife of the painter, Giovanni Antonio Pellegrini (1675–1741).

[7] Probably, *Music (Terpsichore or Erato?), with lyre* (*c.*1732). The sitter was Lucia Panichi, an opera singer who performed in London, 1739–42. She was known as, 'La Muscovita', or, 'Madame Mosokoveti', an opera singer, and was the mistress of Lord Middlesex. See Jeffares, 'Carriera, Rosalba, Mythological Subjects: *Music (Terpsichore or Erato?), with lyre* J.21.1661', *Dictionary of Pastellists before 1800*, http://www.pastellists.com [Accessed 25 Jan 2021].

[8] Chiesa dei Santi Apostoli di Cristo (Church of the Holy Apostles of Christ), most notable for the Cornaro Chapel.

the sprinkling with water, they put salt (Sal Sapientia)[1] into his mouth, spittle with the finger into his ears to open them to receive the word of god, they blew upon him to chase the impure spirit, and anointed him with oyl *Chrismate Salutis*.[2] I observd the women did not follow [f. 23r] the Child to Church as with us, and he was held by the Godfather.[3] in this Church I took notice of an odd inscription over an altar that for every Mass said at that altar there was a soul delivered from Purgatory. I passd the Even[in]g at the place.

Wensday. 15. Fat[h]er Crivelli came and dind with me, in the even[in]g we went out among the booksellers[4] and to the place and from hence to the Palace Giustiniani where we had a very agreeable conversation with a pretty deal of company.

Thursday. 16. Sr Rob. Brown, and his Lady, the Consul and Sig^{r} Gislanzoni dined with me, in the evening I went with the 2 latter to the Place, but not in other company.

Fryday. 17. I went in the morning to Fat[h]er Crivelli who made me some compliments from Father Grandi[5] lately arrivd at Venice from Bologna. Old Marcello had invited himself to dinner with me, and in the afternoon, it being the feast of S^{t} ~~Barbara~~ Marina,[6] we went to the Island of the Gratie,[7] behind St Georges[8] to see the Pilgrims who set out this day upon their ~~to~~ devotion to St Frances of Assise,[9] they go off

[1] In early modern Italy, it was common for the infant to be brought by godparents to be baptized with salt, water and oil. 'Salt was placed in some churches on the tongues of the catechumens, as an emblem of wisdom and an admonition to attain it. With salt, milk and honey were given. In the *Sacramentary* of Gregory the Great, after the benediction and consecration of salt, it is said, "Hac oratione expleta, accipiat sacerdos de eodem sale, et ponat in ore infantis, dicendo, Accipe sal sapientiae in vitam aeternam" ("This benediction being finished, let the priest take a portion of the same salt and put it into the mouth of the infant, saying, Take the salt of wisdom to eternal life").' See 'Salt, Ecclesiastical Use of', *McClintock and Strong Biblical Encyclopaedia*, ed. John McClintock and James Strong, https://www.biblicalcyclopedia.com [Accessed 25 Jan 2021].

[2] The chrisma salutis, or, 'chrism of salvation', a scented oil blessed by the bishop. As Benitez explains: 'Written into the fattest fatness of olive oil by the [Holy] Spirit, identified in the descent of the Spirit on Christ, and mingled with the Spirit in its consecration, the chrism is a creature infused with sanctification'. Benitez, 'Richness, Fatness and Salvation', p. 20.

[3] For more on this custom, see Terpstra, *Religious Refugees in the Early Modern World*, p. 196; Black, *Church, Religion and Society*, pp. 93–4.

[4] Despite his disparaging comments in his diary about Venetian books and booksellers, Folkes visited during the peak of the Venetian book trade. As Walter Panciera indicated, by the 1730s at the height of the growth of the book market, 'the street of the Mercerie could look like an academic road with a large number of bookshops that made it look like a permanent book fair ... in 1735, there were 26 registered print shops with a total of 96 working presses'. Panciera, 'The Industries of Venice', p. 205.

[5] Guido Grandi (1670–1742), historian of the Benedictine Camaldolese order and famous mathematician, friend and correspondent of Newton.

[6] This is the feast day of Santa Marina the Monk (*c*.5th-century Byzantium) whose remains are now in the church of Santa Maria Formosa in Venice, transferred to Venice in 1213 in a reliquary. On her feast day, there was a ritual procession by members of the Venetian Scuola Grande to the parish church of Santa Marina connected to marking the anniversary of the recovery of Padua in 1509 from the troops of the League of Cambrai. See Muir, *Civic Ritual*, p. 213. See also Folda, 'Reliquary of Saint Marina', pp. 496–7; Hahn, *Strange Beauty*, p. 224.

[7] La Grazia or Santa Maria della Grazia, an artificial island between the Guidecca and San Clemente, which took its name from a miraculous Madonna on display in its church. There was also an Hieronymite monastery, a Franciscan convent, and a home for pilgrims to the Holy Land.

[8] San Giorgio Maggiore, an island with a church of the same name designed by Palladio, built between 1566 and 1610.

[9] 'In 1216, while praying in the Portiuncla, Francis had a vision that ordered him to ask the Pope to grant indulgences to anyone who confessed and received the sacraments in the church.' Pope Honorius III subsequently granted a plenary or lifetime indulgence to any pilgrim who received the sacrament in the Portiuncla (a small chapel located within the Papal Basilica of Saint Mary of the Angels approximately 4 miles

in Pilotas (large boats)[1] done out with colours, and most of the company of Venice go to see them in Gondolas, and others in large hired boats calld bataglias, they are the market boats of the city, and from hence comes the University word Battles[2] at Oxford which signifies the same as sizings with us at Cambridg.[3] The Pilgrims set out from this Island ~~are~~ performing this devotion to the Vergine della Gratia before their departure. The Venetians make this a day of masking, without which they have no idea of a festival; I never saw such a throng of boats as this, the Gondoliers striving to show theire dexterity by rowing in the crowd.

I have since I came hither made a pretty many reflexions upon the manner of thinking I have found with relation to religion and I am perswaded there is here unless among the fanaticks I mean such as we should call in england religiously melancholy, much less real belief than in England, nay it often inclines to direct disbelief of the very first cause it self,[4] but in general reveald religion is in no vogue among those that have any degree of learning, nor is it at all wonder full when we consider how very irrational that religion as they are taught by rote, which it cannot but be the first consequence of any knowledg or free conversation to shake off: and never having had any thing more rational put into their heads with it, with the foolerys the whole goes, and it is the opinion of all the wiser sort

[f. 23v] *The Gentleman has formerly been an Inquisitor of Note and is of one of the chief familys, but happening to have some dispute with one that was Inquisitor after him he struck him in the place S[t] Mark upon which he was forced to leave his country and livd some years at the Court of Hanover but has since made his peace.[5]

[f. 24r] that Religion is only a tool of the state to keep the vulgar in awe;[6] this I have met with delivered with great freedom both among seculars and ecclesiasticks I have talkd

(6.4 km) from Assisi) from vespers on 1 August to sunset on 2 August. This indulgence spawned one of many pilgrimages to the saint, most along the St Francis Way from Florence to Rome. Davidson and Gitlitz, eds, 'Assisi', in *Pilgrimage*, pp. 43–5.

[1] Presumably pilot boats. See Martin, *The Art and Archaeology of Venetian Ships*. There were also the *peote*, ornate canopied boats that belonged to the aristocracy, the *tartana* fishing boats, and *burchielle*, heavy cargo boats that carried lumber, salt and sand.

[2] The *OED* indicates for battels: 'In Univ. of Oxford: *(a)* college accounts for board and provisions supplied from the kitchen and buttery; *(b)* (in looser use) the whole college accounts for board and lodgings, rates, tuition, and contribution to various funds, as "My last term's battels came to £40"; also *attributive*, as *battel-bills*.' It is uncertain if Folkes is correct, as 'the etymology of this, with its associated verb, and derivative *batteler*, has been the subject of abundant conjecture'. See *OED*.

[3] Sizings: Cambridge term for the college allowance from the buttery. See *OED*.

[4] Namely, belief in God.

[5] A member of the House of Venier, one of the most powerful dynasties in Venice, dating (fallaciously) their ancestry back to the Roman Emperors Valerian and Gallienus, and they were one of the caste of families listed in Venice's Golden Book of the nobility, serving as doges, procurators, archbishops, consuls, and inquisitors. Their unfinished palace, Palazzo Venier dei Leoni, is now the home of the Peggy Guggenheim Collection.

[6] Although a Freemason, Folkes certainly entertained materialist philosophies before and after his tour of Italy; the Mancunian poet, John Byrom, recorded their conversation when at Folkes's house on Queen Square for dinner on 26 Feb 1725/6, to see his 'books and rarities'. They talked with the surgeon, William Cheseldon, and the watchmaker, George Graham, about Mrs Folkes's very comical pet monkey. They, 'disputed whether he had reason or no', concluding that, 'he was a man without reason'. Byrom, *The Private Journal*, I, part 1, pp. 209–10. Although the precise line that divided deism, natural religion and the revealed religion of the orthodox Christian was subtle in this period, Folkes's travel diary to Italy confirmed his freethinking beliefs.

with. who have not struck in the planest terms to assert every man was obligd to follow the external religion of his prince. It is incredible with what greediness they are fond of any sort of freethinking book, and with what eagerness they speak on the subject and enquire after the ways of thinking of a nation more used to liberty of thought than themselves.[1] The bible is no ways known and I have met with stupendous instances of mistakes about the things contained in it from men otherwise very well acquainted with the common parts of Litterature.[2] I apprehend the first prohibition of reading it, keeps them from enquiry till they are forward in infidel notions and then they think it not worth the looking into it:[3] and it has been with the greatest surprize several have heard me speak of the measure of learning of all sorts that very antient to say no more book contains. but I observd before fanaticism to be exceedingly common, I often go into churches in the afternoon especially, and allways meet numbers of people seeming naild to em, *who depart not from the house of the Lord*, but without any rational devotion seem to dream away their time there in a moping condition: it may I think deserve very serious consideration whether it be not a strong argument of the badness of the Roman ffaith but among the professors of it there seems no medium between, a direct abandon of all religion, and a superstitious adoration accompanyd with direct madness.[4]

I went after my return from the Gratie, to Sigr Giustinianis, where I met Sigr Zanini, Sigr ~~Gran~~ Vanieri*[5] who has been twice in England Father Crivelli etc.

Saturday 18. I waited on the ~~Consul~~ Resident in the morning who is leaving the town for a few days, and in the evening I went with Sigr Gislansony to a conversation at a Lawyers house where the Marquis Poleni[6] was expected, and where he accordingly came; I had a pretty deal of conversation with him and he promisd to come and see me to morrow.

Sunday 19. Early this morning was a very considerable storm of thunder and Lightning. I observd before the most considerable claps, a whistling noise like that of a bullet in the air; and the [f. 26r] Marquis Poleni told me afterwards he observed the same thing and that that noise always accompanies that thunder that does mischief in his country, and

[1] Federico Barbierato's, *The Inquisitor in the Hat Shop*, demonstrates the widespread presence of unbelief in 17th- and 18th-century Venice, based upon his analysis of records of the Venetian Holy Office. Barbierato wrote: 'the clergy started to be thought of as a social group that used bogus concepts such as punishments or rewards after death to keep society within closely controlled limits ... Therefore, even though religion was widely accepted as a necessary surveillance device to maintain the social fabric, it came to be considered above all as an instrument of control, a set of dogma, norms and prescriptions completely removed from any considerations of the afterlife.' Barbierato, *The Inquisitor in the Hat Shop*, pp. 100–101.

[2] Folkes, of course, with Thomas Pellet, had co-edited Newton's *Chronology of Ancient Kingdoms Amended* (1728), and *Observations Upon the Prophecies of Daniel and the Apocalypse of St John* (1733), so he was deeply conversant with the Bible, although a freethinker himself.

[3] Folkes's allusion to the fact Italy was a Catholic country, and the Bible was not in the Italian vernacular.

[4] Folkes's attitude foreshadows that of Ludovico Antonio Muratori (1672–1760), a parish priest in Modena, who attacked popular piety in Italy in his *On the Proper Devotion of the Christian* (*Della regolata divozione de' Cristiani)*, published in Venice in 1747. Muratori: 'sharply condemned the innumberable processions, pilgrimages, cults, festivals, and religious holidays so rooted in popular religious life. Central to his critique was the charge that profane and superstitious practices had thoroughly debased the rituals and practices of popular devotion.' Melton, *Absolutism and the Eighteenth-Century Origins of Compulsory Schooling*, p. 76.

[5] The asterisk refers to the material that Folkes wrote on the verso of f. 23 (p. 70 above).

[6] Giovanni Poleni (1683–1761), FRS, the Italian natural philosopher, physicist and expert in architecture and classical antiquity. For his role in early modern science, see Introduction, pp. 12 and 14 above.

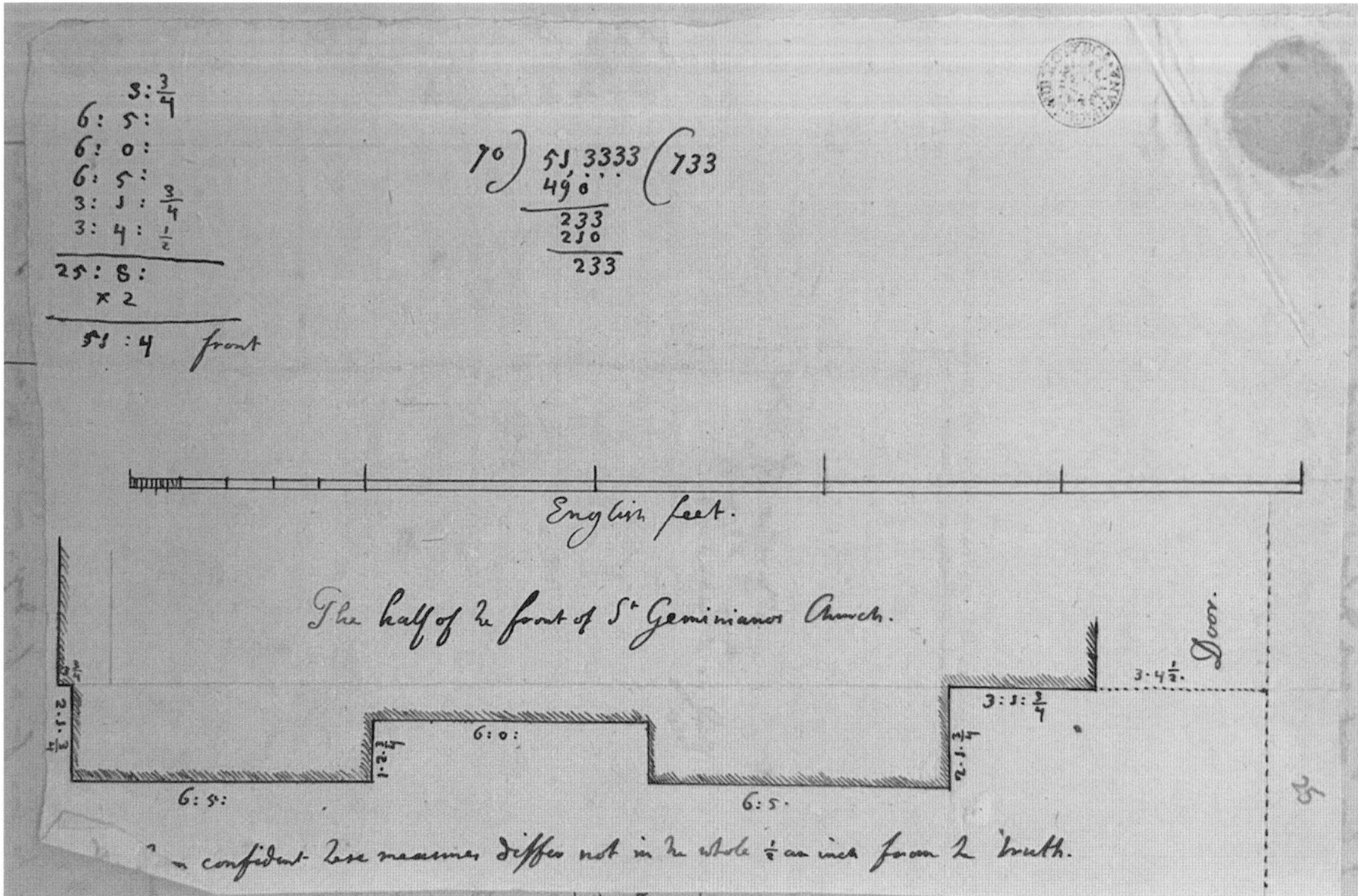

Figure 16. Folkes's sketch of 'half of the front of St Geminianos Church', Venice, in English feet in an attempt to find its standard of measure in construction. He wrote: 'I am confident these measures differ not in the whole ½ an inch from the truth', f. 25r. Courtesy of the Bodleian Library, University of Oxford.

from hence he concluded acidisse fulmina.[1] he came to me about ij a clock and staid most of the morning enquiring about the state of learning and of his friends in England, he returns tomorrow to Padua.[2] I was in the evening at Sigr Giustinianis. Mr Egerton[3] dind with me.

Monday 20. I was not out all day till evening when I went to Wish Sr Rob Brown and his Lady a good Journey they going out of town to Morrow, the consul goes with them for some time he calld upon me in the ~~evening~~ morning. ~~th~~ I after went to take the air and from hence to the place St Mark

Tuesday 21. I was to see Count Bielgne, in the evening going to take the air we met Marshal Schuylenberg returnd from the country. he invited us to his house. I went in the evening to St Marks where I met Sigr Riva, ffather Crivelli, and Sigr Gislansoni.

Wednesday 22. I was but little out only I markt the shadow of the campanile designing to be more exact in my measure of it. in the evening I was with the usual company at Signor Giustinianis.

Thursday 23. I was no otherwise out than to make some enquirys after the original standard of the Venetian measures which I find to be fruitless.

Fryday. 24. I made further enquiry after the measures but to as little purpose as before; I believe there is none they only hold their foot by tradition. I took some measures of St Geminianos Church in order to have made something out by them, but think upon comparing the measures I have taken, it was not built by the Venetian foot but by the Roman Architectonick palm whose length comes out very exactly by it.[4] In the evening I was with the usual company at the Juistinian[5] Palace.

Saturday. 25. I was not out, in the evening was a very great storm of thunder rain and lightning. We have had peaches several days servd at dinner[6] and there are also grapes very common but not yet good.

[1] Poleni's explanation of lightning (*acidisse fulmina*) probably stemmed from Newton's *Opticks*, query 31, pp. 354–5, which itself came from the early modern theory of aerial nitre or atmospheric nitrous acids originating in the works of Paracelsus. Roos, *Salt of the Earth*, pp. 23–4. See also Debus, 'The Paracelsian Aerial Nitre'; Guerlac, 'The Poet's Nitre'.

[2] Poleni was not only a Fellow of the Royal Society and a member of the Académie, but also belonged to the Academies of Sciences of Berlin and St Petersburg, as well as being in touch with many eminent natural philosophers, such as Leonhard Euler, Pierre-Louis Moreau de Maupertuis, Jean Antoine Nollet, Nicolaus I. Bernoulli, Jean Jallabert, Roger Boscovich and Pieter van Musschenbroek. Talas, 'New Light on the Cabinet of Physics of Padua', pp. 49–50.

[3] Samuel Egerton (1711–80) of Tatton Park, Cheshire, who served as an MP from 1754 to 1780. Egerton was apprentice to Joseph Smith in 1729, and by 1730 he was working with Smith and dealing with Canaletto on behalf of his uncle Samuel Hill who commissioned two paintings from the artist. Egerton's Grand Tour portrait by Nazari is still extant at Tatton. On 19 July and 9 Oct 1733, Folkes met Egerton, who was still in Venice in the spring of 1734. In 1738, Egerton returned to England for his succession to his estates Ingamells, *Dictionary*, p. 333.

[4] San Geminiano was in the Piazza San Marco, part of the complete Renaissance remodelling of the piazza and surrounding buildings by Jacopo Sansovino in the classical style. See Figure 16. See also Introduction, p. 25 above.

[5] Giustinian.

[6] Even before the Bellini cocktail, peaches were a long-standing part of Venetian cuisine. Peach stones, radiocarbon dated between AD 650 and AD 700, were found in a sediment core below San Marco's Basilica. Sediment filled in canals when Venice was built on marshland. See Ammerman, et al., 'Beneath the Basilica of San Marco', pp. 1620–29.

Sunday 26. Sig^r^ Alex. Marcello came to dinner with me, and after dinner came with his Interpretor an East Indian of Surat[1] whom I had invited to drink coffe with me, he ~~with~~ and a compagnon he has here with him are sort of factors to a great Indian Merchant that has above 32 Cargo ships of his own trading in several ports [f. 27r] of the East; they come by sea to Moccha,[2] thence with the Caravans thro Arabia to Damascus, and Alexandretta,[3] so by sea to Stanboul,[4] and thence to Smirna,[5] from whence they went to ~~Leghord~~ Leghorn, and so to Lisbon where here business was to get satisfaction for a ship taken by the Portuguese; from Lisbon they returnd to Leghorn and as hit[h]er, whence they go next week to Alexandria in Egypt and so home; this was a good sensible man, a Sonnite Mahometan, and speaking some Italian, they are traders in diamonds and have some very good with them, they say a Grandson of Orungebo[6] is now on the Throne of Industan[7] but they had never seen him nor been at the court which is now kept at Islamabad; they could give no account of the Pagoda of Elora[8] of which I principally wanted to enquire, like true Mahometans having a great contempt of all works of the Gentiles, but these latter they say are still above 2 to one in that part of Industan. I was at night at Sig^r^ Giustinianis.

Monday. 27. I was only out to take the air, but I calld at St Giovanni and Paolo,[9] to

[1] Surat was a city in the Indian state of Gujarat, a large seaport in the 18th century. Folkes was speaking to a diamond trader who was a Sunni Muslim. Traders in Venice and Genoa had controlled the diamond trade until the 15th century, whereupon the Portuguese, with their discovery of the sea route to India, cut out the Italians. The Dutch also began to encroach on the trade in the 17th century, first purchasing diamonds from Surat in 1643, but London replaced Lisbon in the 1670s as the 'main European market for unpolished (or rough) diamonds. Surat itself is still known for processing, polishing and cutting diamonds.' Samuel, 'Diamonds and Pieces of Eight', p. 28. As Ishrat Alam has noted, in India, 'diamonds occur over three wide areas: (1) the eastern side of the Deccan from the Pennar to the Son, (2) Andhra, especially in the Krishna and Godavari basins, and (3) Chota Nagpur and Madhya Pradesh to Bundelkhand'. See Alam, 'Diamond Mining and Trade in South India', p. 311. As Alam commented, India was perhaps the first and possibly the only source of diamonds known initially to the Europeans, but it was not the only place they were mined.

[2] Mocha (or Mukha or Arabic: al-Makhā), a port city on the Red Sea coast of Yemen.

[3] İskenderun, a city in Hatay Province on the Mediterranean coast of Turkey.

[4] Istanbul.

[5] A Greek port on the Aegean coast of Anatolia, now Izmir.

[6] Possibly a corruption of Aurangzeb, or Muhi-ud-Din Muhammad (1618–1707), the 6th Mughal Emperor, who ruled over the Indian subcontinent for 49 years. He spearheaded successful proto-industrialization, and Sharia law. For a recent biography, see Truschke, *Aurangzeb: The Life and Legacy of India's Most Controversial King*. At the time of Folkes's diary, his grandson, Nasir-ud-Din Muhammed Shah or Roshan Akhtar (1702–48), was on the Mughal throne.

[7] Hindustan, the Persian name for the Indian subcontinent.

[8] Kailasa Temple, Ellora. The temples were created from the 6th to the 10th centuries, and were the only site on the Indian subcontinent, to contain major excavations for three of India's indigenous religions: Hinduism, Buddhism, and Jainism. Owen, *Carving Devotion*, p. 1. Antiquaries like Folkes were generally interested in archaeological sites outside Europe. In Folkes's case, this included him collecting evidence for the existence of Rassem the petrified city in the Libyan desert. See Zwierlein, *Imperial Unknown*, p. 289; Roos, *Folkes*, pp. 105–6.

[9] San Giovanni e Paolo, also known as San Zanipolo, a 15th-century church with the status of a minor basilica, in the Castello *sestiere* of Venice. In Folkes's time, the funeral services of the doges were held here, and 25 doges have tomb memorials in the church. The church features notable paintings by Giovanni Bellini of the Saints Vincent Ferrer, Christopher and Sebastian, and had Veronese's famous *The Feast in the House of Levi* (1573) in the refectory, now in the Gallerie d'Accademia in Venice. See Rosand, *Painting in Sixteenth-Century Venice*, chapter 4. See also above, p. 49, n. 5.

look over the monuments etc here is among others that of Bragadenos skin[1] mentiond by Misson.[2] adjoyning this Church is the Scola of St Marco,[3] well describd by Wright[4] to whom I refer, but the pictures are both in an ill light and wretched[l]y damagd.[5] I have bought the great Plan of Venice[6] which seems a very good one, and serves much more to preserve an idea of the town than any description can; I [only] observe Venice is in the main the closest built town I believe in the world there being generally never garden nor yard at all, the canals which answer to the streets are mostly narrower than streets are in great towns, and the land passages are more allys, yet there are some open places which tho not large are more considerable here Misson in general an exact writer seems to think them, but strangers very difficultly get any idea of Venice, all people of any distinction being seldom from their Gondola; and with that the Canals wind so one often comes out in an very different part of the town from what one would expect; however the plan explains that perfectly which I should never have understood without.[7] the number of people I can get no satisfactory account of yet they commonly say 35000[8] but I am told by good hands they fall far short of that computation; If as I have constantly received the bills of Mortality of London contain under a million I should Judg Venice not to exceed a 7 or 8th part of that but I hope to have some further information about it.[9]

[1] The skin of Marco Antonio Bragadin (1523–71), Venetian lawyer and captain general, who was flayed alive by the Ottoman Turks. He was a captain in Cyprus during the war between the Turks and the Venetians, losing the sea port, trading centre and fortress of Famagusta, the only barrier standing between the Ottoman empire and control of the Eastern Mediterranean. After the fortress was besieged, the Turks took revenge on Bragadin by imprisonment, mutilation, and flagellation. His body was then quartered and skin stuffed with straw and sewn, marked with his military insignia, and paraded through Famagusta. Although Amr al-Bahr Mustafa Pasha sent the skin as a war trophy to Sultan Selim II in Constantinople, the Venetians retrieved it in 1577 during the Battle of Lepanto, bringing it to San Zanipolo where it can still be seen today inside an urn. Crowley, 'Bragadin's Defense', pp. 66–75, as well as his *Empires of the Sea,* chapters 16 and 17.

[2] Misson, *A New Voyage to Italy*, I, part 1, p. 303.

[3] Scuola Grande of San Marco, which faces the Campo San Giovanni e Paolo. There is a blot of ink before 'Scuola'.

[4] Folkes is referring to Wright's, *Some Observations*, I, pp. 71–2.

[5] Folkes may have been referring to the works of Tintoretto: *Miracle of the Slave* (1548), *St Mark's Body Brought to Venice* (1562–6) and *Finding of the Body of St Mark* (1562–6). The first two works are now in the Gallerie d'Accademia in Venice, the latter in the Pinacoteca de Brera in Milan. These works were mentioned in Wright's *Some Observations*. He remarked: 'In this School are some of Tintoret's best Performances which are truly fine. The most remarkable of them represents the *Venetians* bearing away the Body of S. *Mark*: which, by some Revelation, they had discovered'. Wright, *Some Observations,* I, pp. 71–2.

[6] Probably Lodovico Ughi's, *Topographical View of Venice* (1729), which was 5 feet × 6 feet (152 × 183 cm), and printed from 20 copper plates on 13 sheets, flanked by 16 views of Venice. It was a bird's eye view at a highly oblique angle, reflecting innovative techniques of surveying. This was probably why the map appealed to Folkes beyond simple navigation. As an antiquarian, Folkes also was a connoisseur of cartography. Item 33 from the Third Night's Sale of the auction of Folkes's prints and engravings notes a map of Venice in his collection. He also had, 'sixteen views in Venice, by Marieski' (item 10). See *A Catalogue of the Prints and Drawings*, pp. 7, 11. Folkes also found Venice confusing and difficult to navigate. See Sweet, *Cities and the Grand Tour*, p. 210. See Introduction above, p. 7, n. 3.

[7] See Misson, *A New Voyage to Italy*, I, part 1, pp. 234–5, for his description of Venice, which Folkes summarizes.

[8] Thirty-five thousand seems quite an underestimate. As Frederic Lane indicated: 'after the plague of 1630–1 had reduced the city's population from roughly 150,000 to about 100,000, it recovered within a decade to 120,000 and fluctuated thereafter between that figure and about 150,000. In 1764–66, the census reported 141,056; in 1790, 137,603 inhabitants.' Lane, *Venice*, p. 424.

[9] Folkes's own estimate was much more likely, ranging from 125,000 to 142,000.

[f. 28r] Tuesday. 28. I am affected with an illness I understand to be very common here with the heat, but it is excessively troublesome. I have very great numbers of eruptions like little heat spots, which give a perpetuall pain like the stinging of wasps every instant and are some beyond expression tho I am otherwise well but very restless and uneasie.[1] I was at night at Sig^{r} Giustinianis.

Wednesd. 29. I got no rest last night with my indisposition and am yet more uneasy with it than yesterday, I have been affected with it about 5 days but I was not excessive troublesome till yesterday. here is to night and so here was last night a concourse of company to Sta Martas[2] where they sup mostly in their boats. I went hither in my boat this evening, and just when onto the Church to which there was a concourse of people and an orchester with musick built upon the occasion, this Church is adornd with pretty good pictures but none of the best hands, Sta Marthas own altar was drest and hung all round with a great quantity of Plate like our milkmaids garlands on mayday. I presume the saint was on a tower of plate *she was busied about many trays* in the scripture,[3] and so they make her their friend by presents of that sort. Alass Dr. Middleton's book[4] falls farr below of the truth of the follys committed in these countrys, the names of the Gods have changed but paganism continues to all purposes.

Thursd. 30. I have had but an indifferent night and am still very uneasie with my indisposition. I was again this morning at the Library of St Mark. the stairs to go up are extremely well adornd but lost as everything here through excessive nastiness.[5] I gave longer this time than before to examin some of the statues in the antechamber to the

[1] Probably prickly heat or sweat rash, called *sudamina* or *sirones*. In his *Practical Medicine* (1634), Daniel Sennert described it 'like little tubercules red and hard, which break out all over the body with an acute itching like the sting of bees, wasps or gnats or like being beaten with nettles'. Sennert, *Praticae medicinae liber quintus*, p. 177, as quoted by Redbourn, 'The History of Sweat', p. 215.

[2] Santa Marta was a 15th-century church on the south-western point of the Island of Venice on the Canale della Guidecca (the building is now owned by the Venetian Port Authority and used for exhibitions and conferences). Folkes is referring to the Feast of Santa Marta, a popular festival celebrated every year on 29 July. Sole was in season (*sfogi zentili* or *Solea vulgaris*), so they served as a large part of the cuisine during the festival. Santa Marta was the patroness of butlers, cooks, housewives and waiters; during the feast, social distinctions were suspended. Canaletto's *La Vigilia di Santa Marta* (1760), now in the Staatliche Museen, Berlin, a night-time scene depicting the festival, shows a distinguished looking foreign visitor being invited to dance by a worker, as well as the feasting and musicians. See Michiel, *Origine delle feste veneziane*, II, pp. 219–28, for a description of the feast of Santa Marta.

[3] In scripture, Martha was known for witnessing Jesus's resurrection of her brother, Lazarus, and for anxiously fretting about the preparation of meals, hence the reference to serving trays in her hagiography.

[4] Folkes is referring to the work by Conyers Middleton, *Letter from Rome, showing an Exact Conformity between Popery and Paganism* (1729). Folkes had a complete edition of Middleton's works in his personal library. See *A catalogue of the entire and valuable library*, no. 92, p. 3. Middleton (1683–1748) is best known for his bitter quarrels with scholar and theologian Richard Bentley (1662–1742). Middleton was made Librarian of Cambridge University in 1722, and Woodwardian Professor of Geology from 1731 to 1734. After his travels in Italy in 1724–5, his *Letter from Rome*, based on ethnographic observation, demonstrated his sarcastic abhorrence of religious 'enthusiasm' and religious mysticism.

[5] Folkes's visit was directly before improvements to the Biblioteca Nazionale Marciana. 'With the election of Lorenzo Tiepolo in 1735 and the nomination of the art historian Antonio Maria Zanetti as *custode* the library underwent a thorough revision and for the first time one can detect a clear cultural policy: the preparation of a modern catalogue, attention to conservation of the collections and their incrementation.' Raines, 'Book Museum or Scholarly Library?', p. 43.

Library. The Ganimede is admirable, and so the Leda with the swan[1] here is also a fine colossal head of Minerva and then 2 statues of falling Gladiators, a bust of Paris or Alys in a Phrygian bonnet, with others of Vespasian, M Aurelius and Severus; and some inscriptons which have been publisht. I am told Sig[r] Gianetti[2] is etching the Antiquitys of this room that extremely well deserve it. In the Library I was shewn a MSS [*sic*] of Ptolemy a very large folio in Greek[3] with Maps which I take notice of it for it being said to be 500 years old, I am not Critic enough in the Age of MSS to determine that, but I observd Cardinal Bessarions arms[4] which are painted in a shield among the Ornaments [f. 29r] of the first leaf are much modern than those ornaments themselves are. The maps are in a coarse stile but like the first printed ones in Ptolemy, the sea is azure colour and so are the rivers very large, and the Mountains represented by a tint of green, no other representation of citys than the names writ. the character is a good round greek letter with accents some at least.[5] I saw besides 2 other MSS of good curiosity tho not antient, they were collections of drawings of Sig[r] Manouchi[6] who wrote the history of Industan,[7] the one are indian ones by the painters of Mogul representing all the Moguls from

[1] For Folkes's comments on the library's decay, see above, p. 66, n. 5. Folkes also described the antiquities earlier in the diary upon his first visit to the library.

[2] In the early 1720s, Antonio Maria Zanetti worked with his cousin on the edition of a two-volume catalogue to depict classical sculptures in the Venetian public collections, the production to which Folkes referred. It was published between 1740 and 1743. See Zanetti, *Delle antiche statue greche e romane*.

[3] This is MS Gr. Z. 388 (=333): https://en.wikipedia.org/wiki/Geography_(Geōgraphikḗ Hyphḗgēsis (Geographical Guidance), by Ptolemy in the Biblioteca Marciana in Venice, containing 27 map projections commissioned by Cardinal Bessarion, and attributed to John Rhosos of Crete (d. 1498). This vellum codex of the Alexandrian Greek, Claudius Ptolemy (*c*. AD 90–*c*. AD 168), embodied the geographic knowledge of the Greco-Roman world. Ptolemy projected the globe onto a two-dimensional plane, and it expressed place coordinates in latitude and longitude in a grid or graticule. Although accurate in its portrayal of Mediterranean geography, Ptolemy erred by representing the Indian Ocean as an enclosed sea. Ptolemy's work was rediscovered by Byzantine scholars in 1300, with 54 manuscript editions prepared, such as the Biblioteca Nazionale Marciana's example. See Burri, 'The Rediscovery of Ptolemy's Geography', pp. 124–50. See also *Mapping Our World*, pp. 18–19; Bardi, 'Islamic Astronomy', p. 342; Rigo, 'Bessarione, Giovanni Regiomontano', pp. 49–110.

[4] The Cardinal's heraldry consisted of a shield with a Cross, clasped by two arms in red and white sleeves to represent the Byzantine and Latin Churches. Above the Cross was a cloud emitting rays, the entire shield surmounted by a cardinal's hat. Bessarion had left for Rome 13 years before the fall of Constantinople, so to some degree he was straddling two cultures. See Monfasini, *Byzantine Scholars*, p. 166; Bolick, 'Culture, Humanism, and Intellect', p. 4.

[5] The map is indeed coloured in this manner, the blue from lapis lazuli, the inscriptions in gilt.

[6] Niccolao Manucci (1639–1717), a Venetian writer, diplomat, artilleryman and cultural intermediary between Europe and Mughal India, best known for his travel account, the so-called, *Storia do Mogor*. Manucci had travelled to the subcontinent in the service of Viscount Bellemont, who died en route to Delhi leading to Manucci's varied career. He not only served as a quack doctor in the Mughal court, but in 1686, he also became a political correspondent in Madras for the English governor, Gyfford. Manucci's collections of drawings, to which Folkes refers, are in the Biblioteca Nazionale Marciana, MS 'The Libro Nero' Codice It. VI. 136 (=8300), a leather-bound volume alternating Indian paintings and sheets of paper inscribed with captions, known as the Libro Nero, due to its black binding. The Libro Nero is analysed in Subrahmanyam, *Three Ways to be Alien*, pp. 163–8, and in Becherini, 'Staging the Foreign', pp. 589–95. The other album of Manucci's collection of Indian paintings and captions, The Libro Rosso (due to its red binding), is now kept in the Bibliothèque Nationale in Paris (Estampes, Rés. Od. 45 pet. fol.). Becherini's analysis demonstrated that some of the paintings, 'were collected within the Deccan region during the period between 1678 and 1685', and others, 'in or around Madras between 1686 and February 1705'. Becherini, 'Staging the Foreign', p. 101.

[7] As Folkes was fluent in French, it is likely he was referring to Catrou's edition of Manucci's work, the *Histoire générale de l'empire du Mogol* (1715). For a modern and abridged edition, see, Irvine, ed., *A Pepys of Mogul India*.

Tamerlane[1] to Orangeb[2] and his sons mostly upon their Elephants,[3] very prettyly done[4] then follow several Raias[5] and other great men of Industan, with some stories of remarkable actions, particularly the presenting Darahs head to Orangeb,[6] and the defeat of [Moradbar][7] as described in Manouchi's history. The other volume was of coarser drawings[8] but of such matters as the Aut[h]or had drawn in his stay in India Idols, temples, priests, habits and manner of the country the whole is curious of any thing I have seen.[9] I omit the cieling of the room as the names of the painters are in the printed description of Venice.[10] ffather Crevelli dind with me to day and being very uneasie all the evening I did not stir out. In the tribune before the Library, is also a Cupid with his bow[11] very like one in the Justinian Galery, here is also a handsome triangular altar of Marble with bassreliefs on the sides and rams heads at the corners.[12] Wright has spoke well of this place.[13]

[1] Timûr Gurkânî or Timur-i-Lang (1336–1405), anglicized as Tamerlane, the Turco-Mongol conqueror and renowned military leader, who founded the Timurid empire, establishing the Timurid dynasty in modern-day Afghanistan, Iran and Central Asia. There is a portrait of him in the Libro Rosso, f. 9r.

[2] Aurangzeb the 6th Mughal Emperor. See above, p.74, n. 6. There is a portrait of him on horseback in the Libro Rosso, f. 29r.

[3] Folkes may have been referring to any of the following, portraying rulers on their elephants in the Libro Rosso: f. 19r, showing Zahir al-Din Muhammad Babur, the founder of the Mughal empire in the Indian subcontinent; f. 23r of Akbar, the 3rd Mughal Emperor; f. 25r of Jahangir, the 4th Mughal Emperor; f. 35r of Murad Bakhsh, the youngest surviving son of Mughal Emperor Shah Jahan; f. 73r of Muhammad Amin Khan, who served in high office during the reign of Mughal Emperor Aurangzeb.

[4] Manuccis's prefatory note to the Libro Rosso noted: 'Before leaving the Mughal kingdom, to satisfy my curiosity I caused to be painted (by means of one of my friends named Mir Muhammad, an official in the service of Prince Shah 'Alam) all these portraits of the kings and princes from Tamerlane up to Aurangzeb, and of the sons and grandsons of the latter, together with the portraits of the kings of Bijapur and Golconda, and with some of the main Gentile princes and other famous generals, all of them drawn from the original portraits of the royal court.' Manucci had, 'the paintings produced at his request during his service at the court of the Mughal prince Shah 'Alam, one of the sons of emperor Aurangzeb Alamgir'. Becherini, 'Staging the Foreign', p. 104. See Plate 9.

[5] Rajas (Rajs).

[6] The scene in which Aurangzeb received the severed head of Prince Dara Shikoh (1615–59), his eldest brother, is portrayed in the Libro Rosso, f. 31r. See Becherini, 'Staging the Foreign', p. 583.

[7] Folkes's reference is unclear. Possibly Murad Bakhsh (1619–61), portrayed in battle in the Libro Rosso, f. 35r. See Becherini, 'Staging the Foreign', p. 383.

[8] The Libro Nero. As Becherini has noted, the paintings were the products of a pictorial tradition developed in the 17th century under the Nayaka rulers of South India. Becherini, 'Staging the Foreign', p. 127.

[9] There are portrayals, for instance, of Krishna subduing the demon Kaliya, and a series of 5 paintings of a Hindu wedding.

[10] The ceiling of the Salon di lettura (reading room) of the Biblioteca has 21 *tondi* (oil on canvas) painted between 1556 and 1559 by seven different artists chosen by Jacopo Sansovino and Titian. 'Biblioteca Marciana', Foundazione Musei Civici Venezia, https://correr.visitmuve.it/en/mostre-en/archivio-mostre-en/veronese-myths-portraits-allegories/2011/10/4609/verone-se-a-venezia-2-marciana-1/ [Accessed 25 March 2021].

[11] The catalogue for the collection of the Statuario Pubblico compiled by Antonio Maria Zanetti in 1736, noted a statue of, 'Cupid stringing a bow', as part of the original Grimani collection. In 1795, the sculptor Antonio Canova, was called upon to suggest a new arrangement for the statuary of the Statuario Pubblico; Canova noted of the Cupid that it was, 'de molta bellezza' [of much beauty]. See Perry, 'Statuario Pubblico', pp. 107, 124.

[12] The catalogue for the collection of the Statuario Pubblico compiled by Antonio Maria Zanetti in 1736, noted a statue of, 'Altar: triangular, to Mars. Rams' heads at corners, dancing putti reliefs', as part of the original Grimani collection. See Perry, 'Statuario Pubblico', p. 124.

[13] Wright wrote: 'Towards the middle of this Part, there is an Ascent to the publick Library, which is a very handsome Room; and, besides the Books and Manuscripts which are its proper Furniture, it is well adorn'd with very good Paintings: there are several Portraits, Histories, and emblematical Pieces, Heads of Philosophers, &c. by the best *Venetian,* and some other Masters, as *Battista Franco, Salviati, &c.* This Library was considerably

Fryday. 31. I had a very uneasie night last night about 12 a clock was a great storm of thunder etc., it seemd to work up with continual lightning and rumbling for near an hour, then came a prodigious flash of Lightning and one of the loudest cracks I have heard over head, with which the storm at once went off. the morning is a little coold after it, they tell me the heat has been for some days more than ordinary. This afternoon we catcht a pretty large Scorpion[1] in our dining room, which I took notice of as the first I have yet seen alive, he was brusht down from behind a curtain, and was somewhat larger than a beetle, or like a small crawfish. I was in the evening at Sig[r] Giustinianis with the usual Company, Prince Castiglion, Sig[r] Vanieri, Sig[r] Mar[co] Ant[onio] Grimani who has been Embassador in England,[2] Sig[r] Quirini[3] etc.

[f. 30r] Aug. 1. Sat. I continue still much out of order with my indisposition, The Resident dind with me to day, as did Sigr Martinelli; I was not out in the evening.

Sunday. 2. I only went out to the Residents who had invited me to dinner being so much indisposd I came strait home. Sig[a] Barbara, the mistress of the house where we are and who is a woman of exceeding good character, told my wife this very odd story that some years since she had a sister then of the age of about 19, and being both troubled with a swelling of the hands common here in cold weather, a gentleman that was to see them perswaded them to rub on the back of their hands some of a liquor he had in a bottle, and which had a strong perfumd smell which Sig[a] Barbara not liking went out and as she found it gave her a ~~sway~~ swimming in her head, she took pains to remove it with warm water and vinegar: but her sister who was not so much affected with the smell kept it upon her hands and kept smelling to them most of the evening, before she went to bed she also complained of the swimming of her head; but soon after she was laid down grew so ill that Sig[a] Barbara who lay with her got up to call her mother to her, but she expired in a very little time her body immediately swelling to be three times its natural size so as

augmented by Cardinal *Bessarione,* as appears by an Inscription upon Marble which is there in Memory of it.' Wright, *Some Observations*, I, p. 60. There is a cross inked in the right-hand side of the margin at this point in the document.

[1] Possibly, *Euscorpius italicus*, common in ruins, buildings and under household furnishings. The largest of the *Euscorpius*, the species has an adult length up to 50 mm. See Crucitti, 'Distribution and Diversity of Italian Scorpions', pp. 281–300.

[2] The Grimani were a prominent aristocratic family in Venice. Not only dominant in trade and politics (there were three Grimani doges), but also notable owners of opera houses (including La Fenice) and theatres. See Ferraro, *Venice: History of a Floating City*, p. 190. Folkes may have confused Marco Antonio Grimani with his younger brother, Pietro Grimani (1677–1752), who was Venetian Ambassador to England from 1710 where he met Isaac Newton, due to his interests in astronomy, and who proposed him as a Fellow of the Royal Society in 1712. During his stay in England until 1714, he attempted to lure English experts in navigation to Italy as part of a planned Venetian nautical school. See Giuseppe Gullino, 'Grimani, Pietro', *Dizionario Biografico degli Italiani*, https://www.treccani.it/enciclopedia/pietro-grimani_(Dizionario-Biografico) [Accessed 14 April 2021]. See also 'Venetian Diplomatic Agents in England', in *Calendar of State Papers*, pp. cxxii–cxxix. *British History Online* http://www.british-history.ac.uk/cal-state-papers/venice/vol1/cxxii-cxxix [Accessed 14 April 2021]. For more on Grimani, see above, p. 57, n. 4.

[3] The Querini were an illustrious and ancient Venetian family. The person Folkes met may have been Giovanni Francesco Querini (b. 1677), or Giovanni Querini (b. 1681), brothers of Cardinal Angelo Maria (born Girolamo) Querini (1680–1755). They were the sons of Paolo Querini and Cecilia Giustiniani. Giovanni and Angelo Maria went on a Grand Tour in 1710, and in 1711 came to England, where they met the historian Bishop Gilbert Burnet and Newton, and also visited the University of Oxford. See Giuseppe Trebbi, 'Angelo Maria (born Girolamo) Querini, *Dizionario Biografico degli Italiani*, https://www.treccani.it/enciclopedia/angelo-maria-querini_(Dizionario-Biografico) [Accessed 14 April 2021].

to be a terrible and frightfull sight to all who saw her. this seeming a very remarkable case, and coming from one of good credit I thought it worth setting down.[1]

Monday. 3. I have had an exceeding ill night and continue much indisposd. so that I have not stirrd out all day; I received an invitation to dinner \for tomorrow/ with my ffamily from Marshall Schuylenbourg. I had had one before but was engaged the same day so could not wait on him; I intend my self that honour now if well enough to go out.

Tuesday. 4. I dind with Marshl Schuylenbourg as I proposd who treated with me with very great kindness and a most noble entertainment, here was beside father Crivelli, Captain Dandolo, a Greek officer[2], Col [*blank*][3] who lives with the Marshal and some others. but I was so much out of order that I returnd home presently after dinner and went out no more that day.

Wednesday. 5. being still very ill I did not go out at all nor rise till 3 this afternoon. ~~here~~ there was a very great storm with Thunder etc. this morning. Mr Gislansoni came and dined with me and in the evening ffather Crivelli[4] calld on me with Sig^{r} Rizetti[5] aut[h]or of a book which he gave me of the nature of light and colours ag^{t}. Sr I. Newton.

[f. 31r] ~~Wednesday.~~ Thursday. 6. being still very much indisposd I have not been out, I did not rise till dinner, the heats are not so much out, but my head aches and I am sick; and most excessively faint and low spirited so that I can no ways amuse my self. The Consul who is just come to town was so kind to come and sat the evening with me.

Fryday. 7. I had some tolerable rest and have been easier to day tho often in pain. the Consul came and dind with me, they call my illness here *scotture* searings, and indeed

[1] As Matthew Lubin has remarked, 'the documentation of state-sanctioned murders [via poison] that the Republic of Venice not only presided over, but wrote down on parchment, set Venice apart from other Italian city-states for its relative openness, and its volume, from the fourteenth to the eighteenth centuries'. Lubin, 'Poisoning as a Means of State Assassination', p. 228. As Venice was an emporium for global spices and *materia medica*, Venetians had a unique specialist knowledge of toxic substances, though in 1410, 8 were outlawed for sale except by two 'tightly-overseen' pharmacies: 'napellum, opium, cantharides, "worms of Caffa", the arsenic "curinum", arsenic sublimate, silver sublimate and red arsenic'. Ibid., p. 228.

[2] Probably a member of the Venetian patrician Dandalo family, whose Gothic palace was the Ca' Farsetti near the Rialto Bridge (now used as Venice's town hall). The family rose to power in the 12th century via Byzantium trade and produced four doges. Although by the early modern era they had declined from their medieval heights, the Dandalos still held several political offices in the Venetian Republic until its fall. See Raines, 'Nobili Scaduti?', p. 157.

[3] Folkes left a blank space here.

[4] Crivelli had written *Elementi di fisica* in 1731, a textbook, which was a reinterpretation of recent discoveries in natural philosophy within an Aristotelian framework. Historian Vincenzo Ferrone considered Crivelli's work to be a 'meaningful record of the attitude of a large segment of society in the Veneto with regard to Newton's works'. Ferrone, *The Intellectual Roots of the Italian Enlightenment*, p. 98. The first volume offered the most complete, if dispassionate, explanation of Newtonian gravity that appeared in Italy (for example, he rejected Newton's concept of a vacuum in space). Crivelli presented Cartesian and Newtonian optics alongside the objections of Rizzetti that 'subtly neutralized them', and he did the same thing in his coverage of astronomy, describing Copernican, Tychonic, and Ptolemaic hypotheses. From reading his manual it is clear there was a wide-ranging knowledge of Newtonian theory in the Veneto, but there was also 'an image of local scholars rejecting what we might define as any sort of Newtonian orthodoxy'. Ferrone, *Intellectual Roots*, p. 99. In this light, Crivelli's visit with the antagonistic Rizzetti in tow may be interpreted as a challenge to the Newtonian Folkes.

[5] Giovanni Rizzetti (1675–1751), was a Paduan nobleman who belonged to the circle of the mathematician Jacopo Riccati (1676–1754), of Castelfranco Veneto. He refuted Newton's ideas of the refrangibility of optics. For a discussion of this dispute and Rizzetti's role in it, see Introduction, pp. 11 and 14 above.

that name represents the sensation like hot needles continually pricking all over especially in the back. I was to night at Mr. Giustinianis.

Saturday. I have had a very bad night and am more uneasie this morning. I went no farther out than just to take the air. and take a turn on the place; my Wife and boy[1] begin to be very uneasie with same indisposition.

Sunday; 8. I had not much rest last night but have had less pain the heats beginning to dry on my back; My Wife is very uneasie with hers, and mine are most troublesome in such parts as did not come out before; The Resident visited me this evening and carried me towards Lido; as we came back there was a storm of thunder and lightning, I was at Sig^{r} Justinianis.

Monday. 9. This being the feast of St Lorenzo[2] the bridge leading to the Church of that name[3] was ~~and~~ adornd with an arch or temple of pasteboard coverd with paper and linnen with images etc and Corinthian Columns in a beautifull manner; as soon as one is off the bridg is the Scola of St Lorenzo where a Priest stood all day with a glass in which is a bone of the Saint[4] which crowds of people were perpetually coming to kiss by him stood a man with a box like a till which he shook at every person to put them in mind of dropping money to it.[5] A little further was the church belonging to the Nunnery in which was very fine musick of the best voices and hands in town,[6] here were seated in 2 gallerys built for the occasion adornd with ornaments of Architecture, Images and festoons made

[1] Also named Martin Folkes (1720–40); Martin Folkes Jr was killed in a riding accident at the Caen Military Academy in Normandy. His short life means that there is not a large amount of archival evidence about him. We do know that Folkes took particular care over his son's education. Just as his father had been, Martin was admitted as Fellow Commoner at Clare Hall, Cambridge, on 29 Oct 1737. Over the years, Folkes had carefully nurtured his son's 'extraordinary taste for medallic [numismatic] knowledge', and frequently had him attend as a visitor to Royal Society meetings, and gatherings at the Mitre Tavern afterwards, to learn to debate and to drink. See Stukeley, *The Family Memoirs*, I, p. 99. In 1739, Folkes travelled with his son to France, 'chiefly with a view of seeing the Academies there, and conversing with the learned men'. BL, Add. MS 4222, ff. 25–26. Folkes wanted Martin to practise his French and attain the discipline and accomplishments belonging to a gentleman of quality, a cosmopolitanism to add polish to his English education. It was also the reason that he took him on his tour of Italy.

[2] The Feast Day of the Saint, known in the Anglophone world as St Lawrence, is now celebrated on 10 August. St Lawrence is the patron saint of both libraries and chefs. He was an early administrative deacon who ran finances and kept records for the early Roman Church, so is connected with archives. More gruesomely, he is associated with chefs for being grilled alive at the order of Emperor Valerian on 10 August 258. Traditionally, the saint is portrayed with his gridiron as a symbol of his martyrdom. See J. P. Kirsch, 'St. Lawrence', *Catholic Encyclopedia*, http://www.newadvent.org/cathen/09089a.htm [Accessed 27 April 2021].

[3] Founded in 812 and located in the *sestiere* of Castello of Venice (at a point approximately midway between San Marco and the Arsenale), San Lorenzo was rebuilt at the end of the 16th century (1592) based upon a design by Simone Sorella. Allegedly the resting place of Marco Polo, the church was deconsecrated in 1810 following the Napoleonic decree; it reopened in 1817 following Napoleon's downfall. San Lorenzo is now used for contemporary art exhibitions. See Ocean Space, https://www.ocean-space.org/about/church-san-lorenzo [Accessed 26 April 2021].

[4] The French closed down the convent as part of the Napoleonic suppressions. Some of these relics were the focus of late-medieval devotional cults to 'new saints at the chiesuola of San Sebastiano at the convent of San Lorenzo under the aegis of the nuns'. The nuns honoured the 'Venetian patrician, ex-bishop, and apparent gardener of their monastery, Leone Bembo, as well as an obscure priest from the parish of San Zan Degolà ('The Beheaded John the Baptist'), simply known as Giovanni Plebiano or John the Priest'. McCluskey, *New Saints in Late-Medieval Venice,* p. 99.

[5] Apart from the saint's traditional hagiography, San Lorenzo's church had a complex sacred history, possessing a fragment of the True Cross that was lost in the Venetian Canals and miraculously restored. Petkov, 'Relics and Society', pp. 267–8.

[6] The music concerts were also a competitive venture between the convents. Wright's *Some Observations* notes: 'The Nuns of S. Lorenzo, and those of S. Maria Celestia, have on their Feast-days, one the 10th, the other

of linnen in a very beautifull manner I have seen several of these sort of things since I came to Venice but none so good as these,[1] the pleating of the linnen is somewhat very curious and represents all sorts of fruits in the festoons of flowers palm branches etc in a manner I could not have believd, the Images have faces and hands of Stucho,[2] the draperys are all linnen in put about them in so gracefull a manner as to represent very pretty statues; all these decorations I am told are the Charge of the Monastery whose nuns were drest[3]

[f. 31v] What I say here of the bank not being to my mind I refer myself to what I have hence noted on that subject in a paper by it self.

[f. 32r] out in their best manner in honour of the feast, the Church was all hung with rich damask, I had not been in it before it is a handsome one, and they have some good pictures,[4] I was very much pleased with one in the Monastery, but which I could herefore only see at some distance through the grate, it is a Crucifix with the Virgin fainting and some other figures, very gracefull and a fine mellow colouring and well preserved there is

the 15th of August, a great Concert of Musick in their several Churches. The Nuns of both these Convents are noble Ladies; and they vie for Superiority with each other, which shall have the best Musick: and therefore each obliges the chief of their Musicians when they engage them to be at their Feast, not to be employed at the other. So that which ever of the two gets the best of the Home-Musicians first for their Feast, puts the other under a Necessity of sending to *Bologna*, or some such distant Place, for others.' See Wright, *Some Observations*, I, p. 99.

[1] An observation confirmed by Wright. At the feast of the San Celestia convent, he noted 'there was an occasional *Portico*, and a Colonnade on the Bridge that leads to the Church, with Extempore-Statues, made up of Pasteboard and stiffen'd Linen Cloth; both without the Church and within'. Ibid.

[2] Stucco, a mixture of lime and white marble powder, which offered a cheap and lightweight alternative to pristine white Carrara marble for sculpture and ornament. As Jones indicated, elaborate stucco relief sculpture was fashionable from the mid-16th century in the Veneto for fireplace overmantels, interior walls and staircases, and it was commonly used in combination with frescoes in both public buildings and private residences. Jones, 'The Business of Sculpture in Venice 1525–1625', p. 20; See also Molli, 'Una "pella come un marmo"', pp. 93–101.

[3] The nuns, often from noble families, were encouraged in the decorative arts and in singing for the public. Venetian nuns were also often patrons of arts and architecture themselves; as historian Benjamin Paul has noted, in this sense, the convent was a 'space for female self-realization in early modern Venice'. Paul, *Nuns and Reform Art in Early Modern Venice*, p. i. Wright wrote of the costume of the nuns of San Celestia in Venice, similar in wealth and size to San Lorenzo: 'Those I saw at the Celestia were dress'd in white; no Veil over their Faces; a small transparent black Covering [called a Whisk] goes round their Shoulders; their Heads were very prettily dres'd; a sort of small thin Coif went round the Crown, and came under their Chin; their Hair was seen at the Forehead, and Nape of the Neck: the Covering on their Neck and Breast was so thin, that 'twas next to nothing at all.' Wright, *Some Observations*, I, pp. 99–100.

[4] Though it is difficult to ascertain the identity of the paintings from Folkes's description, the convent did have a fine collection. The inventory of the church's contents in 1810 is in the Archivio di Stato, Venice (Arch Demanio (1806–13), San Lorenzo, Fasc. iii, 2, no. 21). A panel depicting *The Martyrdom of Saint Lawrence with Two Nuns* by Jacobello del Fiore, (*c.*1425), displayed now in the Rijksmuseum, was perhaps from the San Lorenzo Convent. See https://www.rijksmuseum.nl/en/collection/SK-A-4001, Rijksmuseum, Amsterdam [Accessed 27 April 2021]. San Lorenzo also had in its collections a work by Carlo Crivelli (1430–95), *Virgin and Child (Madonna della passione)* (*c.*1460), now in the Museo Castelvecchio in Verona. https://artsandculture.google.com / asset / madonna-of-the-passion-carlo-crivelli-venice-1430-1437-ascoli-piceno 1495-ca/QAGXX4E3l_YsfQ?hl=en [Accessed 20 March 2025]. An oil painting on panel of *Saint Stephen* (*c.*1520–30), by Girolamo da Santacroce, is now in the Pinacoteca di Brera (Milan), where it arrived from San Lorenzo after the Napoleonic suppression. See 'St Stephen', Pinacoteca di Brera, https://pinacotecabrera.org/en/collezioni/collezione-on-line/st-stephen [Accessed 20 March 2025]; 'Saint Lawrence', The Courtauld Institute of Art, https://gallerycollections.courtauld.ac.uk/object-p-1947-lf-388 [Accessed 27 April 2021]. See also Cotton, 'San Lorenzo', Churches of Venice, http://churchesofvenice.com/castello.htm#sanlor [Accessed 27 April 2021].

also in the Church a good crucifix of Palma.[1] The concourse of people here was prodigious.[2] The Consul dind with me to day and was so good to take some pains to give me an idea of the Bank and Bank money of Venice, better than I have before had. It seems to me in general to be a publick account kept from one man to another therefore more secure than private transactions of which the memorandums may be lost and the witnesses questionable, whereas here the payment made being by a transfer from one mans account to anothers appears in the nature of a publick record; they as I understand neither receive nor pay any money in specie at least not now, but a merchant being to pay another a certain Summ of money does it by a transfer from his account to that of the other, and sums of such bank money are sold and negotiated from one to another like stock among us.[3] and here is this advantage that whereas here their specie money is perpetually changing the bank money is of a more durable sort the same summ of bank money meaning nearly the same at one time as at another whereas the current money does not, therefore as the current money alters the nominal value of so many ducats bank changes when it is to be paid for but the same number of bank ducats is always the same, nearly; nearly I say for the greater or less demand for bank money may they say make some little difference in the price of bank money, this they call the Agio[4] and is to be added to the number of bank duchats to turn them into ducats effective or such as are paid, and this agio rises when the ducat effective raises its nominal value and vice versa. they indeed say there are 2 Agios the one is constant, the other variable, the constant is ½ part added and the variable one plainly sees the reason of. another particular also relating to the bank is they keep their accounts in Livres Sols and Deiners,[5] and their Livre is not the Livre common calld to distinguish it Livre de Piccoli but is worth 10 I think Bank Duccats making very nearly a pound weight of Silver.[6] another particular is

[1] Jacopo Negretti (1548/50–1628) best known as Palma Giovane (young Palma), a Venetian painter.

[2] Wright noted on the San Celestia convent's feast day: 'On their Feast-days the Door of their Convent is flung open, and they stand in Crowds at the Entrance, where I observ'd them talking to their Acquaintance with great Freedom.' Wright, *Some Observations*, I, p. 99. There is a cross inked in the right-hand margin here.

[3] As Frederic Lane has indicated, Venice developed the idea of the giro bank in the medieval era. It was called a giro bank, because there was a demandable bank deposit based on a giro or book entry payment. Lane wrote: 'The main function of a Venetian banker was not making loans but making payments on behalf of his clients. The payor or person making the payment orally instructed the banker to make a transfer to the account of the person being paid (payee). The banker wrote as directed in his book, which was an official notarial record, so that there was no need of receipts ... They were called banche di scritta or del giro because their main function was to write transfers and thus to rotate (girare) credits from one account to another at the command of the merchants.' Lane, *Venice*, p. 147. The total balances of such banks were about 1 million ducats [*ducati*], with as many as 4,000 depositors, or 1 in 30 of the population of Venice. See Ibid., p. 190. The *Negociator's Magazine* of 1777 noted of the Giro Bank, that it: 'is accounted one of the most considerable in *Europe*, having a Fund of some Millions of Ducats deposited in it by the Subjects into the Hands of the Republick, who are Sureties for the same, and pay the Salaries of all the Officers, for whose Care and Honesty they stand engaged, and make good all the Monies to the Bank, which may happen to be lost by Fraudulency, Mismanagement, &c'. Hayes, *Negociator's Magazine*, pp. 345–6.

[4] The *agio* from *agiotage* denoted the difference in exchange between metal of full value and depreciated currency.

[5] Nominal values in Venice were expressed in *denari* (d), *soldi* (s) of 12d, and *lire* (£) of 20s. See Appendix B on Venetian coinage.

[6] Folkes was correct. The *Negociator's Magazine* of 1777 noted that the Venetians, 'keep their Accounts in Livres, Sols, and Deniers Picoli or Current, reckoning 12 Deniers to a Sol, and 20 Sols to a Livre But in the Bank, or the Bankers, keep their Accounts in Livres, Sols, and Grosses, reckoning 12 Gross to a Sol, and 20 Sols to a Livre, and every Livre they value at 10 Ducats Banco or 12 Ducats Current. So that the Ducats are of two

that they keep all their accounts here in the Roman figures which seems an argument of its antiquity. I [f. 33r] hope hereafter to get a further account of it.

Tuesday. 10. I was ~~not~~ at home all day till evening when I went to the place. I saw near the Rialto before an Apothecarys 25 men drest alike with roses in their caps, shaking in boxes the several ingredients for the Venice treacle[1] and as many mortars standing by where they had all been beating the same just before, it seems this is a custom they have to perform all the parts of the operation of making the treacle in publick and the drugs are visited in form by the Physicians before they are put into the cauldron. I had seen ever since I came to Venice the Vipers put in Glasses in the shop windows.[2] this evening I saw a funeral carried through the place there were great numbers of wax light some 6 or 7 inches diameter and 5 foot at least high as large burdens as could well be carried by men in sort of black cases with only holes for their eyes[3] there were also priests alms people and Crucifixes and other reliques carried and the corps laid along on a bier in a religious habit uncoverd it was a women and her hands were joind holding up a Crucifix.

Wednesd. 11. I went in the morning to Sig^{r} Gislansoni to the Zecca where I saw the melting the silver and casting into plates for some new ducats that are going to be coind.[4] The government I learn have bought in the silver for these at 98^{l}:17^{s}:0^{d},16. I after wards was in the office where they test the gold and silver stuffs,[5] no man but the licensed workmen can make them, nor they without a seald thread to be workt in[t]o the beginning of the pieces that nothing may be made privately, and when finisht is[6] test is made of the gold and silver us[e]d in them.[7] I also saw the office for gilt wire all of which

---

Sorts; the Bank Ducats are Par with 52 Pence Sterling; and the Ducats of Picoli or Current, are Par with 40 Pence Sterl. or thereabouts. The Ducat Banco is valued at 6 livres 4 Sols, or 124 Sols Picoli, or 24 Grosses. The Current, or Picoli Money, is what is usually bargained for in buying Goods and Merchandize, and is 20 *per Cent.* Worse than Bank Money.' Hayes, *Negociator's Magazine*, pp. 344–5. See also Appendix B.

[1] Also known as the *panacea theriac* ('treacle' is an English corruption of the word theriac). Therapeutic failures of theriac were commonly attributed to defective manufacture or composition, and so there was strict quality control, and manufacture often was required to take place publicly. Griffin, 'Venetian Treacle', p. 317.

[2] Viper flesh powder was one of the many ingredients in Venetian treacle. As Griffin noted: 'At the final stage the prescribed quantities of 55 herbs previously prepared by various processes, along with the prescribed quantity of squill and viper flesh powder (48 drachms), were added to hedychium, long pepper and poppy juice (all at 24 drachms); eight herbs including cinnamon and opobalsam (all at 12 drachms); 18 herbs including myrrh, black and white pepper and turpentine resin (at 6 drachms); 22 others and then Lemnian earth and roasted copper (at 4 drachms each); bitumen and castoreum (the secretion of beaver); 150 drachms of honey and 80 drachms of vetch meal.' Griffin, 'Venetian Treacle', p. 318.

[3] Members of a religious confraternity or scuole, voluntary lay associations, organized to provide charity for the living and the dead. D'Andrea, 'Charity and Confraternities', p. 424. The costume Folkes describes indicates they were probably members of the Scuola Grande di Santa Maria della Misericordia. See Bamji, 'The Materiality of Death', pp. 119–36.

[4] The most common type of Venetian gold coin was the *ducati* featuring the doge's image receiving the standard from San Marco. See Appendix B on Venetian coinage.

[5] Venice was famous for goldbeating and wire making. Goldbeating came to Venice *c.* AD 1000 from Byzantium; hence the Ca' d'Oro Palace on the Grand Canal, the 'House of Gold' that once shone with 22,000 sheets of goldleaf. The drawers of gold wire for fabric and jewellery were called *tiraori*. See Zwingle, 'The World's Greatest Goldbeater', https://craftsmanship.net/the-worlds-greatest-goldbeater/ [Accessed 21 June 2021].

[6] Folkes may have meant *a* test is made.

[7] Purity tests were important as the gold *ducati* were mostly used in foreign trade and not for domestic circulation. For the rarity of gold coins in circulation in the Venetian Republic, see Viale, 'Consumer Behaviour', p. 40, n. 21. See also Appendix B on Venetian coinage.

is made in Zecca, they are by ordinance to lay 25 carats[1] of Gold on a Mark of silver; I saw here round barrs of 27 inches long weighing 25 Marci each[2] which were covered with a plate of ~~silver~~ \gold/ weighing 625 Carats, they heat the bar then wrap the plate about it after another heat it is spread close and burnisht on then it again receives a strong heat in the fire and this fixes it, then it is drawn through lesser and lesser holes, and some of it is brought to such a degree of fineness that by some [haires?][3] as they say make 17 bracci[4] in length of 1 carat of metal.[5] I was this evening at the Giustiniani Palace.

[f. 34r] Thursday 12. ffather Crivelli dind with me and I went with him after dinner to Sig^r Zendrinis[6] who is the Mathematician of the Republic who is chiefly employd about their rivers, he showd me some instruments for levelling etc but in no ways comparable to what we have in England. I was also at another persons who is a practical Mathematician to see some instruments but nothing was particular what he had best were some English things but none of those neither of the very best sort. there was one good workman here called Bernard Facini;[7] of whom I saw a diagonal scale well divided and a good compass box with a contrivance to take the needle off the pin when it was laid by etc.

Fryday 13. I was not out till evening when I went to Sig^r Giustinianis where I staied till supper and then came home. I am better but not yet easie of my indisposition. My Wife is very uneasie with it.

Saturday. 14 I visited in the morning the Resident and the Consul, the latter came home to dinner with me and sat with me the best part of the afternoon, I was not out in the evening:

[1] A carato or carat, was a unit of weight for gold, silver and pearls, and in Florence and Venice 144 carats = 1 oz. See Edler, *Glossary of Medieval Terms*, p. 321.

[2] Gold and silver were measured by the marco, subdivided as follows: 1 marco=8 oncia=192 denaro=4608 grano. See Homans, *A Cyclopedia of Commerce*, p. 1946; Edler, *Glossary of Medieval Terms*, p. 171.

[3] This word is unclear, possibly 'haires' for the hair-fine wire, or just simply 'wires'.

[4] One Venetian braccio is .683 metre. See Cardarelli, *Encyclopaedia of Scientific Units, Weights and Measures*, p. 88.

[5] This is wire drawing through a draw plate which had a series of tapering holes. Metalworkers would grip the wire with strong pliers and pull it through the next smallest hole. As Andrew Oddy commented: 'A steady pull is essential, and in the most primitive technique the artisan gripped the end of the wire projecting through the die and then either walked steadily backwards across the workshop or sat on a swing with his feet on the anvil on which was clamped the draw-plate, so that he could move backwards and forwards by straightening or bending his legs.' Oddy, 'The Production of Gold Wire', p. 79.

[6] Bernardo Zendrini (1679–1747), the 'public mathematician' of Venice responsible for hydraulics. The Museo Galileo in Florence notes: 'To him we owe the replacement, around 1738, of the weak banks of palafitte and stones, erected to protect the lagoon of Venice from high seas, with the famous "murazzi", barriers constructed along the shore between Pellestrina and the port of Chioggia with blocks of Istrian stone cemented with pozzolana mortar.' 'Bernardo Zendrini', Scientific Itineraries in Tuscany, https://brunelleschi.imss.fi.it/itineraries/biography/BernardinoZendrini.html [Accessed 21 June 2021]. Folkes may have underestimated Venetian ingenuity and engineering. The *murazzi* were not just dikes but featured stairways whose, 'lowest steps sank deep under the water in order to break the waves, dissipating their violent energy before they struck the upper part of the wall'. They were constructed from 1744 to 1782, only partially collapsing 'with the high tides that submerged Venice on November 4, 1966'. Gambolati and Teatini, *Venice Shall Rise Again*, p. 38.

[7] Bernardo Facini (1665–1731), a Venetian astronomer, mathematician, and scientific-instrument maker, and author of various inventions such as the surveying protractor with magnetic compass (1694), now in the Naval History Museum of Venice; Museo Galileo, 'Bernardo Facini', https://catalogue.museogalileo.it/biography/BernardoFacini.html [Accessed 21 June 2021]. Facini also designed a spiral-scale slide rule with the inclusion of vernier-like markings that run on a band outside the scale; the only copy is in the Adler Planetarium and Astronomy Museum in Chicago. See Chamberlain, 'Long-Scale Slide Rules Revisited', p. 25. See also Bedini, *Clockwork Cosmos*; Bedini, 'A Marine Quadrant by Bernardo Facini', pp. 597–9.

Sunday. 15. being the feast of St Roch[1] was a day of funct[i]on of the Doge when he went to the Church and Schola of the Saint, and the Church of the Frari[2] in the neighbourhood; the Consul calld upon me in the morning and we went together first the ffrari, where I had been with Mr. Mace on the 23 June this church was drest out on the occasion, and I was again well entertaind with the sight of some of the Monuments and the Pictures, there is besides a fine Altarpiece ~~what~~ where the Putta is Marble in relief finely cut by Victoria.[3] The Virgin was drest out in her best cloaks and a Silver Image of St Antonie was ~~but~~ set out all Goldsmiths work. from hence I went to the Church of St Roch which was set out in its best manner with decorations of pillars festoon etc of pleated linnen and plate above the chief altar, it is but a small church but adornd with good pictures by Tintoret of the Miracles of the Saint;[4] adjoyning is the Scola Grande of St Roch, which was also nobly drest out, and a parade made of their plate which is magnificent and in prodigious quantity. here seem to be some of the greatest works of Tintoret the whole having been in a manner painted by him, in the ground hall are many others, the Annunciation, flight into Egypt, Adoration of the Magi, and Slaughter of the Innocents all of his famous works.[5] there is a Noble staircase[6] of 2 ffigures half way at the

[1] The feast day of San Rocco commemorated the end of the plague in 1576. A contemporary painting by Canaletto, *Venice: the Feast Day of San Rocco* (*c.*1735), in the National Gallery, London, depicts the procession outside the church. See 'Canaletto: The Feast Day of St Roch, Venice', https://www.nationalgallery.org.uk/paintings/canaletto-venice-the-feast-day-of-saint-roch [Accessed 12 Jan 2021]. See Plate 10. In 1478, the saint's remains were actually recognized legally by the Council of Ten to bring help to the sick. See Valcanover, *Jacopo Tintoretto*, p. 9.

[2] Santa Maria Gloriosa, familiarly known as the 'Frari', or Ca' Grande for its size, a Gothic Franciscan church and principal seat of the Conventual Franciscan order in Venice, was begun *c.*1340 and finished in the 15th century. The building is 102 metres long × 48 wide × 28 high, making it one of Venice's most imposing buildings. Lugato, *The Frari Basilica*, p. 1. The church is almost a museum of Venetian Renaissance art, the most famous piece being Titian's *Assumption of the Virgin* (1518) for the main altar, 'framed by the triumphal arch of the stone tramezzo', as well as Titian's *Madonna di Ca' Pesaro* altarpiece, commissed in 1518 by Jacopo Pesaro, commander of the papal fleet, for his family's worship and burial. See Howard and Cosato, 'Santa Maria Gloriosa dei Frari', p. xii.

[3] Folkes was probably referring to the putti on the Zane family altar (*c.*1568) sculpted by Alessandro Vittoria (1525–1608) that are above the full-length statues of St Peter and St Andrew, though the relief was largely in stucco. His observation is complicated by the fact, as noted by Victoria Avery, that the altar has only partially survived, having, 'lost the central stucco relief of the *Assumption of the Virgin* and the crowning *putto*'. Avery, '"Belo et onorato per onor ancho suo"', pp. 223–33. Lugato noted that in 1753 the altarpiece was replaced by Giuseppe Nogari's glorification of the Franciscan Joseph of Cupertino. Lugato, *The Frari Basilica*, p. 57. See also Howard and Cosato, 'Santa Maria Gloriosa dei Frari', p. xvi.

[4] This was a painting cycle (oil on canvas) by Jacopo Robusti, known as Tintoretto, conceived as early as 1528 as a decorative scheme for the lateral walls of the presbytery of San Rocco. The works, 'narrate the final episodes in the life of the titular saint' and include: *St Roch Healing the Animals* (1567); *St Roch in the Wilderness* (*c.*1580s); *St Roch Cures the Plague Victims* (1549); *The Arrest of St Roch* (*c.*1580s). The *Christ at the Pool of Betheseda* (1559) was painted by Tintoretto for a cabinet that housed the liturgical silver. When the cabinet was eliminated, the two canvases were 'sewn together in 1674 and reinstalled on the right nave wall'. Costa et al., *Tintoretto in Venice,* pp. 95–7.

[5] Folkes was correct. The ground floor hall or *Sala Terrena* was not only an entrance hall, but was used for liturgical services, and features a series of 8 large canvases completed by Tintoretto between 1582 and 1587, which show episodes from the life of the Virgin and the childhood of Christ. Valcanover, *Tintoretto,* p. 98.

[6] The *scalene* or grand staircase was designed by architect Antonio Scarpagnino, appointed in 1527 and serving as *proto* until 1549. Scuola Grande di San Rocco, Scuola Grande, http://www.scuolagrandesanrocco.org/home-en/not-only-tintoretto/the-grand-staircase/ [Accessed 1 July 2021].

landing place over 2 pictures one the annunciation by Titian[1] the other the visitation by Tintoret;[2] on either side the great ascent of the [f. 35r] stairs beyond, are 2 large and good pictures of Zanchi[3] \and Negri/[4] one references the plague of Venice ~~with~~ in 1630, with the ~~Virgin~~ prayers made to the Virgin and St Roch on that occasion, here are represented the sick dying of the plague and the dead bodies throwing off one of the bridges into boats a piece of great horrour and well disposd: over against are the Virgin and St Roch stopping the Plague, Venice representd by a woman with attributes, returning thanksgiving, a ray of light shines from heaven upon the sick whose countenances are turnd to hope and joy I lookt a great while with pleasure on these 2 pictures tho of a more modern master than the others, which are indeed like most of the pictures in the churchs in a manner spoild with the smoke of the candles they continually keep burning before them. above these stairs is the chief hall[5] painted above the wanscoat and then cieling by Tintoret[6] but all much damagd. the wainscote of this room is very curious carvd by Francisco Planta[7] there are sort of herns[8] from distance to distance representing allegorical subjects, by beggars and cripples St Roch being the patron of such[9] and over against the door is a representation of a Library in a curious manner the books carvd in

[1] Titian, *The Annunciation* (*c.*1535). The painting was acquired by the Scuola in 1555 via the legacy of the jurist, Melio da Cortona, a member of the confraternity of San Rocco. It is placed above one of the arches on the landing of the staircase, and contains traditional iconography: the lily of chastity carried by the angel to Mary's feet that erases the sin of Eve. Valcanover, *Tintoretto*, p. 105.

[2] Jacopo Tintoretto, *The Visitation* (1558). Originally made for the church of San Rocco, the painting is now installed in the Scuola. See Valcanover, *Tintoretto,* pp. 105–6.

[3] Antonio Zanchi, *The Virgin Appearing to the Plague Victims* (1666). commissioned in commemoration of the 1630 plague by the 'Guardian Grande', Bernardo Briolo. The Grand Staircase, Scuola Grande di San Rocco, http://www.scuolagrandesanrocco.org/home-en/not-only-tintoretto/the-grand-staircase/ [Accessed 1 July 2021]; See also Valcanover, *Tintoretto*, p. 107.

[4] Pietro Negri, *The Virgin Saving Venice from the Plague*, completed on 20 August 1673 by commission from the Scuola's 'Guardian Grande', Angelo Acquisti. It is the same size as Zanchi's work. The two canvases are mounted on the opposite wall of the staircase, the paintings accommodating the pilasters along the stairwell. See Valcanover, *Tintoretto*, p. 107.

[5] The *Sala Capitolare*, or the chapter room, also called the *Sala Superiore*, used as a meeting room for the brothers of the Scuola, completed in 1567.

[6] In 1576, Tintoretto completed his painting of the large central panel, *Lifting up of the Bronze Serpent*, for the *Sala Capitolare* or the *Sala Superiore* in the Scuola Grande of San Rocco. Subsequently, on 13 Jan 1577, Tintoretto, 'offered to paint the other two large pictures for the ceiling requesting only payment for materials and colours', and he left it to the Scuola to determine the renumeration for his work, an offer they readily accepted. After beginning on the two paintings, *The Gathering of the Manna* and *Moses Striking the Rock*, Tintoretto, 'declared himself ready to paint all the ceiling paintings under the same conditions he had indicated on 13 January', and this latter offer was also accepted. See Valcanover, *Tintoretto*, p. 45. He also completed 10 paintings on the walls illustrating episodes from the New Testament and the altarpiece in the *Sala Capitolare,* completed in the summer of 1581. *Sala Capitolare* – Chapter Room, Scuola Grande di San Rocco, http://www.scuolagrandesanrocco.org/home-en/tintoretto/chapter-room/ [Accessed 1 July 2021].

[7] Folkes was referring to Francesco Pianta di Giovane, who created 12 allegorical sculptures in walnut for the *Sala Capitolare* between 1657 and 1676. These pieces constituted a parade of virtues and vices, 'ending with a homage to the arts of sculpture and painting'. Pianta, Scuola Grande di San Rocco, http://www.scuolagrandesanrocco.org/home-en/not-only-tintoretto/pianta/ [Accessed 7 July 2021]. See also Valcanover, *Tintoretto,* p. 11.

[8] A sculpture with a head and sometimes a torso, usually carved above a squared lower section.

[9] Folkes is referring to 24 wooden rococo reliefs depicting episodes from the life of St Roch, realized between 1741 and 1743, and signed by Giovanni Marchiori. See Rossi, *Giovanni Marchiori*, pp. 56–87.

wood,[1] these are some of the best things of the sort that I have seen. on one side is another room calld the Albergo,[2] where is at one end the fine Large Crucifix of Tintoret and in are and in better preservation than most of the other pieces.[3] also Christ before Pilate of the same[4] a noble picture, the cieling painted as the great hall. The floor of this chamber is very rich inlaid with porphyry verd[5] antique and other rich marbles and here their chief relicks were set out. here were also some set out under canopys in the great hall and in the room or hall below was a St Roch as big as the life drest out in cloaks with his dog a living dog provided for the occasion, who certainly partook of some sanctity after that honour.[6] These Scole are I understand something like the Halls in London, and belong to certain fraternitys or congregations of tradesmen that have given and gave money for charitable uses, etc to marry poor maids, or provide for orphans of a certain sort etc, and they are governd in their charitys by a number of the fraternity like the court of assistants[7] in the companys of London. I was this evening at Sigr: Giustinianis: where was a concert of some of the most famous people particularly Sigr Tartini of Padua said to be the first Violin in the World,[8] I understand nothing but that I liked the instrument better than I have done and they told me he plaid and was the only one that could what they call a terza corda, that is I think that he can stop 3 strings[9] [f. 36r] at once and so make one instrument do as it were 3 parts. there were also others they told me extraordinary performers, and Sigr Giustiniani plaid himself on the Violin which he does very well, being a considerable master of musick.

Monday 16. I was not out till evening when I went to take the air ~~who~~ with the Consul who gave ~~me the honour~~ of the favour of his company.

Tuesday. 17. I was not out at all the Consul dind with me, hence was a great storm of

[1] The bibliophile Folkes was correct in his estimation of the sculpture. 'Between the winged *Fury*, chained and blindfolded, one of the liveliest sculptures of the series, and *Curiosity* with its covered face, opens the bookshelf with its five shelves stuffed with books (64 of them!), an absolute masterpiece of wooden sculpture for its detailed realism in the rendering of each element: the books, the reading glasses propped on the book at the bottom, the quill with its inkwell.' Pianta, Scuola Grande di San Rocco, http://www.scuolagrandesanrocco.org/home-en/not-only-tintoretto/pianta/ [Accessed 7 July 2021]. See Plate 11.

[2] The *Sala dell'Albergo* was the room in which the governing body met, containing the bank vault or Scrigni for the safe keeping of the Scuola's documents, relics, money and candles. See Sala dell'Albergo, Scuola Grande di San Rocco, http://www.scuolagrandesanrocco.org/home-en/tintoretto/sala-dell-albergo/ [Accessed 7 July 2021].

[3] Tintoretto, *Crucifixion* (1565), still in its original position at installation.

[4] Tintoretto, *Christ before Pilate* (1566/67).

[5] Verd, a green ornamental marble consisting of serpentine with calcite and dolomite.

[6] After St Roch contracted the plague, he was expelled from Piacenza, where he was treating patients. His faithful dog brought him bread, and an angel healed his wounds from the disease; sometimes the saint was portrayed with a plague scar on his left leg.

[7] A court of assistants was a governing council of senior members of professional, trade, or livery companies; the analogy Folkes makes with the structure of governance of the Scuola is apt.

[8] Giuseppe Tartini (1692–1770), Venetian violinist and composer. His most famous composition was the fiendishly difficult 'Devil's Trill Sonata' with double stop trills. Tartini's *Treatise on Ornamentation,* composed prior to 1750, was translated into French in 1771, and was, 'probably seen by C. P. E. Bach and most certainly studied by Leopold Mozart'. For a translation into English and commentary, see Babitz, 'Treatise on Ornamentation: Giuseppe Tartini', p. 75.

[9] A triple stop is playing three notes simultaneously, usually by sounding the lower notes briefly and allowing them to ring, and then the bow plays the upper notes.

Thunder wind and rain in the evening, the wind and rain came on in an instant a perfect little Hurricane without any gradual beginning.

Wednesday. 18. I staid at home except taking the air with the Consul in the evening: he is the most conversable man I meet here; he is going to Paluello[1] tomorrow for some ~~two~~ days. He lent me a gold ducat of the republick[2] on which I have made some remarks.[3]

Thursday. 19. I had a bad night being yet much out of order with my late illness, in the evening I visited Mr. Smith: otherwise I was not out.

Fryday. 20. Mr. Gislansoni was to see me, I was not out all day till evening that I went to the Giustiniani Palace.

Saturday. 21. I was only in the evening to wait on the Resident who is ~~well~~ \ill/ of the Gout. Mr. Smith was to see me in the forenoon. we had a storm of thunder and lightning in the afternoon. I have employd my self some days in understanding the Venetian money, and the changes that have been made in it as also the affairs of their Bank.

Sunday. 22. I was in 2 churches I had not before been in St Marina and St Maria Formosa.[4] but saw nothing very remarkable. There was a great storm of thunder and lightning this morning, it is reported a jew[5] was killd in the town, as also that the thunder a saetta[6] they call it here struck a priest that was going to say mass in the Church of the Scolas[7] but did him no real hurt. I was the residents this evening who is ill, I met there the Lord Radnor[8] who has been some time at venice. talking with the Resident about the trade of Venice he told me he was informd their revenues were yearly about 800,000£ Sterling[9] out of which they paid the interest of their debt which he promisd to get me an account of also of the several branches of their revenue.[10]

[1] Paluello (Stra), a village between Venice and Padua, about 30 km from Venice and the location of the Villa Pisani.

[2] Venice first minted the gold *ducat* in 1284 (also known as the *zecchino*), with a weight and fineness of 3.5 grammes of pure gold (.986). See Appendix B for more information about Venetian coinage.

[3] Folkes was doing experiments to weigh coinage and establish standards of measure. Joseph Harris noted: 'the late learned and curious Martin Folkes, Esq; found by a nice examination when he was at Venice, that 150 of those [Venetian] carats do make pretty exactly one ounce Troy'. Harris, *An Essay Upon Money and Coins*, p. 55.

[4] Santa Marina, founded in 1130, was noted for its relic of the right hand of St Margaret of Antioch; the church was demolished in 1820. Santa Maria Formosa was the first church in Venice that was dedicated to the Virgin Mary, built in the 7th century by the Tribuno family. John Ruskin described the church as an example of what he terms the 'Grotesque Renaissance'. Ruskin, *Stones of Venice*, III, pp. 124–6.

[5] For the status of Jews in Italy, see below, p. 143, n. 4.

[6] *Saetta* means 'arrow', 'lightning' or a 'thunderbolt'.

[7] One of the churches of the confraternities of Venice.

[8] Henry Robartes, 3rd Earl of Radnor (*c.* 1695–1741), who spent much of his life after 1726 living in Italy with a singer from Naples. See Black, *Italy and the Grand Tour*, p. 122. Robartes was a patron of the opera, having for instance Baldassare Galuppi's *La ambizione depressa,* which premiered at the Teatro Sant'Angelo in Venice, dedicated to him, as well as Geminiano Giacomelli's *Epaminonda* (performed at San Giovanni Grisostomo, Venice in 1731). Radnor had in his entourage Charles Stanhope, a younger brother of the Earl of Chesterfield, a mathematician and a colleague of Folkes. Selfridge-Field, *A New Chronology of Venetian Opera*, pp. 427, 431.

[9] As Pezzolo has indicated: 'the Venetian state drew its financial resources first of all from indirect taxes, that is to say, from those regarding the transit of goods, consumption, and economic activities in general. The part of government revenue to its domain (property and land rentals, minting rights for coinage, etc.) was quite modest.' Pezzolo, 'The Venetian Economy', p. 270.

[10] 'By 1207, Venice had adopted the *prestiti,* a system of forced loans ... levied according to the citizen's ability to pay based on the value of his property and assets recorded in the census register.' Al-Bawwab, 'The Zecca Mint', p. 7. As Pezzolo has indicated: 'it is quite probable that the Venetian patriciate was the most important lender to the state, and thus to its own government institutions ... in 1787 Venetians held 79.7 per cent of the Venetian debt, foreigners had 19.5 per cent, and just .8 percent was in the hands of subjects of the *terraferma.*' Pezzolo, 'The Venetian Economy', p. 272.

[f. 37r] Monday 23. Abbe Conti[1] who is just come to town came to see me with the Consul and Abbè —.[2] I had known the Abbè very well in England above 16 years ago, I had a very agreeable conversation with him as he is a Gentleman of great knowledg and politeness, and indeed the most knowing man of all the Italians of his side the country. this being Bartolemy day[3] the Gunners went out the Bombardiers they call them, to perform an excersize at Lido, they came home in several boats with musick etc I was at night at Sigr Giustinians where it was proposed to make some of Sr I. Newton's experiments on fryday next.[4] I was desired to look for a convenient place in the house to morrow morning, Sigr Conti has prisms some very good from England.

Tuesday 24. I saw a room at Sigr Giustinians where I breakfastd with Abbè Conti, we had a pretty deal of conversation on several points of literature. I then waited on Ld Radnor. and in the evening father Crivelli coming to me I was again at Sigr Giustinianis, where there was musick a cantata on the death of hector etc the words by Abbe Conti and the music by Sigr Marcello[5] who is most famous for a proper adapting musick to the nature of the subject. Mr. Gislansoni was also with me this morning.

Wensday: 25. I was not at all out Abbe Conti The Consul and ffather Crivelli dind with me and I passed the day in a very long and agreable conversation.

Thursday: 26: I was at the Zecca they have begun to coin, the plates are first cut into square pieces somewhat ~~longer~~ heavier than they should be which are after hammerd with a pretty large hammer and then blanckt they are after that brought to a weight by clipping off the corners and shearing what is necessary after which, being still very ill shaped about 60 at a time are pinchd one upon another in an iron like a wide pair of pinchers, and in that state are hammerd on the edges with a large hammer till they are somewhat like round; then they are badly stampt, by one man with a hammer no bigger than he can easily manage with his right hand whilst he holds the upper die with his left. I observd they usually gave about 12 strokes to a half Ducat and 14 to a ducat, and that of the former they made about 4 in a minute and of the latter 3 or 3 and a half. but the money is very ill stampt.[6]

[1] Abbé Antonio Schinella Conti (1667–1749). Conti was born and educated in the Veneto but spent several years abroad before returning to Venice in 1726. He was in London between 1715 and 1718 to become part of Newton's intellectual circle, where he met Folkes. Conti would also serve as a go-between for Leibniz and Newton in the dispute over the priority of discovery of the calculus. Through Conti's offices, Newton thought he could lessen Leibniz's prestige in Italy, which had grown during the calculus dispute. See Casini, 'The Reception of Newton's Opticks', p. 217. On 9 July 1733, Conti wrote to Folkes concerning Rizzetti's attacks on Newton's theory of the refrangibility of light. See RS MS/790/28, Letter from Antonio Conti, Venice, to Martin Folkes, Royal Society Library, London.

[2] It appears Folkes could not remember the name of Conti's colleague, as he leaves a dash.

[3] The Feast Day of St Bartholomew (25 Aug) was an autumn ritual of eating and drinking. 'St Bartholomew, who was said to have been flayed alive, was an appropriate if grisly patron for butchers.' Burke, *Popular Culture*, p. 277. In Venice, it was a 'Festa Del Porchetta' and 'Carnival Crowds in Venice looked forward to the ritual slaughter of twelve pigs and a bull in the piazza outside the doge's palace'. Terpstra, *Cultures of Charity*, p. 34.

[4] Venice was a favourite haunt of Newtonian mathematicians and their intellectual circles. Part of the reason Folkes went there was to convince their virtuosi of the truth of Newton's work in optics. For more on Folkes's demonstrations, see the Introduction, pp. 11–15 above.

[5] Benedetto Marcello (see above, p. 56, n. 1) brother of Alessandro Ignazio (see above, p. 56, n. 1).

[6] The stamping of the blanks with the zecca's dies certified them as official Venetian coins. As Ives has noted, the design of the Venetian *ducat* was unchanged for centuries. Despite Folkes's disparaging comments, 'the weight and intrinsic quality of the coin were as important as its external appearance'. Ives, 'Introduction', *Venetian Gold Ducat*, p. 1 in ANS Digital Library, http://numismatics.org/digitallibrary/ark:/53695/nnan28786 [Accessed 13 July 2021]. See Plate 12. Folkes, of course, as an acolyte of Newton, was interested in minting techniques; in

[f. 38r] Fryday: 27. I was at the Palace Justinian in the morning, to have[1] made some of Sir Isaac Newtons optical experiments but the Sun not stirring I could no more than get things in order for another day. The Apparatus belonged to Abbe Conti who was there, and he had brought some very good prisms from England.[2] I was again at his house in the evening with the usual company, besides the Consul and Abbe Muazzo[3] a very worthy ingenious man who was to visit me the first time with Sig[r]. Conti.

Saturd. 28. I was again at the Giustinian Palace and made ~~of~~ most of the Experiments of the first book of Opticks[4] with very good success and to the satisfaction of all that were here which were mostly men of the first people of Venice, Sig[r]. Rizetti was also here who made several objections but none of any moment. I was afterwards with the Consul at the Residents, the former dind with me and showed me after dinner a very ingenious cipher, I was also visited this afternoon by Count Bielgne; the Consul at night went with Mr. Conti and me to Sig[r] Giustinianis. Sig[r] Justinian set me a present of spanish Chocolate.[5]

Sunday 29. I was with the Consul at Sigr Contis, and we went all three to visit M[arsha]l Schuilenberg, and Coll. Stratico[6] a very ingenious and agreeable gentleman that twas with him the Marchal kept us to dinner. The Coll there told me he had been to all the Islands of the Archipelago which are many of them still full of Antiquitys but in very bad condition. there is a piece he told me of the wall of Constantinople next the sea

the 1690s, Newton had introduced 'time and motion' studies to make the minting of coins more efficient. See Iliffe and Mandelbrote, 'Newton and the Mint', University of Oxford, https://newtonandthemint.history.ox.ac.uk/ [Accesed 21 March 2022]. See Iliffe and Smith, 'Introduction', p. 25; and Appendix B on Venetian coinage.

[1] There is a cross inked in the right-hand margin at this point.

[2] For the importance of prisms in Folkes's demonstrations to the Venetian nobility, see the Introduction, pp. 12–15 above.

[3] Abbé Antonio Muazzo was a public Professor of Mathematics and Architecture associated with the Venetian Arsenal. See Frascari, 'The Particolareggiamento', pp. 3–12. Muazzo and Conti convinced Algarotti to perform prism experiments in Venice to refute Rizzetti's work. See *Mémoires concernant la vie et les écrits du Comte François Algarotti*, pp. 16–17.

[4] Newton, in Book I of the *Opticks*, discusses several phenomena including the colours that make up white light, and doing a series of experiments with prisms, which Folkes reproduced, probably imitating Desaguliers's simplified experimental demonstrations to the Royal Society of Newtonian optics in 1714. See Desaguliers, 'An Account of Some Experiments of Light and Colours', pp. 433–47. As Alan Shapiro has analysed, in Book II 'Newton set forth his observations on the colours of thick films and thick plates', including the phenomenon of Newton's rings. See Shapiro, *Fits, Passions, and Paroxysms,* p. 3. In the *Opticks*, Newton also gives an explanation of double refraction, as he sought priority over the explanation given by Christiaan Huygens. In Folkes's opinion, his experimental performance had the required effect of promoting Newtonianism in the Veneto. When he left Venice, Folkes noted that his experimental success of Newton's optics showed that he had, 'conversd with many of them [the Venetians] in a manner no English man almost has done before me'. See Folkes's diary entry for 12 July 1733 (p. 99 below).

[5] A gift that would have appealed to the epicurean Folkes. 'By the eighteenth century, coffee and particularly tea would overtake chocolate in terms of mass quantities imported, yet the latter was the first stimulant beverage consumed in significant quantities by Europeans.' Norton, 'Tasting Empire', p. 666.

[6] Possibly Giuseppe Battista Stratico or Stratikos (1690–1770) of the noble Stratico or Stratigo family, Venetians of Greek descent from Zadar, then capital of Venetian Dalmatia (now in Croatia). His grandson, Conte Simone Stratico (1733–1824), a Professor in the University of Padua, was made Fellow of the Royal

that is built of fragments of Antient Marbles particularly an immense quantity of verd Antique.[1] going to take ~~tair~~ the air this evening, I calld to look at the Athenian lyons before the gate of the Arsenal, ~~here~~ there are 2 very large one on each side the door that on the right hands is seiant[2] and is about 11 foot high. The other is about the same size but couchant.[3] The snout of the first and above half the head of the second is modern, they have some spirit but I cannot say I very greatly admire them. at some distance on the ~~right~~ left hand is a lyoness standing much less than the Lyons and not I think so good and further still to the left another lion yet less couchant, they are all of white marble and were brought from Athens hither. I was at night at Sig^{r} Justinianis.

[f. 39r] Monday. 30 I went with the Consul to visit Abbe Muazzo who had been to see me with Sig^{r} Conti he is a very worthy man and of good learning he received me in a very pretty Library stord with usefull books, and talkt with me with great ffreedom upon religion it self showing me several protestant books for which he showed a great esteem, in particular he showed me a text of the old Testament I had not taken notice of where it is said speaking of the Pascal lamb[4] this is the Lamb your ffathers eat when they come out of Egypt. he promisd to come and see me often and seemd delighted with an opportunity of conversing without reserve. I then went back with the Consul where I met Abbe Conti and sat with him till the consul and he went out of town which they did at one a clock the Abbe promised to see me again before I left Venice, he is indeed a most agreeable man and of very good knowledg, and has a thorough contempt for the ridiculous things of his own people.[5] after dinner I took the air and passd the evening at the Residents who is still laid up with the gout.

Tuesday 31 I was not any where but at the Giustiniani palace in the evening. I had however an adventure a Lady in a Gondola askt for me calling her self the Countess of Valvasoni,[6] my servants carried her up but when I came to her I found her a errant beggar and one that has been in the same manner with every body of any fashion in Venice. she also seemd well disposd to have banded[7] or anything else she could have been of service in the end she cost me a pistole[8] to get rid of her.[9]

[1] Venice conquered Constantinople in 1204, and thus acquired great quantities of statuary and precious marbles from the civic and sacred buildings of the capital of the Eastern Roman Empire. Some of these, of course, decorated San Marco. Folkes here was referring to verd, a green ornamental marble.

[2] In heraldry, seiant or sejant means sitting upright. Folkes is describing the Piraeus Lion. As Rhys Griffiths has indicated: 'At the Arsenale in Venice stands a white marble lion whose story stretches from the classical world to the decline of the Ottoman empire, via the Vikings. Around three metres in height, the Piraeus Lion was one of a pair of statues that stood guard – probably as a fountain – at the Athenian port of Piraeus, having been sculpted circa 360 BC. It was removed from the harbour in 1697 by the invading Doge of Venice, Francesco Morosini, while he had Athens under siege during the Great Turkish War of 1683–99.' Griffiths, 'The Piraeus Lion', p. 70. Morosini wished to, 'carry off to Venice some memorial of Athens which could vie with the four bronze horses, carried thither after the capture of Constantinople'. Miller, *Essays on the Latin Orient*, p. 412.

[3] In heraldry, of an animal lying with the body resting on the legs and the head raised. This is another lion from Morosini's war booty, which had stood near the temple of Theseus and was added to the Venetian Arsenal in 1687.

[4] A lamb sacrificed at Passover.

[5] Folkes's reference to Catholic religious ritual.

[6] There is a town called Valvasone with a medieval castle in the Friuli Venezia Giulia region, approximately 100 km away, so close enough to be probable, but far enough away to allow deception.

[7] To 'band' or be 'banded' is a shortened form of bandy or bandied. See *OED.*

[8] A Venetian gold coin resembling a Spanish gold double *escudo*. The coin originated in Spain in 1566, and it was to become the ancestor of many European issues.

[9] For the state of beggars in Venice, see above, p. 57, n. 5.

Sep. 1. Wednesday. Mr. Martinelli brought me a scheme from the Resident of the Popes Lottery[1] which he desired me to tell him the disadvantage of. it is rather a game than a Lottery but farr worse than the royal oak[2] or ace of hearts, there is 26 percent in the best cases against the player and in the others 40 and 70 per Cent yet are all the people at Rome we are told mad after it. Dr. Michaelotti was to see me.

Thursday. 2. I was again to prepare some of the Experiments of the Opticks at Sig^r^ Giustinianis, our Experiments there have made a good deal of noise and many of the chief Nobility are very desirous to see some if them which I have promised to repeat on ~~saturd~~ tuesday next. tho it is but in a gross manner I can make them want of a convenient apparatus.[3] they are very satisfactory, and carry demonstration to all that are capable of comprehending them.

[f. 40r] Fryday. 3 I was no ways out but to take the air; Mr. Gislansoni was with me all the morning.

Saturday. 4. I was at the place in the morning. there was a show of thunder in the Evening. I was at night at Sig^r^ Giustinianis. with the usual company.

Sunday. 5. I was at the Residents, and afterwards with Sigr Giustiniani.

Monday. 6. I prepared some more of Sr I.N. optical Experiments, and In the afternoon I was with the Resident.

Tuesday. 7. I made the Experiment I had intended at the Justinian Palace I have now shewn all that are considerable in the book of opticks in the first book, and tho my apparatus was none of the best I was sufficiently satisfyd with them and am told it is what was never done with any success here before, there were several of the Chief nobility and some others here present. and every one that has any sense seems thoroughly satisfyed as far as he understands. tho I was a little surprised at some mistakes Crivelli fell into and was pretty obstinate in a great while which showed me he is not so perfect master of the Philosophy as I thought before.[4] Sigr Giustiniani had before asked me to dine with him this day and gave me a very noble entertainment, with a sort of Princely magnificence for the quantity of his plate, and choice of his things he set in our way of eating. here were at dinner beside himself and Lady. Sig^r^ Loredano,[5] Sigr

[1] Pope Clement XII, ruler of the Papal States from 1730 until 1740, revived the papal lottery in 1731 which had been forbidden under his predecessor Benedict XIII, raising a half million *scudi* for architectural projects in Rome, including initiating the Trevi Fountain and paving the streets. See James Loughlin, 'Pope Clement XII', *Catholic Encyclopedia*, https://www.newadvent.org/cathen/04030a.htm [Accessed 2 Aug 2021].

[2] The 'Royal Oak Lottery – a clear pun on Italian *occa*, or goose (translation of Catalan *auca*) – which was introduced in Britain by the Italian guitar player and composer Francesco Corbetta in 1661.' It was a game in which there was a cabled column, set up on a board with stakes laid out. 'A die or multi-sided ball was dropped into the column, so that it falls down to the bottom and rolls out onto the layout.' Depaulis, 'Bingo! A Material History of Modern Gaming', p. 43.

[3] There may be a word missing here: [for] want of a convenient apparatus.

[4] Considering Crivelli's neutral stance in presenting Newton's optics in his *Elementi*, his obstinate behaviour during the demonstration was not surprising. Rizzetti also continued his anti-Newtonian campaign until the 1740s, with diminishing success.

[5] A member of the illustrious Loredan family, a Venetian noble family eminent in the republic since the early 11th century, who amassed great wealth in the spice and silk trade, and produced several doges, statesmen, military commanders and writers. Folkes may have had dinner with Francesco Loredan (1685–1762), who became the 116th Doge of Venice in 1752, serving until his death in 1762. However, Francesco had several brothers, so this identification is uncertain.

Quirini,[1] Sig^r Barbarigo,[2] noble venetians, and Father Crivelli, Father Stevini[3] and Father Picenti[4] Somaschi religious,[5] all which were part of the company at the Experiments, and the 2 last mentioned Priests men of good learning and very great modesty whom I like as well as any I have conversd with. Sig^r Quirini is also a man of very good taste and great ingenuity in short Sigr Giustinianis company in general is all that is of Learning or good taste in Venice, and I have had through his acquaintance opportunity of conversing, and being upon a better foot with all the most considerable nobility that any stranger has been here for many years. when I came home Count Bielgne came to see me and I went with him to take the air in the garden of Sigr Molini[6] which tho small is famous as being in Venice where ground is scarce. here we met Madame Bavella[7] a Venetian Lady of merit who was come also to take the air in the same, and we walked about 2 hours. the garden has a fine prospect towards Murano.

[f. 41r] Wednesday. 8. I was not out other than to take the air.

Thursday. 9. Mr Gislansoni was with me, and dind I made some calculation for him of the value of the fflorence money. and afterwards went to a gold beaters[8] to see the manner of their working with him. I have made some remarks on it in a paper by it self.

Fryday. 10. Father Crivelli calld to go with me to the Soranzo Library.[9] it is pretty large tho not half what they say I judg here may be about 15000 volumes.[10] I saw the Mentz[11]

[1] A member of the Querini or Quirini family. Like the Loredans, he was a representative of the *Case Vecchie*, or old houses/families that were in Venice from its earliest days. Ca'Querini is on the left bank of the Grand Canal, proximal to Campo Santa Maria Formosa, and is now the Pinacoteca Querini Stampalia, an art collection and museum and home to the Fondazione Querini Stampalia.

[2] Gregorio Barbarigo (1711–1801), scion of an aristocratic family of Istrian origin. See above, p. 68, n. 1.

[3] Just possibly Jacopo Stellini (1699–1770), an Italian Somascan abbot, polymath and philosopher, and friend of violinist Giuseppe Tartini, whom Folkes heard perform. In Venice, Stellini was a tutor to the family of Giovanni Emo, and due to their patronage in 1739 became a professor at Padua teaching moral philosophy. See Stefano Perini, 'Jacopo Stellini', *Dizionario Biografico dei Friulani*, https://www.dizionariobiograficodeifriulani.it/stellini-iacopo [Accessed 7 Dec 2022].

[4] Probably Giovan Bernardo Pisenti (1707–42), a Venetian Somascan cleric educated by the Jesuits, who also studied with Eustachio Manfredi in Bologna, and himself served as a personal tutor to Girolamo Ascanio Giustinian's son of the same name. Folkes's diary shows that he spent much of his time in Venice at the Giustiniani Palace with Girolamo's family, so it is likely that he met Pisenti there. Pisenti made the first translation of George Berkeley's *New Theory of Vision* (1709), issued with the introduction to the *Treatise concerning the Principles of Human Knowledge* (1710), the only translation of any part of the book until a German edition in 1869. In scholar Antonella Barzazi's opinion, the initiative to publish Berkeley's work was probably born in the wake of the Newtonian optics controversy aroused in the Venetian world by the attacks of Rizzetti. See Barzazi, *Gli affanni dell'erudizione*, p. 181.

[5] Order of the Somascans in Venice. See below, p. 38, n. 1.

[6] A scion of the Venetian Molin family, which produced Doge Francesco Molin and, in the 18th century, the Bishop of Brescia, Giovanni Molin.

[7] This surname is unclear.

[8] Goldbeaters or *battiloros* made gold leaf by hammering. Gold was also beaten to be cut into strips and used to cover thread for spun gold. Venice introduced the art to the rest of Europe. See above, p. 84, n. 5.

[9] The Biblioteca Soranzo at Venice, the library of Senator Giacomo Soranzo (1686–1761), which was dispersed after his death.

[10] Soranzo collected 4,000 manuscripts and 20,000 printed books. See 'Giocomo Soranzo', Archivio dei possessori, https://archiviopossessori.it/archivio/51-soranzo-giacomo [Accessed 9 Dec 2022].

[11] A copy of the Mainz Bible published in 1462, subsequently purchased from the Soranzo Library by the Parisian bookseller Giovan Claudio Molini (1724–1812) in 1762, and then sold in Paris in 1763 to Louis Jean Gaignat (1697–1768). After a series of owners, the Bible was sold to Quaritch in 1884 at the Syston Park sale for £4,950. See 'Giacomo Savani', Owners of Incunabula; 'Hebrew Manuscripts from the Canonici

bible on vellum and they told me they had but did not shew me the Psalms in 1457[1] and the Rationale. I also saw a pretty many more books but the generality were not in fine care. nor do I take the Library to have been at all judiciously collected. an immense quantity of Fathers schoolmen councils etc. they have some M.S. but those I know nothing of having been shewn none but some of old Provencal Poets.[2] There are some [[~~xxxx~~]] marbles about and some pictures, and a large collection of the medals of the heads of great men mostly modern and many only casts. on the whole this library no ways answered the character I had heard of it. the books are in glass cases of Cypress but very ill made and inconvenient. I was at Sig[r] Giustinianis at night. it has been sultry hot all day and has thunderd this evening with a little rain

Saturd. 11. the weather is quite altered to cold. in the mornings so that there is no more heat than is agreeable enough.

Sunday. 12. I was hardly out.

Monday. 13. Count Bielgne was to see me and I visited the Resident:

Tuesday. 14. One Mr Wills an English Gentleman and his Lady being come to town I visited them, they came from England by sea to Leghorn and hence by fflorence Bologna an[d] Ferrara hither.[3] I was in the evening as usual at Mr Giustianianis, when Prince Castiglione made me complements from Sigr Marini,[4] the Sardinian Envoy in such a manner as I think my self obligd to go and see him.[5]

Wednesd. 15. I waited on the Resident and Count Bielgne, after dinner Mr Wills returnd my visit. I was at night at the Giustinian palace as usual.

Thursday. 16. I was at Cav[r6] Marinis but did not find him. Mr Gislansoni and Father Crivelli dind with me and after dinner I was with them at St Georges,[7] where I received

Collection', Digital Bodleian, https://digital.bodleian.ox.ac.uk/collections/canonici/ [Accessed 9 Dec 2022]; Ricci, *Catalogue raisonné,* p. 98.

[1] Possibly the Mainz Psalter of 14 August 1457, the earliest European book with a colophon that states the date of its publication and the names of its makers, Johann Fust and Peter Schoeffer of Gernsheim. Its most famous feature is its innovative use of decorative initials printed in red and blue ink. It is exceedingly rare, with only ten integral copies remaining. See White and Needham, *Peter Schoeffer,* pp. 16–17; 'Psalterium'. Incunabula Short Title Catalogue, https://data.cerl.org/istc/ip01036000?style=expanded [Accessed 9 Dec 2022].

[2] Folkes was entirely too dismissive of this renowned library, which contained, for example, the earliest illuminated copies of Fiore dei Liberi's treatise on fencing and martial arts entitled, *Fior di battaglia* (The Flower of Battle). The copy in the Morgan Library and Museum in New York was from the Sorenzo Library. See 'Il flor di battaglia/Fiore dei Liberi', MS M.383, *c.* 1410–20, https://www.themorgan.org/manuscript/77302 [Accessed 12 Dec 2022].

[3] There is not a good deal known about Mr or Mrs Wills. Ingamells merely cites this diary. See Ingamells, *Dictionary,* p. 1006.

[4] Cosimo Giulio Cesare Francesco Marini (d. 1739), Sardinian Envoy to Venice. He was the great grandchild of Claudio Marini, Ambassador to Louis XIII of France. See Cantaluppi, 'Le carte del genovese Claudio Marini', p. 319.

[5] 'Good relations with Sardinia were sought actively in the 1720s, 1730s, and 1740s, and an alliance with it was a major plank of British policy during the War of the Austrian Sucession.' Black, 'Eighteenth-Century Foreign Policy', p. 35. See above, p. 38, n. 1.

[6] An abbreviation for 'Cavalier'.

[7] Probably the Benedictine monastery San Giorgio Maggiore, located on the island of San Giorgio Maggiore next to the church of the same name. In 1782, the library began to be dispersed during the dissolutions, and in 1797, the 'French entered the city and started taking away to Paris the best books according to a selection prepared by the commissaries Berthollet and Monge'. McLeish, 'Monastic Provenances', The Conveyor,

civilities from the Abbot and Father [Ruola?][1] a Gentleman of good learning. I was at one of the ffathers cell that [f. 42r] has a good loadstone, it lifts about 12 pound and is spherical about 3 inches Diameter, but miserably armd. he also showd me some watches, a binocle telescope and some Microscopes but nothing of any goodness.

Fryday 17. I was with Mr Mills at the Church of St Daniel[2] near the Arsenal where I had not yet been, there is over the altar a good picture of Daniel in the Lyons Den by Peter of Cortona,[3] and an annunciation of Luca Jordano.[4] It is a nunnery and we saw some of the Ladys in the parlor. ~~Afterwards~~ After dinner I met at the place Cavalier Marin, with whom I had a pretty long conversation he is a Gentleman of Letters and great politeness, but dreadfully afflicted with the gout. He understands English and has a taste of most sciences, and treated me with great distinction and civility. I also met Procurator Quirini with whom I had a pretty deal of talk on the article of money, he is in the Magistracy to whom that [province][5] belongs.

Saturday 18. Dr Michaelotti was to see me with 2 Swedish Gentlemen. Mr Schelsius[6] Prof. of Astronomy at Upsal, and a young Gentleman who travels at the Kings expense to study Architecture.[7] they are very well bred understanding men and make a handsome figure. Mr Scelsius presented me with a book he has published on the Aurorae boreales, and brough[t] me compliments from Mr Dopelmeyer of Nuremberg.[8] the Marquis Poleni

Bodleian Library, https://blogs.bodleian.ox.ac.uk/theconveyor/monastic-provenances-of-early-printed-books-in-bodleian-collections-case-1/ [Accessed 7 July 2023].

[1] This word is unclear.

[2] San Daniele Profeta, founded in 820 by the Bragadin family, at the time of Folkes's visit a convent for Augustinian nuns. It became a French barracks in 1806 and an Austrian one in 1815. When the church was demolished in 1839 the remaining buildings became the college of the Royal Imperial Navy, and it is now used as a barracks. Cotton, 'San Daniele Profeta', Churches of Venice, http://churchesofvenice.com/demolished.htm#sandanprof [Accessed 9 Dec 2022].

[3] Pietro da Cortona (Pietro Berettini), *Daniel in the Lions' Den* (1663/4), Gallerie Accademia, Venice.

[4] Luca Giordano (1634–1705), Baroque painter and printmaker. See above, p. 59, n. 1.

[5] This word is unclear.

[6] Anders Celsius (1701–44). His work on the aurora borealis, the *Nova methodus distantiam solis a terra determinandi* (*New Method for Determining the Distance from the Earth to the Sun*), was first read to the Royal Society on 8 April 1736 and printed in *Philosophical Transactions*. The record of the reading of Celsius's book is in the RSL. RBO/19/57. See also Celsius, 'Observations of the Aurora Borealis', pp. 241–4. His research also involved the study of auroral phenomena, which he conducted with his assistant, Olof Hiorter (1696–1750), and he was the first to suggest a connection between them and changes in the magnetic field of the Earth. When Celsius later came to London, Folkes would help arrange for him to obtain precision scientific instruments used in his expedition with Pierre Louis Maupertuis (1698–1759) to Lapland to measure the shape of the Earth. In December 1736, Celsius wrote to Folkes, stating he was, 'one thousand times grateful for the courtesies you did for me during my stay in London'. RSL, Library, RS/MS/790/21, Letter from Celsius to Folkes, 3 December 1736. See Roos, *Folkes*, pp. 12, 167–76.

[7] Probably Jonas Meldercreutz (1714–85), a young mathematician who would be appointed to the chair of mathematics at Uppsala in 1751.

[8] When he was in Nuremberg on his tour before reaching Italy, Folkes met with a local salon headed by mathematician, astronomer, and translator Johann Gabriel Doppelmayr (1677–1750) and gathered information about the local coinage. Doppelmayr had written the *Historische Nachricht* (1730), a history of the mathematicians and artists of Nuremberg, which included illustrations of medallion portraits of his countrymen, a topic of conversation he may have had with Folkes. Doppelmayr would later correspond with Folkes concerning his atlas of cosmological maps known as the *Atlas coelestis* (1742), and his writing of the first German work on electricity. While on tour, Folkes wrote to Sir Hans Sloane, the Royal Society President, to recommend Doppelmayr for election to the Royal Society. See RSL, RS/MS/250/26, Martin Folkes; RS/MS/790, Letters 36–9; BL, MS Sloane 4058, f. 60, Folkes to Sloane, 1 October 1733.

was also to see me who is come upon some business to Venice. Mr Gislansoni dind with me

Sunday.19. Mr Wills and his Lady dind with me, after dinner we were at the Oratorio of the Incurabili[1] and in the evening I was at Mr Giustinianis.

Monday. 20. I was with Mr Wills at the Great Council where we were admitted to see them ballot;[2] In that all the Nobles set together, they chuse by ballot 36 Electors, this is done by 60 drawing gold balls out of 2 urns placed on either side the seats of the Doge and Council of 10,[3] for greater Expedition then those 60 draw again from a 3^d Urn in which are only 36 golden balls; this done, they are divided by tickets given by the Chancellor acording to the order of their drawing into 4 chambers 9 into each; there they have 9 Offices to elect unto, and in each chamber the ballot is repeated, for the nine Offices so as one to draw each upon which he nominates to that Offices such a person as he thinks proper, and the persons thus nominated in the 4 chambers [f. 43r] which may prove 4,3,2,1 or only one to each office, are again balloted for by the Grand Council: this last ballot is performd by putting a ball into one of a considerable number of balloting boxes carried about by Children for that purpose, to make the thing quicker. Some other Offices are nominated to the ballot by the Senate ~~only~~, first who go for that purpose into the next room where they are all rangd in order, and each takes an oath by touching a paper that is handed to him by one of the Avocadores.[4] I saw every thing with great conveniency and was treated with very great civility among the noblemen, several doing me the honour of coming to talk to me with great kindness, among which were Proc^r. Quirani, Sig^r Valieri,[5] Sig^r Loredano, Sig^r Grimani the Procurator that was Ambassador in England,[6] Sig^r Marc

[1] For the Incurabili, see above, p. 47, n. 3.

[2] The Great Council of Venice met in a hall seating more than fifteen hundred every Sunday, in which an elaborate balloting procedure for government office took place, governed by several procedures to prevent cheating, including the use of a boy, the *ballottino*, who looked after the balls (*ballotta*) and urns used in the lottery. See Lane, *Venice*, p. 255. As Coggins and Perrin indicated, the Venetians uniquely exploited 'the virtue of lotteries – their tendency to restrain corruption – while negating through the use of election their leading drawback – a clearly inferior person may be elevated to a post above his level of ability'. Coggins and Perali, '64% Majority Rule', p. 712.

[3] The Council of Ten, one of the major bodies of government in Venice, in operation from 1310 to 1797. The officials, elected to a one-year term, had a broad mandate over matters of state security and were empowered to punish members of the nobility for infractions, including banishment and capital punishment. It also employed a system of anonymous informers and police powers; *bocche dei leoni* (lion's mouths) were placed around Venice allowing Venetians to report bad behaviour by written notes which, when credible, led to state investigation. See Lane, *Venice*, pp. 400–403; Madden, *Venice: A New History*, pp. 188–9, 207.

[4] Avogadori de Comùn, the public prosecutors, who oversaw the compliance of councils and organs of the state to the Venetian constitution.

[5] Member of the Valier family, Silvestro Valier (1630–1700), served as the 109th Doge, his wife Elisabetta Querini, the last dogaressa to be elected. During his reign, the Morean War between Venice and the Ottoman empire came to an end. His father, Bertuccio Valier, served as the 102nd Doge from 1656 to 1658. The noteworthy Valier funeral monument by Andrea Tirali is in the church of Santi Giovanni e Paolo

[6] Pietro Grimani (1677–1752), who served as Ambassador to England from 1710 to 1714. The post was not particularly demanding, given the neutrality of the Venetian Republic during the War of the Spanish Succession. During his time in England, Grimani became acquainted with Newton due to his interest in astronomical studies, and in 1712, he was elected Fellow of the Royal Society, proposed by John Keill. Pietro Grimani also served as 115th Doge of Venice from 1741 until his death. Favilla and Rugolo, *A Brief History*, p. 13. See also 'Venetian Diplomatic Agents in England', in *Calendar of State Papers* pp. cxxii–cxxix; *British History Online* http://www.british-history.ac.uk/cal-state-papers/venice/vol1/cxxii-cxxix [Accessed 12 Dec 2022]. Giuseppe Gullino, 'Grimani, Pietro', *Dizionario Biografico degli Italiani*, https://www.treccani.it/enciclopedia/pietro-grimani_%28Dizionario-Biografico%29/ [Accessed 12 Dec 2022]. For more on the Grimani, see above, p. 67, n. 4.

Antonio his brother,[1] Sig Giustiniani, [[xx]][2] Sigr Priuli,[3] Sig^{r} Barbarigo, and others all of the first note. and coming out I was complimented with the greatest civility by a gentleman then a stranger to my self but whom I afterwards understood to be Sigr Papafava, a person of distinction as a man of letters, and one of the first houses.[4] After dinner I again saw Sig^{r} Marini and sat with him a pretty while, he has something applyd himself to the doctrine of chances; and told me among other things a remarkable particular said to have happened in the choice of a person into a place by the ballot that had actually but 4 friends in the whole council. but those 4 drew 4 of the 36 balls so as to be divided into the 4 different chambers, where again each of them drew the same number so as to nominate to the same place by which means he was alone referred to the ballot of the great council, where it is very unusual, and unprecedented to reject any person so recommended. I was at night at Sigr Giustianinis who explained to me many particulars of the forms observed in the Council and Senate.

~~Monday~~ Tuesday. 23. I had invited the 2 Swedish Gentlemen to dinner, and I passd the day with them, they gave me several curious accounts both of their own country and their observations in Germany, they also gave me the measure and weights of their own country in which they have been both been very curious, and I am so well satisfied of their acquaintance that tis a satifaction they are going the same way and will be of service to me observing the same things as my self.[5]

[f. 44r] Wednesd. 22. I was at the Residents, but otherwise not out till I went in the evening to Sigr Giustinianis, who gave me a letter he was procured from Sig. Pasqualigo[6] to recommend me to Cardinal Quirini[7] at Rome, I have been treated by this Gentleman every way with great distinction, and indeed here from ~~all~~ \many of/ the Nobility ~~be~~

[1] Marcantonio Grimani (1675–1752), the heir via primogeniture to the family fortune. He was married in 1697 to Paolina Querini di Giovanni.

[2] This word that Folkes struck out is unclear.

[3] A member of the Venetian patrician family, who were in Venice from the 11th century, achieving great wealth and eminence by the 16th century. The family had a prominent place in the life of the Republic, both in politics with three doges – Lorenzo (1556–59), his brother Girolamo (1559–67), and Antonio (1618–23) – and in the military with the Admiral, Francesco (about 1430–90), who is known for the purchase of Cyprus. The Priuli also were prominent in the arts and sciences, with Alvise (1497–1560), learned orientalist and diplomat, and in the Church, with cardinals Lorenzo (1537–1600), Matteo (1583–1634), Alvise (1651–1720), Pietro (1669–1728), and Antonio Mario (1707–72). See 'Priuli', *Enciclopedia on line*, https://www.treccani.it/enciclopedia/priuli/ [Accessed 12 Dec 2022].

[4] Venetian patrician family, a branch of the Carraresi family originally from Padua, admitted to the Great Council in the 17th century due to their financial support of the Venetian War of Candia against the Ottoman Turks.

[5] Celsius would prove to be a companion for Folkes as he travelled to Rome to investigate weights and measures. The diary of Sir Andrew Mitchell (1708–71), politician and fellow traveller in Rome, recorded that, he, Celsius and Folkes went to see the Marquess Caponi, whose palace courtyard in Rome had a standard stone for the Roman foot. See Introduction, p. 23 above.

[6] Probably Bishop Domenico Pasqualigo (1698–1766), who was Bishop of Scardona, consecrated on 1 June 1750.

[7] Girolamo Querini (1680–1755), the youngest of the three sons of Paolo Querini and Cecilia Giustiniani. In 1696, he entered the Order of St Benedict Cassinese, Florence, and took the name Angelo Maria, after which he was ordained, becoming Professor of Sacred Scripture in his monastery in 1705, and then abbot. On 22 Nov 1723, he was elected Archbishop of Corfu. He was created a Cardinal in 1726 and the next year was transferred to the diocese of Brescia, with personal title of Archbishop. He also served as librarian of the Holy Roman Church (in the Vatican), from 4 Sept 1730 until his death. Miranda, 'The Cardinals of the Holy Roman Church', https://cardinals.fiu.edu/bios1726-ii.htm [Accessed 12 Dec 2022].

received extraordinary civilitys, and I believe may say I have conversd with many of them in a manner no English man almost has done before me. there are but few here that have much of letters but such as have are extraordinarily fond of any man of fashion that has any thing of the same taste.

Wednesd. 29. It has been very cold all this week, which has occasiond my going very little out further than to Sig^{r} Giustinianis at night and to make some necessary visits. the cold is far more intense than at this season in England and is come all at once I see the tops of the mountains of the ffriul[1] all coverd with snow from whence I suppose the cold wind comes. I was however one day to see the Certosa[2] on an Island between Muran and Lido and about a mile from Venice it is a handsome convent with a good Church,[3] the cells of the monks are so many distinct houses joined by the Cloisters and each has below a sort of hall or parlor and above stairs a small dining room, a bed chamber, a little study and an Oratory, and a small garden of his own without. they have no library in common but each a few books of their own. a brother who was Sacristan as such had leave to speak with strangers and he showd me about seemingly very glad to see any body, he talkt some miserable Latin, and walkd all over a large vineyard they have, with me. they speak to each other he told me on 3 days of the week and all holy days besides when they dine together in the refectory. they eat fish and drink wine but eat no meat. eggs are also allowd them. the Cell I was in was exceedingly neat. I thought they might not have touchd money, but I askt him if it may be allowable to ~~buy any little~~ pray him to buy any little book to remember me, and I perceived by his answer he might; and he was excessively thankfull for a Ducat I put into this hand. they come not to Venice but take the air at Lido and in the neighbouring islands.

[f. 44v] # I was this evening to have taken leave of Sgr. Giustiniani to whom I have have been very generally obligd at Venice, he going out of town to morrow, ~~so~~ but had not the fortune to meet with him, so are oblig'd only to thank him by letter for his favours.

[f. 45r] Thursday. Oct. 3 I was at the Churcher of the Servites[4] where I had not been before. there are some good pictures, and it is a fine large Church. Mr Smith came and set with me most of the afternoon.

Fryday. 2. I was at Mr Smiths to see his drawings and some pictures, etc that he has very fine Mr Wills was with me, the rest of the day I spent in writing of letters at Night I was at Sigr Giustinianis.

Saturday. 3. designing to leave Venice the next week I visited Count Bielk, the Sardinian Minister Cav^{r} Mariani, and Prince Castiglione and prayd their company at

[1] Mountains of the Friuli Venezia Giulia, or the Dolomites.

[2] The island known as La Certosa, guarding the northernmost entrance to the lagoon. The island is out beyond San Pietro di Castello and close inside the Porto di Lido, between Sant'Elena and Le Vignole.

[3] La Certosa was settled in 1422 by Carthusian monks from Florence after an earlier Augustinian community had left. La Certosa was an important last resting place for the Venetian oligarchy, particularly the church of Sant'Andrea. The church was occupied by the Carthusians for nearly four centuries, until the Napoleonic suppression of the monasteries in 1806 and 1810. The monastery's complex was methodically dismembered and used for the storage of gunpowder. Mid-17th-century drawings confirm Folkes's description of separate cells for the monks. See Figure 17. See McAndrew, 'Sant'Andrea della Cortosa', pp. 18, 20, 21, 28.

[4] Santa Maria dei Servi, a monastery of the Servite order founded in the Cannaregio District of Venice in 1316, with the majestic Gothic church completed by 1474. Much of it was demolished in the Napoleonic suppressions. It was the home of Paolo Sarpi (1552–1623), a friend and patron of Galileo.

Figure 17. 'Isola S. Spirito – La Certosa di Venetia', Table 182, in Vincenzo Coronelli, *Singolarità di Venezia. Vedute, chiese, isole*, Venice, 1710. Biblioteca Nazionale Marciana, Venice. This illustration confirms Folkes's description of separate cells and gardens for the monks. Courtesy of the Biblioteca Nazionale Marciana.

dinner to morrow. Sig^{r} Martinelli dind with me and after dinner Abbe Conti came to see me who is returnd to Venice to see me before my journey, he staid with me all the evening till we went together to Sig^{r} Giustinianis. he cannot dine with me to morrow being to be at Marshal Schuylenburgs where I should also be if I was not engagd.

Sunday. 4. I visited the Resident who has been at Sr R. Browns but is returnd with a fit of the gout, we disport48ed away an hour or 2 as usual, when I came home and found my company who staid with me all the afternoon, after dinner. Abbè Conti came to us and I had a day of very agreeable conversation.

Monday. 5. Abbè Conti and Mr Bodin[1] dind with me and the Consul was with us in the best part of the day who was so kind to come to see me before my journey as I am preparing to leave Venice; I was also visited to day by ffather Crivelli and Cav^{r} Marini. # [2]

Tuesday. 6. Prince Castiglion visited me in the morning; the Consul dined with me, and in the evening I was at the Comedy at the Theater of St Luke, this is the first that have been since I came to Venice. the house is large and ~~capable~~ convenient but the entertainment lower than any thing I have seen at Bartolomew fair.[3] Count Bielgne who had visited me this afternoon carried me into his box, my own being full with My ffamily, Mrs Wills and Sig^{r} Martinelli who went with me. Sig^{r} Conti sat wth us.

Wednesd. 7. I put up my things for my journey so was not out in the morning, the Consul dined with me and in the evening carried us all to a play at St Lukes the entertainment was much like what we had the last night. their plays begin [f. 46r] late here not ending till eleven a clock

Thursday. 8. I was visited by Abbe Conti in the morning with Mr Bodin. then the Consul went with us to the Arsenal[4] which is the best worth seeing of any thing in Venice; they generally say it is about 3 miles round,[5] encompassd with a high wall and only one entrance by \water through/ a Canal shut with great gates and barricades, and[6] deep enough for their largest ships to come in that are built and deposited here. the entrance from land is one also close by the other and with a very magnificent gate and the Athenian Lyons lying before it. this arsenal is like the Kings Yard at Chatham Portsmouth etc in England,[7] but tho it is not equal in stores with them in some particulars, it exceed them in this that here is every manufacture relating to their shipping in the same place, the foundery of the Canon both brass and Iron, the Anchor forges, the docks both wet and

[1] It is not known who this man is; he does not appear in Ingamells's *Dictionary*.

[2] The hashtag-shaped symbol that Folkes wrote here is a reference to material he noted on f. 44v (see above, p. 99).

[3] St Luke's was a theatre devoted to comedy, often *Commedia dell'arte* performances, like the comedies of Four Masques. Its season was in the autumn from the Monday before St Luke's day (18 Oct) to the eve of St Martin's Day (10 Nov). It was coincident with autumn *villeggiatura* or holiday season for the aristocrats in their country villas. This comedy period (which opened after 1725 with special entertainments) was not used much for opera until the 18th century. See Selfridge-Field, *A New Chronology*, p. 26.

[4] Venice's complex of public shipyards and armouries was considered by Francesco Sansovino 'the base and the foundation of the greatness of the Republic'. Davis, *Shipbuilders*, p. 2. The Arsenal is often considered by historians as an example of proto-industrialization, particularly in shipbuilding, but whatever the interpretation, it certainly was the heart of the Venetian naval empire. In the early modern period, its great storehouse of munitions astonished travellers on the Grand Tour, like Folkes.

[5] Folkes was not far off; it was about 3.2 km.

[6] Folkes wrote 'and' over 'but' here.

[7] Folkes referred to Chatham Royal Dockyard on the river Medway in Kent, which built numerous ships for the Royal Navy.

dry for all their great ships and a very long water in the middle into which they launch from the Stocks all round; one thing very peculiar is all their docks are under cover,[1] so that a ship is quite filled within doors; and laid up in the same manner. they had several building when we were there both men of war and Gallies; the longrooms for the cables and other stores are very handsomely fitted up with dorick pillars so as to make a very beautifull as well as Grand appearance.[2] The Consul here had provided us a very handsome refreshment of Oysters just taken before us out of the sea, with several sorts of wine etc in an exceedingly genteel manner. their largest ships here are of 70 guns and the Admiral carries 80 but is but little bigger they told me than our 70 gun ships. we saw here in a house on purpose the bucentor[3] wherein the Doge marries the sea, it is the most beautifull vessel for shape and richness of ornament all gilt and carvd on by an excellent artist that can be imagined. it gave me much more pleasure and admiration than I could have thought, being indeed a masterpiece of beauty in its kind it was built about 4 years since in the time of the Doge Mocenigo.[4] We saw three old ones very fine also they had been but no ways comparable to this. They keep it in a house where it is drawn out of the water with capstans,[5] it carries 46 oars of four men to an oar and a fifth to overlook them. The state [f. 47r] room for the Doge and Senate is of a vast size, his seat opens behind to the sea when he turns about and flings 'em the ring with the words sponso te, etc.[6] there is also one fine barge for the other functions at this time a building.[7] there are here also armorys of small arms but those not comparable to which is in the towr of London. upon the whole I never saw a finer sight. after dinner I was again visited by the Abbe Conti, and also by Mr Smith and Sig^r^ Marcello. I was at night at a play at St Samuels[8] a handsome theater but not so frequented as the other.

[1] The undercover water areas are called *squeri*.

[2] Folkes is describing the Ropewalk (*Corderia de la Tana*) originating in 1579, having been built by Doge Nicolò da Ponte (r. 1578–85). The Ropewalk is 346 yards (316 metres) long, and it is supported by 92 Doric pillars.

[3] The *Bucintoro*, the gilded state barge of the doges of Venice, propelled by 168 rowers. Until 1798, it was used in the annual ceremony on Ascension Day (*Festa della Sensa*) 'marrying' the Doge to the Adriatic Sea, the Sposalizio del Mar. 'In the ceremony the *Bucintoro* was rowed in procession out of the lagoon into the Adriatic, where the Doge cast a ring blessed by the Patriarch into the waters, symbolising the marriage of Venice to the sea.' Folkes saw the last *Bucintoro*, constructed in 1728–9 with its gilded wooden allegorical sculptures by Antonio Corradini. After the fall of Venice to Napoleon in 1797, the *Bucintoro* was burnt to recover the gold and in 1824, broken up in its birthplace, the Arsenale. 'The Bacino di San Marco on Ascension Day', Royal Collection Trust, https://www.rct.uk/collection/404417/the-bacino-di-san-marco-on-ascension-day [Accessed 15 Dec 2022].

[4] Alvise III Sebastiano Mocenigo (1662–1732), the 112th Doge of Venice 1722–32.

[5] Capstan: 'A piece of mechanism, working on the principle of the wheel and axle, on a vertical axis, the power being applied by movable bars or levers inserted in horizontal sockets made round the top, and pushed by men walking round, whereby the apparatus is made to revolve and wind up a cable round its cylinder or barrel; it is used especially on board ship for weighing the anchor, also for hoisting heavy sails, etc., and for raising weights out of quarries, mines, coal-pits ... and the like.' See *OED*.

[6] *Desponsamus te, mare, in signum veri perpetuique dominii* [We espouse thee, O sea, as a sign of true and perpetual dominion]. Muir, *Civic Ritual*, p. 122.

[7] This vessel may have been the ship for the patriarch of Venice, who is taken on a separate boat to the *Bucintoro* to perform the ceremony. At the time of Folkes's visit, the Patriarch was Francesco Antonio Correr (served 1734–41), a priest of the Capuchin Order.

[8] Teatro San Samuele, an opera house and theatre on the Rio del Duca in Venice. The theatre belonged to the Grimani Family. Michele Grimani (1696–1775), was the most active in its management. The Teatro San Samuele was used especially for Ascensiontide operas, and it stood from 1656 until 1747 when the building was destroyed by fire. A new building was opened in 1748, and it continued under different owners until the

fryday. 9. being to leave Venice to morrow I went in the morning to take leave of Marshal Schuylenburg, Prince Castigilione Count Bielk and Cavalier Marini \and Mr Smith/; and then came home where Abbè Conti The Consul and Mr Bodin dined with me; after dinner I was visited by, Comt Bielk Cav^{r} Marini, Mr Wills and Mr Martinelli \Dr Michelotti/ and Mr Smith with Mr Egerton who gave me my letters and settled my accounts, I went in the Evening to the Resident who is confined with the Gout. and so having finisht my affairs I set out to morrow morning in a Burcello[1] which I have orderd for 7 o clock. The Consul was so kind to give me several more papers relating to things here I had a mind to enquire after.

Figure 18. 'Burcello etc. a Conveyance from Padua to Venice', in Edward Wright, *Some Observations Made in Travelling through France, Italy, etc, in the years 1720, 1721 and 1722*, 2 vols, London, 1730, I between pages 42 and 43. Wellcome Images, London. Folkes took a similar conveyance when leaving Venice for Padua.

---

theatre was demolished in 1894. The impresario and dramatic poet, Carlo Goldoni, had his work performed at the theatre, as did Antonio Vivaldi in Autumn 1735, namely, his operas *Aristide* and *Griselda*. See Talbot, *The Vivaldi Companion*, pp. 12, 92.

[1] As noted by Wright in the mid-18th century: 'the Burcello is a large handsome boat; the middle part of which is a pretty room, generally adorn'd with carving, gilding and painting. 'Tis drawn down the Brenta with one horse to Fusino, the entrance into the Lagune; and from thence to Venice, 'tis hawl'd along by another boat, which they call a Remulcio, with four or six rowers. The passage down the Brenta is very pleasant, being enliven'd on each side with pretty villages, and with palaces, many of them built by Palladio, which are villa's to the noble Venetians.' Wright, *Some Observations*, I, p. 43.

Saturday. 10. I set out from Venice in a Burcello between 8 and 9 a clock, the Consul kindly to his other favours adding that of bearing me company to Padua. we crosst the Lagoon and went up the Brenta,[1] the same views I had taken notice of coming with the perpetual fine seats of the venetians on both sides the river, make a prospect I have never seen equald. on the outside of a house at [*blank*][2] belonging to the ffamily of Bembo, are 2 sham balconys with 2 figures looking out in each painted by Paul Veronese[3] and in good preservation and strength of colours still. one might well pass a long time in seeing and

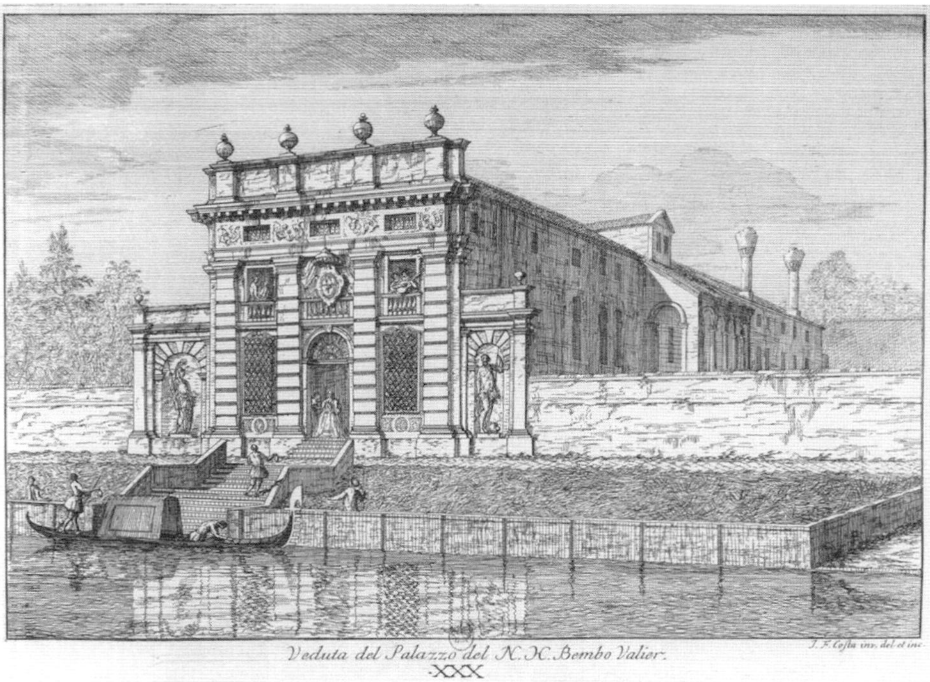

Figure 19. Giovanni Francesco Costa, 'Veduta del Palazzo del N. H. Vembo Falier', plate 30, in *Delle delicie del fiume Brenta espresse ne'palazzi e casini situati sopra sue sponde dalla sboccatura nella laguna di Venezia fino alla citta di Padova*, Venice, 1750. Bibliothèque nationale de France, Paris. The trompe l'oeil paintings of the 'sham balconies' that Folkes observed can be seen in this engraving. More famously, Veronese also painted the interior of the piano nobile of Villa Barbaro with illusionistic frescoes. Courtesy of the Bibliothèque nationale de France, Paris.

[1] The Brenta canal, constructed in the 16th century, directly linked Padua and the Venetian lagoon and carried goods and travellers via Strà, Dolo, and Mira, to Fusina, at the margins of the lagoon directly opposite Venice. The banks of the canal became a favoured area for Venetian nobles to build their villas and hold humanist salons. Moretti, 'Barbaro brothers', https://www.thebritishacademy.ac.uk/blog/how-barbaro-brothers-created-perfect-renaissance-villa/ [Accessed 20 Dec 2022].

[2] Folkes left a blank space here as he could not remember the name of the village.

[3] This is the Villa Bembo Valier, nicknamed, 'la Chitarra', or the guitar. As Sapienza has noted: 'The frescoes of Villa Bembo Valier known as "La Chitarra" decorated the monumental façade on the river. The figures looked out from a balustrade, simulating false windows'. Though Folkes thought the frescoes were by Veronese, their attribution is disputed. Sapienza, 'Notice sur le cycle', pp. 21–4.

describing what is upon this River whose sluices also are considerable. I was surprised to see the trees here bearing still a good summer look notwithstanding the excessive cold we have had but to day if [*sic*] a fine warm day. we came by Paluello where Sr Robert Brown and his lady very kindly givest us to have staied some time in as pleasant a seat as ever I have seen; but as our journey was fixt we only drank tea, and they were so good to [f. 48r] give us their company in the boat as far as the furthest sluice ~~wh~~ next Padua,[1] where their coach attended them and we took leave. Abbe Conti has a little seat also next over against Sr Roberts. I leave this part of the world with regret from the many distinguishing marks of favour and esteem I have received both from the English and the most remarkable, both for quality and accomplishments of the Venetians. I arrived at Sigr Barbaros at Padua about 8 a clock. having been favoured with an order to have the gates kept open for me from the Podesta.[2] We landed just without the town where I had ordered a Coach for my wife. I am much fatigued having been about all day yesterday and having had excessive bad rest last night. I am also in great pain of a broken skin I have got on board the Burcello.

Sunday. 11. Dr Mengoni who is Bidello è Bibliothecario della natione Britannica[3] came to me; and I went out with him and the Consul to look about the town. I saw the pavilion at the Padua cornaro, a celebrated piece of Architecture of Falconetti.[4] I was

[1] 'Politically, Padua belonged to the Venetian mainland state, having come under Venetian sway upon the overthrow of the Carraresi in 1405. To its conquered cities of the terraforma Venice allowed a degree of political autonomy that was intended to perpetuate the semblance, if not the reality, of the *status quo ante*.' Schulz, *Giammaria Mosca*, p. 13.

[2] The podestà was the holder of the highest civil office. Girolamo Bolani, was Podestà in Padua from 12 May 1734 until 12 May 1735, when he died. See Gloria, *I podestà*, p. 38. For the role of the podestà, see above, pp. 37–8, n. 3.

[3] Domenico Mingoni, who served as beadle of the Library of the English Nation at the University of Padua, established on the model of the Natio Germanica in 1649. It was a library for Scots, Irish, and English. The University of Padua itself was founded in 1222 by a group of professors and students from Bologna who wished for more academic freedom. 'Initially, the University was organized as a free corporation of students and organized according to ethno-geographic criteria, the "nationes" that, in turn, referred to two large groups: the citramontani (Italians) and the oltramontani (non-Italians). Therefore, it was the students who approved the statutes and elected their professors, paying them with the money they were able to raise.' 'The University of Padua', Padova medieval, http://www.padovamedievale.it/info/university-of-padua/en [Accessed 22 Dec 2022]. The 'Nations' had designated representatives who formed the University council, headed by a syndic and pro-rector, elected annually at the beginning of August. See Villani, 'Between Anatomy and Politics', p. 153. The University of Padua had international fame in the early modern period, largely because of the freedom of thought permitted by the Republic of Venice, of which Padua was a subject from 1405 to 1797. The Universities of Padua and Bologna were the most influential sites for the study of natural philosophy (science) in the late 14th and 15th centuries. By the 18th century, Padua had declined, and Leiden and its iatrochymical curriculum was considered the major attraction for English medical students abroad. See Park and Pennuto, 'Science and Medicine', pp. 368–9, 375. Other contemporary guidebooks also indicate the decline of the university; in 1707, De Blainville reported of the University of Padua, 'At present it is entirely stript of all its former Lustre'. De Blainville, *Travels*, I, p. 455.

[4] Loggia e Odèo Cornaro, a theatre and music room built in 1524 and conceived of by Alvise Cornaro (1480/4–1566) a writer, art patron, and hydraulics expert. The designer was Giovanni Maria Falconetto from Verona. The Loggia stage was designed as a *scaenae frons* to emulate an open-air Roman theatre, and it was decorated with classical motifs in fresco and stucco. See 'Le arti alla "corte" di Alvise', Latorlonga, https://www.latorlonga.it/loggia-e-odeo-cornaro [Accessed 20 Dec 2022]; Moretti, 'Quivi si esercitaranno', pp. 136–8.

then at St Anthonies Church[1] where I again took notice of the bass reliefs of brass and figures of the same metal and great candlestick all which I am informd are works of Donatelli.[2] The round building also behind the great Altar which I before took for a Chapter house,[3] is designd to contain the treasury that is to say the relicks of this church it is one of the finest rooms I have seen very richly adorned with marble and carvings the figures are of Parodis[4] who did some also at the high altar and about the Saints shrine, in this rotunda fronting the entrance are to be 5 repositorys shut with folding doors. those of the middlemost are only finishst and are all of silver and gold chasd, in a fine and magnificent manner the work by Sig^r Ridolfi who is dead but has left one pair more they say unfinisht. on the right hand overagainst the shrine is a pretty chappel all painted as they say by Giotto.[5] it is the historys of the crucifixion and passion of our Lord. the shrine it self of St. Antonio and the chappel enclosing it is fine beyond what can well imagen. the history of his life in all relief some of which are by

[f. 48v]Plus Potuit B[eata] V[irgo]. facere de Deo, quam Deus de se ipso. St Bernardini[6] Per ipsam, et in ipsâ, et cum ipsa, et ab ipsâ habet mundus et habiturus est omne bonum.[7] Velocior est Salus nonnumquam memorato nomine mariae, quam invocato nomine filii Dei. St Anselm.[8]

[f. 49r] Sansovino[9] are masterpieces all of white marble. the figures also of Marble and

[1] Basilica Pontificia di Sant'Antonio di Padova, known as Il Santo. The Basilica housed the tomb of St Anthony (d. 1231), and was one of the most prestigious churches for the Franciscan order and a major destination for pilgrims, native and foreign (the *forestieri*). It was also a 'key civic symbol for the city of Padua'. Johnson, 'Approaching the Altar', p. 627; see also pp. 629, 636, 647. The material was gilded bronze, a reflection of local wealth, as well as the aesthetic tastes of the Conventual Franciscan friars, which favoured rich decoration. Johnson. Ibid., p. 640.

[2] Folkes means Donatello, who worked on the high altar for Il Santo in the 1440s. He executed the high altar in the Basilica Pontificia di Sant'Antonio, called the *Altare del Santo*, between 1447 and 1450. There are seven lifesize statues (three Franciscans, including Anthony, and 3 local saints), as well as four large reliefs in bronze with the saint's miracles. Donatello also sculpted a limestone relief of *Christ in the Emtombment*, and a monumental bronze *Crucifix* (1444–7). Paoletti and Radke, *Art in Renaissance Italy*, p. 264. The bronze candlestick was not made by Donatello, but by Andrea Biosco or 'Riccio' in 1515. 'The Presbytery and Main Altar', Saint Anthony of Padua, https://www.santantonio.org/en/content/presbitery-and-main-altar [Accessed 21 Dec 2022].

[3] The Chapel of the Relics (Treasury Chapel), erected towards the end of the 17th century, according to the architectural plans of the sculptor and architect, Filippo Parodi. 'The Chapels of the Relics (Treasury Chapel)', Saint Anthony of Padua, https://www.santantonio.org/en/content/chapel-relics-treasury-chapel [Accessed 21 Dec 2022].

[4] Filippo Parodi (1630–1702) from Genoa was responsible for the six marble sculptures of saints and the *Glory of Saint Anthony* (1689–97) in the Capella del Tesoro (Chapel of the Treasure) in Il Santo.

[5] Giotto painted frescoes in the Benediction Chapel in Il Santo (under the patronage of the Scrovegni family) before moving on to complete his more famous cycle in the Scrovegni Chapel which was connected to the Palazzo Scrovegni. He also did wall paintings in the chapter house, currently being restored. See Di Giovanni, 'I Dipinti Murali', pp. 355–91.

[6] From San Bernardino of Siena, *Opera omnia*, 2:375. This should read: 'Plus enim potuit beata Virgo facere de Deo, quam Deus de se ipso' [The Blessed Virgin was able to do more for God than God could do for himself].

[7] 'Through her, and with her, and from her the world has and will have every good.'

[8] St Anselm [pseudo; Eadmerus Cantuariensis], *Liber de Excellentia Beatae Mariae*, cap. 6. 'Salvation is sometimes swifter through recalling the name of Mary, than through invoking the name of the Son of God.' The text is not by Anselm as Folkes indicates, but by Eadmer (*c.*1060–*c.*1124), a Benedictine monk and contemporary biographer of Anselm. The absolute perfection of Mary was the subject of Eadmer's *De excellentia*. See Westphall, 'A Borrowing', p. 2.

[9] There are nine marble reliefs with scenes from the life of St Anthony, indeed some by Sansovino, as well as other regional artists such as Antonio Minello. See 'The Chapel of Saint Anthony', Saint Anthony of Padua Basilica, https://www.santantonio.org/en/content/chapel-saint-anthony [Accessed 22 Dec 2022].

bronze with ~~with~~ some of silver are no less curious. the number of lamps burning in large silver vases[1] and one as they say of pure Gold and which cost ~~measure~~ \above/ 2000 gives an air of surprizing grandeur. here are numberless things more than can be obsevd by a traveller in the little time he can allot ~~it~~ himself. but the whole of this church is not pleasing it is dark and crowded with things and very dirty. I was next at the Schola of the Saint which I had not seen before, it is all paintd in fresco with historys of his miracles.[2] 4 of the pieces are of Titian[3] and I think in as grand a manner as any of his works I have seen. particularly that where a jealous husband has just killd his wife. suspecting his child to ~~have~~ \be/ none of his own. and the regret of the same where the mother being restored to life the infant names his father by the command of the Saint.[4] the other frescoes are of other hands Campagnoli etc[5] there is also in the same a picture of Paulo Veronese on canvass. I went from here to the fine Church of S[t] Justina[6] which gave me new entertainment and admiration tho I had been so often here before. besides the things I had before noted the Consul made me observe some inscriptions I had not before read in a little chappel dedicated to the Virgin and St Prosdocimus[7] between the Church and Convent.[8] The wall is painted with figures of angels holding tablets on which are inscribed sentences to her honour. 3 of which I thought worth transcribing for the oddness of them. having thus spent the morning I came to my inn where I found S[r] Robert Browne and his lady had done us the honour of coming to dinner with us, and staying best part of the afternoon I was not any more out than a little about the street just in the evening. the town is extreamly large but excessive ugly all but the

[1] On the sides of the altar-tomb of St Anthony, built by Tiziano Aspetti in 1594, are two 17th-century silver candelabra on marble bases.

[2] The Scuola del Santo or Scoletta (little school), which was the headquarters of the Archconfraternity of St Anthony of Padua, built in 1427, and expanded in 1504. Wilk, 'Titian's Padua Experience', p. 51.

[3] Titian was called to Padua in December 1510 to paint three frescoes for the *Sala Priorale* or meeting hall of the Scuola. The artist finished all three frescoes in December 1511: the *Miracle of the Newborn Child*; *The Miracle of the Jealous Husband*; and the *Miracle of the Reattached Leg*. These are Titian's earliest, 'securely documented paintings and his first independent commissions after the death of Giorgione'. Ibid., pp. 51, 56.

[4] Folkes was clearly impressed by these works, which were innovative in composition. If Folkes had looked more closely, he may have realized that the raised right arm of the resurrected woman is actually a plaster relief that was painted by Titian, casting its own shadow. This was the only instance of three-dimensional work by the artist, who in his later work achieved spatial illusionism via painting alone. Morosini, 'New Findings', pp. 163–4.

[5] 'Campagnoli etc', was written in the margin. This was a reference to *Risurrezione di una ragazza annegata* by Domenico Campagnola (*c*.1500–64), who was associated with Titian's workshop. Wilk, 'Titian's Padua Experience', p. 51.

[6] Basilica of Santa Giustina, adjacent to the 10th-century Benedictine Abbey, which was initially built in the 5th century, originating as a funerary basilica in the cemetery outside the city walls, and reconstructed several times, more completely in the 15th and 16th centuries. It was suppressed during the Napoleonic invasions of 1797, stripped of its art, which was sent to Paris, and the buildings were only returned to the Catholic Church in 1917. It is now an autonomous priory. It was dedicated to Justina, a local princess, who was thought to have died in the persecutions of Emperor Maximian, also called Herculius (*c*.250–310). Mackie, *Early Christian Chapels*, p. 40.

[7] The sanctuary of Prosdocimus of Padua, also known as the Sanctuary of Santa Maria, dedicated to the first Bishop of Padua, St Prosdocime of Padua (d. *c*. AD 100), and the site of his burial place. The chapel also held a relic of the Virgin purportedly painted by St Luke, the medieval dedication to the Virgin probably commemorating the icon's presence in the chapel. See Mackie, *Early Christian Chapels*, pp. 40–41

[8] These are the inscriptions Folkes noted on f. 48v (p. 106 above).

Campo Sancto where stand S[t] Justinas, and the piazza or open place[1] before the palace of the Podesta which is very grand but out of repair.[2] the town hall I before took notice of. Dr Mengoni tells me the University is nothing in the flourishing condition it has been there is however about 600 students.[3] which are divided into about 11 nations each of which have some particular officers. they now [f. 50r] are much more orderly than they have been the streets are said to have been very dangerous formerly but now are as safe as other places.

Monday.12. I bargain with Giovanni Sardella to carry me[4] to Rome in my own berlin[5] and find another for my servants and baggage. being informed it is very difficult for me to go by \Bologna/ fflorence. so he is to take the road by ferrara Ravenna and Loretto. he cannot have his things ready till wednesday morning and I am to give him 130 Cequini[6] to keepe my self wife and the children on the road, and is to provide us license to eat gras. the price is great but he is said to be the best provided of all those sort of people and having had trouble with my former veturrini was willing to go with a man of Character, which Sardella has very great from the Resident S[r] Robert Brown M[r] Consul and Mr Smith that have all usd him. having then settled my journey I went out to see the house of Count Orsetti a Noble Lucchese[7] that lives here. he has a fine collection of pictures, one called Raphael a madonna and Christ but in the stiff manner of perugino.[8] some very good ones of John Bellino.[9] a fine Christ carrying the cross by Corregio, some parmigianinis[10] and a good madonna of Holben[11] with several others. he has also a great number of groupes of marble and brass, by Carvers whom he has kept

[1] All part of the building complex now called the Palazzo Communale, the civic centre.

[2] The Palazzo del Podestà, also known as Palazzo Moroni, built between 1539 and 1601, was after the design of architect Andrea Moroni (1500–1560). The palace replaced a pre-existing medieval building and was commissioned under the mandate of Marcantonio Contarini – Padua's podestà, or city Mayor, between 1538 and 1540. Courtyard Elevation of Pallazzo del Podestà, Padua, https://collections.vam.ac.uk/item/O617612/ courtyard-elevation-of-palazzo-del-drawing-unknown-circle-of/ [Accesssed 22 Dec 2022].

[3] For the University of Padua and its 'nations', see above, p. 105, n. 3.

[4] A *vetturino* or coachman. A cross is inked in the right-hand margin here.

[5] A four-wheeled carriage with a separate sheltered seat behind the body.

[6] The Venetian *ducat* or a *zecchino*, named after the Zecca. Women stitched *zecchini* on their clothes as decoration; hence the moniker sequin. Each coin weighed 3.5 grams of nearly pure gold, so this was 455 grams of gold, an enormous sum indeed. See Appendix B on Venetian coinage.

[7] Native of Lucca in Tuscany. The Orsetti family originated in San Donnino di Marlia, a rural village located near the Tuscan city of Lucca, where they relocated at the beginning of the 15th century. Lucca was a centre for silk production and trade. By the mid-17th century, Orsetti family members owned the second-largest textile workshop in Lucca, with 95 looms. Their companies thrived in Italy, as well as in Germany and Eastern Europe, especially Poland. Their silk trading company, Filippo Orsetti e Compagnia, flourished from 1695 to at least 1744. 'Keeping the Books: the Rubinstein Collection of the Orsetti Business Archive', University of Kansas Special Collections, https://blogs.lib.ku.edu/spencer/fall-exhibit-2022-keeping-the-books-the-rubinstein-collection-of-the-orsetti-family-business-archive/ [Accessed 26 Dec 2022].

[8] The artist Perugino [Pietro di Cristoforo Vannucci] (*c.*1450–1523).

[9] The artist Giovanni Bellini (1430–*c.*1516). For other discussions of his work, see above, p. 54, n. 6.

[10] The Mannerist painter Girolamo Francesco Maria Mazzola (1503–40), also known as Francesco Mazzola or, more commonly, as Parmigianino.

[11] Possibly Holbein.

Plate 1. *Fra Mauro Map* (Il Mappamondo di Fra' Mauro) (*c.*1450), ink on parchment on panel support, *c.*230 × 230 cm. Biblioteca Nazionale Marciana. Alamy Images. Folkes had contemplated doing an edition of the work.

Plate 2. Canaletto (Giovanni Antonio Canal), *Piazza San Marco* (late 1720s), oil on canvas, 68.6 × 112.4 cm. Metropolitan Museum of Art, New York, Purchase, Mrs Charles Wrightsman Gift, 1988. Courtesy of the Metropolitan Museum of Art, New York. The view is largely contemporaneous with Folkes's visit to Venice.

Plate 3. Pietro Longhi (Pietro Falca), *The Temptation* (1746), oil on canvas, 61 × 49.5 cm. Metropolitan Museum of Art, New York. This is an example of the genre scenes Longhi painted after his failure at history paintings, such as the monumental ceiling painting *Fall of the Giants* (1734) in the Palazzo Sagredo that Folkes denigrated. Courtesy of the Metropolitan Museum of Art, New York.

Plate 4. Salvator Rosa, *Democritus in Meditation* (1640s), oil on canvas, 96 × 154 cm. Statens Museum for Kunst, Copenhagen. The tenebrous and melancholy quality of this work is pervasive, causing Folkes to call the work 'black and unpleasant', when he saw it in the Palazzo Sagredo. Courtesy of Statens Museum for Kunst, Copenhagen.

Plate 5. Attributed to Raphael, *The Last Supper* (*c.*1514), pen and ink with stylus lines and pointing, 30.6 × 46.5 cm (sheet of paper). Courtesy of the Royal Collection Trust / © His Majesty King Charles III 2023. Folkes recorded seeing this drawing in the Sagredo Library when the collection was intact.

Plate 6. Paolo Veronese (Paolo Caliari), *Mystical Marriage of Saint Catherine* (*c.* 1565–70), oil on canvas, 377 × 241 cm. Gallerie dell'Accademia di Venezia, Italy. G.A.VE – Archivio fotografico. Courtesy of the Italian Ministry of Culture. Folkes recorded seeing the painting in Santa Caterina in Venice, its original location.

Plate 7. Francesco Villamena, *Presentation of Christ at the Temple*, after Paolo Veronese, (1597), engraving, 36 × 44.2 cm. Rijksmuseum, Amsterdam. Folkes noted this engraving of the original painting that he saw in San Sebastiano. Courtesy of the Rijksmuseum, Amsterdam.

Plate 8. Tommaso Lombardo, *Madonna and Child and Saint John* (*c.*1535–1560), Melio da Cortona Altar in San Sebastiano, Venice. Author's photograph.

Plate 9. Anonymous seventeenth-century Indian artist for Nicolò Manucci, *Princess Palanquin on an Elephant at the Court of Shah Alem Aurangzeb* (*c.*1680), watercolour on paper with gilded inserts, 38.5 × 25.5 cm. Bibliothèque Nationale de France, Cabinet des Estampes, Paris, Codex CD 45 (*Libro Rosso*), f. 13r. Bridgeman Images. Folkes recorded seeing this manuscript during his visit to the Library of San Marco, Venice.

Plate 10. Canaletto, *Venice: The Feast Day of Saint Roch* (*c.*1735), oil on canvas, 147.7 × 199.4 cm. The National Gallery, London. This painting by Canaletto of the Feast Day of Saint Roch is nearly contemporaneous with Folkes's visit to Venice. The Scuola Grande di San Rocco hung an exhibition of paintings on the Scuola's facade on their feast day. Courtesy of the National Gallery.

Plate 11. Francesco Pianta di Giovane, *Curiosity* (1657–76). Allegorical wood sculpture of a library much admired by the bibliophile Folkes, carved as part of a walnut panelling scheme installed in the *Sala Capitolare* of the Scuola Grande di San Rocco, Venice. Alamy Images.

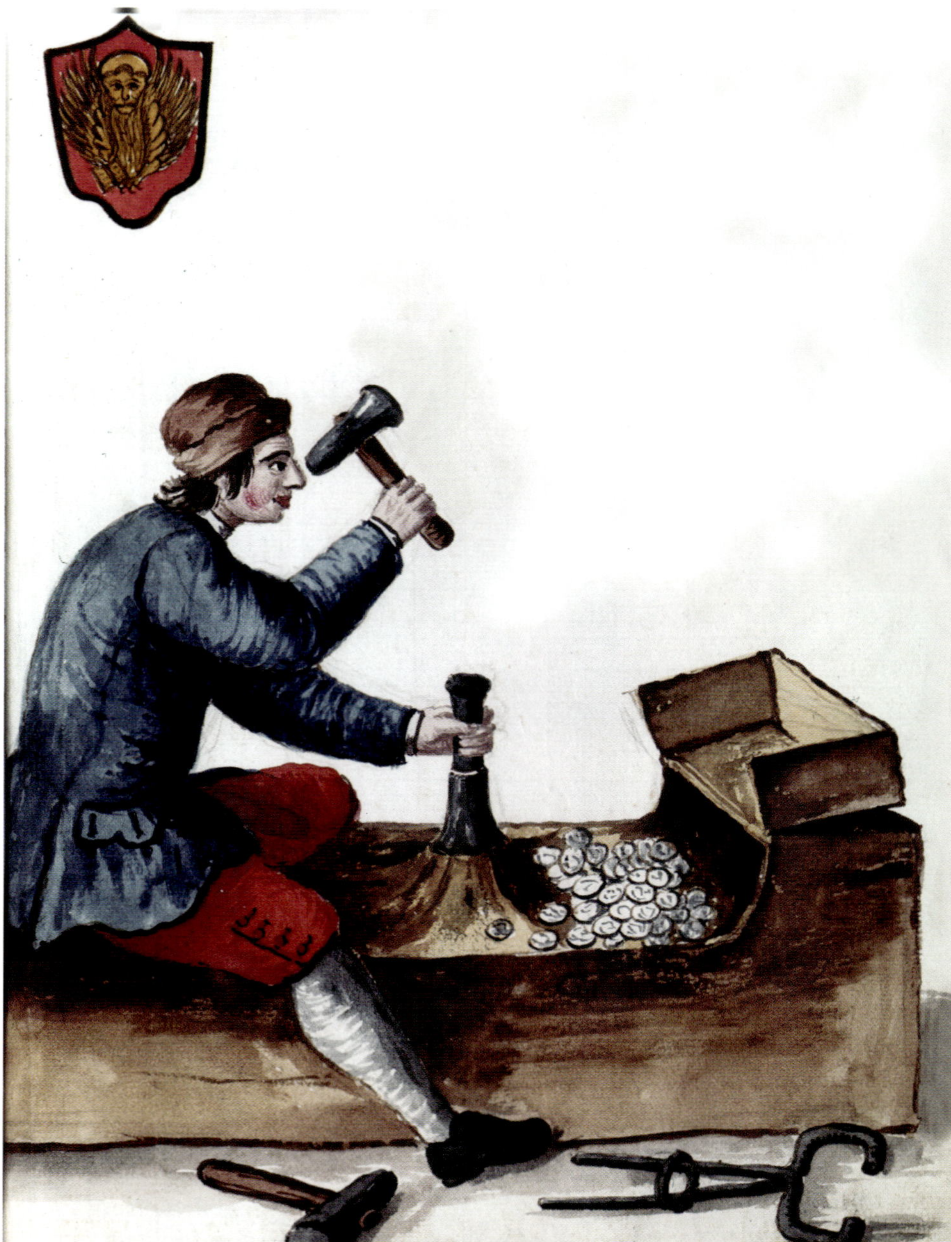

Plate 12. Giovanni Grevembroch (Jan van Grevenbroeck), *Venetian Coiner at Work* from the manuscript, *Gli abiti de veneziani di quasi ogni età con diligenza raccolti e dipinti nel secolo XVIII*, or, *Customs of the Venetians* (eighteenth century), watercolour and ink on paper. Museo Correr, Venice, Gradenigo Dolfin 49 (=191). Bridgeman Images. Folkes described the process of coining via hammering that took place in the Zecca or Mint of Venice.

entirely at his service for many years they are fine things in their sort[1] but have something not great in their taste which made them far less entertaining to me than some others that have much less labour there are commonly 8 or 9 figures in a small groupe. and they value them much upon being all of a piece to the very loose links of [[~~*one word*~~]] a marble chain that is in one of them. they say 2 of the groupes ~~who~~ were sold to Sigr Pisani[2] for antique at an immense rate. if it is true it only proves the want of judgment of the Sigr. for they are in no sort like antique. The Count receivd us as he does all strangers with great civility and showd us his things himself, he has travelld and been in England in K. Williams time.[3] When we had taken leave of him we went with Dr Mengoni to see the library of the English Nation which is kept in his house. There are 3 nations of the University that have Librarys the English, German and Greek, this is meerly a name for I think there are not above 300 vols. [f. 51r] and those very ill chose and in bad condition; the English are one of the Nations in this University priviledgd above the rest; and which performing the forms may have their degrees immediately upon their admission whereas those that are not so priviledgd must wait four years upon matriculation before they can commence. the discipline of the University is excessive bad or rather none. every scholar has a vote in the making their officers by which it has frequently happend that a rakish young fellow by being popular among those of his own sort has been chose Syndic or (vice Syndic)[4] when they have no Syndic, which is their chief Magistrate, and has ruled accordingly having a great power, and entirely protected the scholars and maintained them in the most infamous crimes and irregularitys.[5] several instances of this were given me. Dr Mengoni keeps a book wherein most of the English that go by ~~what~~ write their names and give him a pistole or at least a Cequin for his attendance.[6] I read here those of all of my ffriends that have been in these parts. The

[1] The Villa Reale di Marlia in Lucca was bought by Oliveri and Lelio Orsetti in 1651, who recreated a Baroque garden park with extensive statuary, so it is possible their descendant employed workmen to decorate his house and gardens similarly in Padua.

[2] The Pisani were a Venetian patrician family. Alvise Pisani (1664–1741), served as the 114th Doge of Venice (1735–41), and his brother Andrea Pisani (1662–1718) was captain-general of the Venetian navy. The family demolished their villa in the Stra, and in 1720, engaged the family architect, Girolamo Figimelica Roberti, to design a new villa and garden. The garden structures and maze were begun before the main building. It is possible that the statuary was bought from Orsetti for these purposes. See Tieto, *La meravigliosa Riviera,* p. 288.

[3] Possibly Bartolomeo Orsetti. The *State Papers Domestic* in the reign of William III recorded on 7 Nov 1692: 'Passes for Mr. Bartholomy Orsetty, and Jacques Giominiani, his servant, to go to Harwich, and embark for Holland.' 'William and Mary: November 1692', in *Calendar of State Papers Domestic: William and Mary, 1691–2*, p. 495.

[4] For the University of Padua and its 'nations', see above, p. 105, n. 3.

[5] Paul Grendler asserted that the violence in student-universities in Italy was because they did not exercise very much disciplinary authority. Local governments were hesitant to confiscate weapons or punish student violence. Grendler, *The Universities*, pp. 503, 506. However, as Davies has argued, although there was leniency, this was not the whole picture, and he has provided evidence that there were a variety of responses to student crimes, including homicides being punished by execution, or by fines if they occurred in brawls, explusion, or in lesser cases, public apologies. See Davies, 'Violence and Italian Universities', pp. 512–14.

[6] As Wright indicated, Mingoni kept a 'Register of Strangers he attends upon; his fee is a Pistole'. Wright, *Some Observations*, I, p. 43. There are 950 signatures in the Visitors Book, 'well maintained until 1730 and petered out in 1765'. Ingamells, *Dictionary,* p. xvi.

D of Richm[on]d when Lord March.[1] Mr Hill,[2] Lord Herbert, first lord Hervey,[3] Mr Stanley,[4] Mr Sloane,[5] Mr Wray,[6] Lord Darcy[7] and very many more of my acquaintance. from Dr Mengonis I went to the gardens of Papafava, and in the way went into the new church not yet finisht of St Maria de Piangi (of complants).[8] I took notice of it as something singular it is quite round only with sort of circular niches for 2 altars and the Organ. The chief altar is in the center and surrounded with pillars at some distance enclosing a sort of tempietto, whose dome ioyns that of the church it self. The Gardens

[1] Folkes's dear friend, Charles Lennox, 2nd Duke of Richmond and Duke of Aubigny (1701–50). He was Earl of March from his birth, as he was the heir to his father's dukedom, and he went on Grand Tour for nearly three years following his marriage to the 13-year-old Sarah Cadogan in 1719. Lennox signed the Paduan register book on 10 January 1721. See Ingamells, *Dictionary*, p. 639. Lennox provided Folkes with several letters of introduction to Venetian nobility. See Introduction, p. 26, n. 3, above.

[2] Probably Thomas Hill, Lennox's tutor, who accompanied him on the Grand Tour. Hill later played an important role in commissioning Canaletto to execute a pair of paintings for the Duke, when the artist came to England seeking work. In May 1746, Tom Hill wrote to Richmond: 'The only news I know to send you, is what I had this day from Swiney [Owen McSwiny] at the Duke of Montagu's [Richmond's neighbour], where we dined, and he, I think, got almost drunk. Canales, alias Canaletti, is come over with a letter of recommendation from our old acquaintance the Consul of Venice ... I told him the best service I thought you could do him w'd be to let him draw a view of the river from y'r dining-room, which in my opinion would give him as much reputation as any of his Venetian prospects.' The result of this letter was Richmond commissioning two views of London from Richmond House: *Whitehall and the Privy Garden* and *The Thames and the City of London*.

[3] The Whig politician and Freemason John, Lord Hervey (1696–1743), a friend of Folkes and Lennox. Like Folkes, Hervey went to Clare College, Cambridge.

[4] Possibly George Stanley (1692–1733/4), who was at Clare College, Cambridge, with Folkes (admitted 1710), and whom Folkes nominated as a Fellow of the Royal Society in 1719. Stanley married Sara, daughter of Sir Hans Sloane. He unsuccessfully contested the parliamentary seat of Lymington in 1729 and committed suicide in 1734. See 'George Stanley', Royal Society Catalogue of Fellows, https://catalogues.royalsociety.org/CalmView/Record.aspx?src=CalmView.Persons&id=NA5361&pos=6 [Accessed 29 Dec 2022]. There was a George Stanley that visited Padua and signed the Visitors book on 26 Aug 1717. See Ingamells, *Dictionary*, p. 889.

[5] Presumably William Sloane (d. 1767), who signed the register book in Padua on 29 January 1717. See Ingamells, *Dictionary*, p. 864. It is likely this was Hans Sloane's nephew. He was made Fellow of the Royal Society in 1722, and he was a member of the Freemason's Lodge that met at the Dolphin in Tower Street in London. William Sloane and Folkes often met in the St Paul's Churchyard Tavern with other Masons. See Roos, *Folkes*, p. 80.

[6] Daniel Wray (1701–83), FRS (1729), FSA (1741), antiquary, writer and friend of Folkes's, who signed the register book in Padua on 20 March 1726. See Ingamells, *Dictionary*, p. 1021. Wray travelled in France and Italy with James Douglas, later 14th Earl of Morton (1722–28). He also served as deputy teller of the Exchequer (1745–82) for Philip Yorke, Lord Hardwicke, and as a trustee of the British Museum (1765). See 'Daniel Wray', Royal Society Catalogue of Fellows, https://catalogues.royalsociety.org/CalmView/Record.aspx?src=CalmView.Persons&id=NA1094&pos=2 [Accessed 30 Dec 2022].

[7] Possibly James (Jessop) Darcy (*c.*1707–33), 2nd Baronet Darcy of Navan, who was admitted as a fellow commoner to Clare College, Cambridge (Folkes's college), in 1722, and matriculated in 1723. Darcy signed the register book in Padua in September 1727. See Ingamells, *Dictionary*, p. 276. Folkes, along with his brother William Folkes, and Sir Hans Sloane, proposed Darcy for membership of the Royal Society in 1729. See 'James Darcy', Royal Society Catalogue of Fellows, https://catalogues.royalsociety.org/CalmView/Record.aspx?src=CalmView.Persons&id=NA7346&pos=1 [Accessed 30 Dec 2022].

[8] Santa Maria del Torresino also known as Santa Maria del Pianto, or, Our Lady of Sorrows or Laments (Folkes wrote 'Piangi' and 'complants' instead of 'complaints'). The odd shape results from an 'oratory here adjacent to a torresino (turret) on fortifications'. See Figure 20. The current church was built in 1718–26 to designs by librettist and architect Gerolamo Frigimelica Roberti (1653–1732), and it was completed by Sante Bonato. The church is called to this day 'Torresino' because of Frigimelica's odd central crenellated tower. Cotton, 'Santa Maria del Torresino', Churches of Venice, http://churchesofvenice.com/padua2.htm#margh [Accessed 31 Dec 2022].

of Papafava are much talked of here, tho there is very little in them to those that have seen gardens in our parts, there are tall yew trees hedges and Arbours, and what they show as curiositys are a maze of green hedges of which I have a plan;[1] and a *Palatio de Herbe*. a set of Arbours laid out into apartments of a house, there is the portico salons and bed chambers, with beds made of mossy banks; there are also some very fine orange trees but standing in tubbs the climate not being here quite good enough to let them stand quite in the ground and open. I then [f. 52r]came here to my inn where I was visited by Sig^r Riva the Professor of Astronomy here[2] and went not out the rest of the day. They are making wine in the yard of the inn they bring the grapes home in a cart, in which also they tread them and the wine runs out into tubs by 2 holes in the bottom of the cart. it is red and looks thick and ropy like blood from a butchers stall as they carry it away in pails. no ways tempting to drink.[3]

Tuesday. 13. I was at Marquis Polenis but he is out of town. I was also at the convent of St Giovanni di Verdara,[4] where 3 of the ffryers (who are all of noble familys shewd me their convent and walkd about with us with great civility. there is in the Refectory a fine picture of the Marriage of Cana by Padovanino.[5] we then took a walk about the town which is near 4 miles round but miserably inhabited. over a coffee house in the Piazza is a large stone of infering [*sic*] to the memory of some Scholars killd a few years since by the Sbiri;[6] and who were here said to have been exemplarily punisht. this fact I was told however was a quarrel that happend between the Podesta and the Syndic of the University, in which the latter was greatly to blame, but the Sbiri in executing the podestas command either exceeded their orders or those orders were ~~simply~~ not thought fit afterwards to be justifyed and this was set up to satisfie the scholars. I went not out this afternoon staying to see my things put on the coaches for my journey in the morning. I leave the Consul with concern who has shown me uncommon civilitys and is one of the best sort of men in the world.

[1] Folkes's assessment is typical of Grand Tourists. The Frenchman, De Blainville, visited the Papafava garden near Padua in 1707, 'looked upon as a little Miracle', and thought it, 'indeed a handsome enough Garden for a private Gentleman but the only Singularity in it is a Labyrinth, out of which a Man would have some Difficulty to extricate himself without a Guide'. De Blainville, *Travels*, p. 143; See also Hunt, *Garden and Grove*, p. 104. There was a summer house in the middle of the maze that afforded a view of Padua.

[2] Ludovico Riva (1698–1746), Professor of Astronomy in Padua.

[3] Folkes was a connossieur of French wine. When Royal Society President, he often retreated with his friends to dinner at Pontack's Head, the usual destination after the St Andrew's Day elections at the Society. Pontack's Head was known for the best Bordeaux wine in London, as the owners possessed the Haut-Brion vineyard. A 1739 newspaper clipping noted: 'To-morrow being St. Andrew's Day, the Royal Society will hold their Annual Meeting ... for the Election of a President and other Officers ... after they will proceed to Pontack's where a splendid Entertainment will be provided for them.' RSL, RS/MM/20.

[4] San Giovanni di Verdara, a Benedictine monastery, founded in 1221. By 1436, it was granted to the Augustine Canons Regular of the Lateran, and in the 16th century became an abbey, accumulating a renowned library. After the order of canons was supressed in 1784, the library was dispersed and the church deconsecrated. It now serves as a military hospital. Several notable manuscripts from the library, including manuscript copies of works by Galen and Hippocrates, are now in the collections of MS Holkham, Bodleian Library, Oxford. See 'Padua, Monastery of San Giovanni di Verdara', Bodleian Library, Oxford, https://medieval.bodleian.ox.ac.uk/catalog/org_243863172 [Accessed 31 Dec 2022].

[5] Alessandro Varotari, called Il Padovanino (1588–1648), an Italian painter of the late Mannerist and early Baroque Venetian School, and a follower of Titian. 'His Marriage of Cana, in the Academy [Accademia] at Venice, is generally considered his principal work; it was formerly in the monastery of San Giovanni di Verdara.' Farquhar, *Biographical Catalogue*, p. 192. Folkes did not provide a closing bracket in this diary entry.

[6] *Sbirri* or university police.

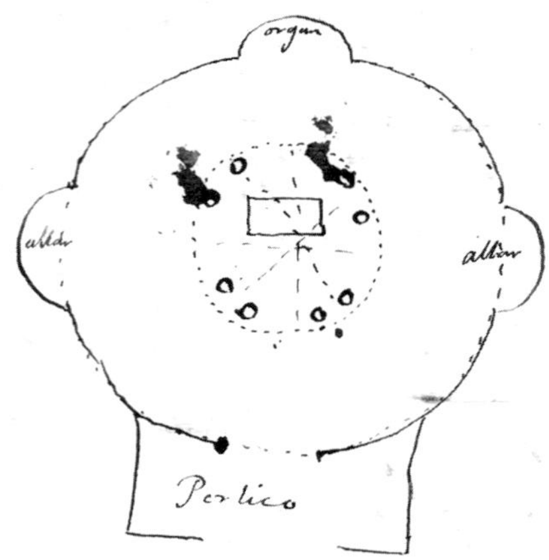

Figure 20. Folkes's sketch of Santa Maria del Torresino on f. 50v of his diary. Courtesy of the Bodleian Library, University of Oxford.

This evening Sardella came to me to desire a pair of his horses having been kept longer out than he intended, that I might not go out till about noon to morrow, and lie where we should have dind which he promises shall make the journey rather more easie to me than otherwise, so I have settled it accordingly.

Wednesd. 14. not being to set out till after eating a bit of dinner I walkt again into the town; and was at the Sala grande[1] where is the Monument erected to the Memory of T. Livius and some others,[2] it is the largest room I have seen Westminster hall excepted. The sides have been pa[i]nted ~~up~~ all over with old history paintings.[3] I was then at the Dome, an unfinisht Church, wherein is the tomb of Cardinal Barbarigo[4] B[isho]p of[5]

[1] The Palazzo della Ragione, the town hall, medieval market, and hall of justice known colloquially as 'Il Salone', the big hall.

[2] In the 13th century: 'an inscription thought to be from the tomb of Livy (64 or 59 BC–AD 17), the Paduan-born historian of ancient Rome, was unearthed. A discovery of what were thought to be his bones was made in 1413.' Venice had conquered Padua a decade earlier, and this discovery was probably an expression of local *campanilismo* (sense of pride for one's hometown). Two reliefs of Livy were subsequently 'installed on the exterior of the Palazzo della Ragione by the mid-15th century', to which Folkes refers in his diary. See McHam, 'Renaissance Monuments', pp. 464, 466, 469–70.

[3] There was a fresco cycle in the Palazzo della Raggione, originally executed by Giotto and his school, from 1306 until 1309. The frescoes were destroyed by a fire in 1420, and the roof was replaced by a large vault, 'reportedly the largest vaulted room in fifteenth-century Europe. Giovanni Nicolò Miretto (fl. 1423–*c.* 1440) and his assistants began repainting the *Sala* before 1425 and completed more than 244 scenes depicting the professions and emotions influenced by the planets'. Prokop, 'The Allegorical Frescoes', Frick Collection, https://www.frick.org/blogs/photoarchive/allegorical_frescoes_palazzo_della_ragione_padua [Accessed 3 Jan 2022].

[4] Gregorio Giovanni Gaspare Barbarigo (1625–97), Bishop of Padua from 1664 until his death. His beatification was celebrated in 1761, and in 1960 he was subsequently canonized by Pope John XXIII.

[5] This passage continues on f. 53r (p. 114 below). The discussion on f. 52v was an addition by Folkes on the facing page.

[f. 52v] I also saw this morning the Monte de Pieta,[1] the Lumber house, or Municable corporation house of this city. a very grand and magnificent stone building adornd with statues and other ornaments by falconetti. They take in pawn here as I was told, at 5 percent interest for one year. after which they are upon proclamation for 6 months and the proprietors pay half a Sol per lira for each month. at the end of which they are sold.

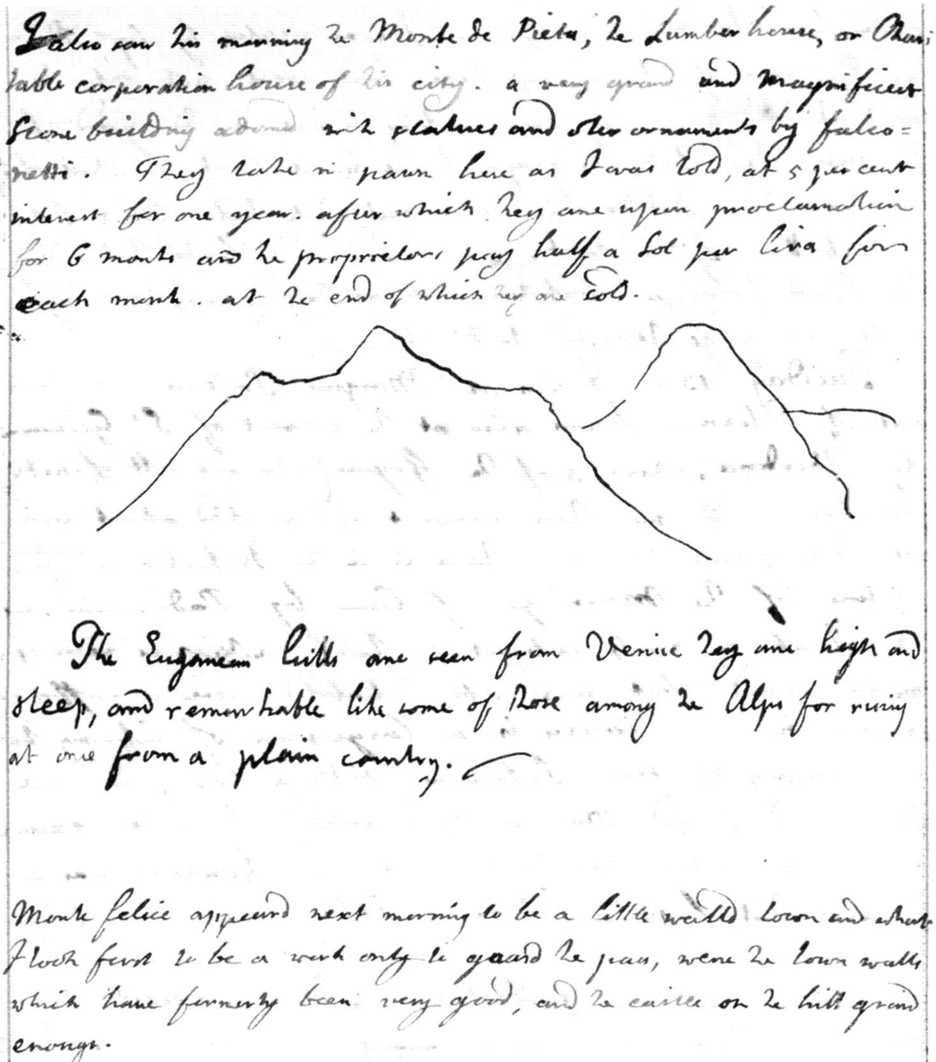

I also saw this morning the Monte de Pieta, the Lumber house, or Municable corporation house of this city. a very grand and magnificent stone building adornd with statues and other ornaments by falconetti. They take in pawn here as I was told, at 5 per cent interest for one year. after which they are upon proclamation for 6 months and the proprietors pay half a Sol per lira for each month. at the end of which they are sold.

The Euganean hills are seen from Venice they are high and steep, and remarkable like some of those among the Alps for rising at once from a plain country.

Monte felice appeard next morning to be a little walld town and what I took first to be a work only to guard the pass, were the town walls which have formerly been very good, and the castle on the hill grand enough.

Figure 21. Folkes's sketch of the Euganean Hills on f. 52v of his diary. Courtesy of the Bodleian Library, University of Oxford.

[1] The Monte di Pietà were 'charitable institutions of credit that lend money at low rates of interest, or without interest at all, upon the security of objects left in pawn, with a view to protecting persons in want from usurers'. Umberto Benigni, 'Montes Pietatis', *The Catholic Encyclopedia*, https://newadvent.org/cathen/10534d.htm [Accessed 29 Dec 2023]. Folkes also referred to Veronese architect, Giovanni Maria Falconetto (*c.*1468–1535), who renovated the facade of the Monte di Pietà. Jacobi, *The Architecture of Banking*, pp. 136–7.

The Euganean hills[1] are seen from Venice they are high and steep, and remarkable like some of those among the Alps for rising at once from a plain country.

Monte felice[2] appeared next morning to be a little walld town and what I took first to be a work only to guard the pass, were the town walls which have formally been very good, and the castle on the hill grand enough.

[f. 53r] Padua, and several good pictures in the Sacristy, particularly one a Madona and Christ of Titian finely presentd. I was then at the house calld Titus Livius[3] what the aut[h]ority of that story is I could not hear, but there are in the walls several antique altars and monumental inscriptions built in which makes it worth seeing as a repositarity of antiquitys of that sort, but I observd never a one very remarkable. they are all in print. after dinner I immediately set out my wife the Children and self in my own Berlin,[4] and the servants in another, we came upon a bank road by the side of a river which we kept to our right hand or rather 2 for they both joynd in about our midway and crosd our road to run towards the sea on our left. the first of these which came from Padua is ~~calld the~~ as I take it part of the Brenta, and the other is called fossa de la Restara,[5] they both joyn at a place called Bataglia[6] and go off to sea as I just mentiond taking the name of Bachaglionè.[7] the ground to our left was all low but very rich full of plantations of vines india corn[8] and trees intermixd. ~~at~~ at some distance on our right were the Euganean Hills, which we came up to at Montefelice where we lay, and where the pass between them and Montefelice[9] is guarded by a wall and a castle on the latter which has formerly been a considerable work but is now much ruinous however standing on a high steep hill makes a good appearance the village of Montfelice is at the foot just of the hill where we met a tollerable inn. this place is about

[1] The hills are a few km south of Padua, of volcanic origin, celebrated for their beauty in poems such as Shelley's, 'Lines written among the Auganean Hills', in answer to Petrarch's earlier poem about this landscape. See Weinberg, *Shelley's Italian Experience*, p. 22. The terroir produces the Montecchia DOC wines, and there are hot springs serving the spa towns near Padua: Battaglia terme, Montegrotto Terme, and Abano Terme.

[2] Monselice, which is on the road from Padua to Ferrara (*c.*20 km from Padua), and which possesses an impressive Castello, built in 1405 by the Marcello family. See 'Castello de Monselice', https://www.castellodimonselice.it/visita-il-castello-di-monselice-padova-veneto/#storia-castello [Accessed 2 Jan 2023]. There is also a ruined Byazntine fortification called the Rocca (fortress).

[3] This is the Casa degli Specchi, which was formerly known as the Casa di Tito Livio, the supposed site of Livy's birth (the present address is 79 Via del Vescovado). The facade, however, is of the late Venetian Renaissance, and it was built at the beginning of the 16th century by Annibale Maggi da Bassano. 'It was already a tourist sight in the sixteenth century. An allegorical fresco by Domenico Campagnola in the house includes a likeness of Livy in the act of writing, with history (or Rome) at his side and a river god, probably the Tiber, below.' Trapp, *Studies of Petrarch*, p. 303. See also Joost-Gaugier, 'The casa degli specchi', pp. 113–24. Folkes's idiosyncratic 'repositarity' derives from repository.

[4] A cross was drawn here in the margin, probably referring to the notes Folkes wrote on f. 52v.

[5] This name is unclear.

[6] The town of Battaglia Terme is a: 'fundamental intersection in the Paduan canal network. It is the meeting place of the Bisato Canal (1139) from Monselice and Este, and the Battaglia Canal (1189–1201), an artificial canal fed by the Bacchiglione River, at Padua. The timely regulation of water levels is performed right in the centre of Battaglia by means of an important hydraulic structure, the monumental Arco di Mezzo ...' Water Museum of Venice, https://www.watermuseumofvenice.com/network-en/the-padua-network/battaglia-terme-canal-village-en/ [Accessed 4 Jan 2022].

[7] Folkes means the river Bacchiglione, which empties into the river Brenta near Chiogga, and flows through Vicenza and Padua.

[8] Maize, specifically flint corn (*Zea mays*, var. *indurata*).

[9] For Monselice, see above, n. 2.

14 mile from Padua and all the way we had very pretty houses on the ~~wat~~ road and some fine ones one of Sigr Delfins as grand as almost any thing I have seen.[1] there were also some pretty villages and one little town called Bataglia[2] prettyly buildt on each side the river with a good bridg over it. we came in a little after 5 a clock.

Thursday. 15. I had a good supper last night and lay clean, I set out this morning between 5 and 6, and immediately left the hills behind me travelling through a rich enclosd country very pleasant and villas by the way as yesterday. the soil is so heavy I should have to plow with 6 oxen to each plow, their plows have wheels as in Norfolk.[3] I passd a little village calld deserto,[4] and soon after over a bridge a large drain calld fossa[5] Monfeliana, then to a little town called Solesino[6] where Sigr Pisani has a very fine seat, this is a little market town chiefly for poultry. we then passd a navigable river calld fossa novana on a bridg

[f. 54r] it is a navigable river, and falls just below into a small lake calld Lago de Anguillara, which I did not see but was I judg only prevented by the hazyness of the morning. at some distance from hence we passd the Adige.[7] on a flying bridg[8] it has exceeding high banks but was now low, I think of about the size of the Thames at Putney.[9] on the other side we met a little country funeral of a child, the body was laid barefaced on a bier with a garland on its head and strewd about with flowers; and the children that attended were also drest in garlands. A little further we crossd the litle river [Ceverolo?],[10] and another (that was quite dry) over bridges; and came beween 9 and 10 to Rovigo where we bait.[11] Rovigo[12] is a pretty walld market town, there is in it

[1] The Dolfin patrician family of Venice owned several villas, but based on Folkes's location, this is probably the Villa Dolfin – Dal Martello, known as La Mincana, a 16th-century Venetian villa situated in Due Carrare, in the province of Padua. It was placed along the public road. It is now a vineyard. See 'Villa Dolfin dal Martello known as La Mincana', https://villevenetetour.com/en/venetian-villas/villa-dolfin-dal-martello-known-as-la-mincana [Accessed 4 Jan 2022].

[2] Battaglia terme. The Venetian bridge, the Ponte del Scaini, connects the two banks. The inn was probably salubrious as this was and is a spa town.

[3] Folkes's stately home was Hillington Hall in Norfolk, and his account papers in the Norfolk Archives show his deep familiarity with farming techniques and yields.

[4] Deserto d'Este is a small village in the municipality of Esta in the province of Padua, formed around the rural estate of the Valerio nobles.

[5] A ditch or drain.

[6] Solesino in the province of Padua, about 50 km south-west of Venice. It is near the Villa Pisani Vescovana, built in the 16th century by Cardinal Francesco Pisani, a patrician and Bishop of Padova. The house is still extant. See Villa Pisani Vescovana, https://www.visitabanomontegrotto.com/en/venetian-villas/villa-pisani-vescovana-en/ [Accessed 7 July 2023].

[7] The river Adige, the second longest river in Italy.

[8] A flying bridge was a temporary bridge for military purposes. See *OED*.

[9] In 1729, the Putney Bridge was built across the Thames, a wooden structure 240 metres long and 7 metres wide that provided the only dry crossing of the Thames between the City and Kingston. As Folkes was in London during the 1720s, it probably seemed to him an apt comparison. See Stone, 'Putney Bridge', The History of London, https://www.thehistoryoflondon.co.uk/putney-bridge/ [Accessed 5 Jan 2023].

[10] This word is unclear.

[11] Of travellers: 'To stop at an inn, originally to feed the horses, but later also to rest and refresh themselves; *hence*, to make a brief stay or sojourn'. See *OED*.

[12] Rovigo, capital of the eponymous province, standing on the low ground known as the Polesine, about 40 km south-southwest of Padua, on the Adigetto canal. As it was on the mouth of the river Po, it was in a strategic location, as it could control northern Italy's river ports. Hence, it was subject to invasions; its brick walls were built in the 1130s by the Viscounts Este, who controlled the territory. In 1482, Rovigo came under Venetian control. See 'Rovigo', in *Encyclopaedia Britannica*, https://www.britannica.com/place/Rovigo-Italy [Accessed 29 Dec 2023].

a good church[1] now fitting up the outside is quite rough but the inside is much on the model of St Pauls[2] tho much less, there are several Altar pieces very handsome of marbles of several colours with good pictures over them. this town is calld in Latin Rhodiginum [*sic*]. I set out again from hence a little before 12. and traveld in much the same sort of road as in the morning for some miles till I came to a small pretty town called Arqua,[3] after which the country became more low and less enclosd it put me several times much in mind of the Smeeth in Marshland,[4] a little forward I crossd what they call Canale bianco, a tollerable navigable River[5] in a ferry, and proceeding some miles further at a place called *Hospidaletto*,[6] passd from the Venetian territorys into the Dukedom of Ferrara[7] belonging to the Church. I soon after came to the side of the Po, it is enclosd between very high banks as indeed all the waters are here as swelling greatly on the melting of the Snow from the mountains; we kept better than a mile on the bank with the River to the left, and then crossd it in a flying bridg at a place called ponte de Lagoscura,[8] where is a village calld Pepoli; the River here is very fine I believe near as wide as the Thames at Lambeth between the banks, tho all the waters are now excessive low, and several canals that seem to have large beds are quite dry. this country abounds very much in Willows poplars, and Larixes,[9] and has been as flat all this day as holland or the Isle of Ely, the churches are all extreamly pretty, and here are vast numbers of fine houses, every thing is built with some taste and there is hardly to the least farm that has not a good deal of elegance in it. about 3 or 4 miles from the place I passed the Po is the City of Ferrara where I lie tonight at the Lione-[f. 55r] Bianco seemingly a good inn. The City of ffferrara is walld with a rampart facd with brick and bastions and a very broad wet ditch.[10] it appears very large but from the outside so little of building is seen and so much of open ground that I could hardly think at first sight it was a City. it is said to

[1] The cathedral of Rovigo, rebuilt from 1696 to 1791 on top of an 11th-century church. The facade was never completed. 'Duomo di Rovigo', Religiana, https://religiana.com/duomo-di-rovigo [Accessed 5 Jan 2022].

[2] The interior of the Duomo di Santo Stefano in Rodrigo has some similarity to St Paul's, London without the ornate decoration.

[3] Arquà Polesine, about 7 km south-west of Rovigo.

[4] A 'smeeth' in East Anglian dialect is a level space or marshland. Folkes was clearly thinking of his home county of Norfolk.

[5] Castelbianco is a tributary of the river Tartaro-Canalbianco-Po di Levante.The Castelbianco is the second part of the river's course, its length is 78 km from Torretta to Volta Grimana, flowing in the province of Rovigo.

[6] Probably the Ospitale Bondeno. The term was written over by Folkes, who also wrote it again in the right-hand margin on this page.

[7] Ferrara was a fief of both the Holy Roman Empire and the Papal States, under the control of the House of Este until 1597, whereupon it devolved to the Papacy (Pope Clement VIII Aldrobrandini) after an earthquake in 1570 that devastated the local economy. The former duchy was then overseen by a papal legate. See Tuohy, *Herculean Ferrara*, p. i. As Fosi has noted: 'the legate's function in Ferrara was ... eminently political, but it could flip into a delicate diplomatic mission, especially when international politics menaced the papal borders. Relations with the Republic of Venice, with the Duchy of Milan – and thus with Spain – and even with the House of Savoy put the reins of policy in legates' hands.' Papal legates also were involved in policing and keeping an eye on the nobles' deviant behaviour, 'watching over their vendettas and their alliances for faction's sake, which might undercut buon governo'. Fosi, *Papal Justice*, pp. 16, 18.

[8] The village of Pontelagoscuro.

[9] Larch trees, genus *Larix*, of the family *Pinaceae*.

[10] The Castello Estense is a moated medieval castle in the centre of Ferrara, with a large block and four corner towers. The water extended all the way to the castle, and it ran through a system of locks all around the city in a ditch. The city itself was surrounded by fortified walls, built by members of the ruling d'Este family.

be about 5 miles round. the streets are wide and mostly strait, and there are many very good houses, yet the appearance of the whole seems poor and uninhabited particulars I know none yet this is the coup d œil,[1] and I came in but just with light enough to form a general idea; I hope to see something of it in the morning, not going away before 11 a clock. We have travelld today a pretty good days journey for the manner we are in of going with a Veturino, it being I believe about 35 English miles, considerably more by the maps of Maginus[2] but those seem to me tho in general pretty good to give the measures between places too large in all this part of the country: it is the same with us in Marshland and I suppose is commonly the practise in draind lands. I think I have observed some faults in those maps yet in the main they seem better than what is commonly found in works of that sort.

Fryday 16. We lay clean and well, I got up to go about the town presently after 7 a clock. and find it to be a better town much than I before took it to be, the streets are not only strait but full of very fine houses tho the ordinary ones are mostly mean, the close walks[3] are lesse off that [*sic*] I saw in all the towns of the Venetian territorys and what is quite new in this part of the world fferrara is mostly very well paved and paths of stone to defend the people on foot. yet what I observd last night of the thinness of people is very true except just about the Market place where there is some air of business and it being market-day I saw a tolerable concourse of pe[o]ple from the country. The Cathedral[4] is a very large church and very Gothic on the outside but within it has been entirely put into the modern form and is not at all unlike the inside of St Pauls at London tho less, at the upper end of the choir behind a great Altar which stands single, is a pretty good copy but a little altar of Michael Angelos last judgment.[5] The other altars are well ornamented with rich marbles and most of them with good pictures, some of Dosso[6] and Scarselino[7] both fferranese. I was pleased particularly with 2 the one a St. Laurence[8] and the other as I take it the death

[1] *Coup d'œil* or a glance that takes in a comprehensive view.

[2] Giovanni Antonio Magini (or in Latin, Maginus) (1555–1617), Italian astronomer, mathematician (Professor of Mathematics at the University of Bologna), and cartographer. He was best known for his *Atlante geografico d'Italia* (Geographic Atlas of Italy), printed posthumously by his son in 1620, with maps of every Italian region. He also did a smaller-size format printing of Ptolemy's *Geographia*. 'Giovanni Antonio Magini', MacTutor, https://mathshistory.st-andrews.ac.uk/Biographies/Magini/ [Accessed 6 Jan 2022].

[3] Probably courtyards or enclosures about or beside a building.

[4] Folkes was correct that the Cathedral of San Giorgio Martire was much altered in the inside. It was consecrated in 1185, but the interior was completely remodelled in the 17th century in a grandiose classical style. Verzar, 'Ferrara Cathedral', Grove Art Online, https://doi.org/10.1093/oao/9781884446054.013.90000369817 [Accessed 12 Jan 2022].

[5] 'The bowl-shaped apse is vaulted with the grandiose "Last Judgment" by Bastianino (16th century). In what is his masterpiece the artist displays great dramatic force, imagination and unconventionality in this work clearly inspired by Michelangelo.' See 'Interior', Ferrara terra e acqua, https://www.ferraraterraeacqua.it/en/ferrara/discover-the-area/art-and-culture/churches-baptisteries/cathedral/interno-della-cattedrale [Accessed 6 Jan 2022].

[6] Dosso Dossi (?1486–1542), who spent his career as a court painter to two successive dukes of Ferrara, Alfonso I d'Este (r. 1505–34) and his son Ercole II (r. 1534–59). Humfrey, 'Dosso Dossi: His Life and Works', p. 3.

[7] Scarsellino or Ippolito Scarsella (1550/51–1620), an easel painter and fresco artist who was an important representative of the School of Ferrara.

[8] The artist's given name was Giovanni Francesco Barbieri (1591–1666), and with the 'exception of an interlude in Rome', he spent most of his life not far from Ferrara in Cento. Perlove, 'Power and Religious Authority', p. 19. This particular painting to which Folkes refers was commissioned by Cardinal Lorenzo Magalotti from Guercino in 1628 for his private chapel in Rome. 'Martirio de San Lorenzo, DiPinto', Catalogo

[f. 55v] Pope Urban 3. Is buried in the Cathedral[1]

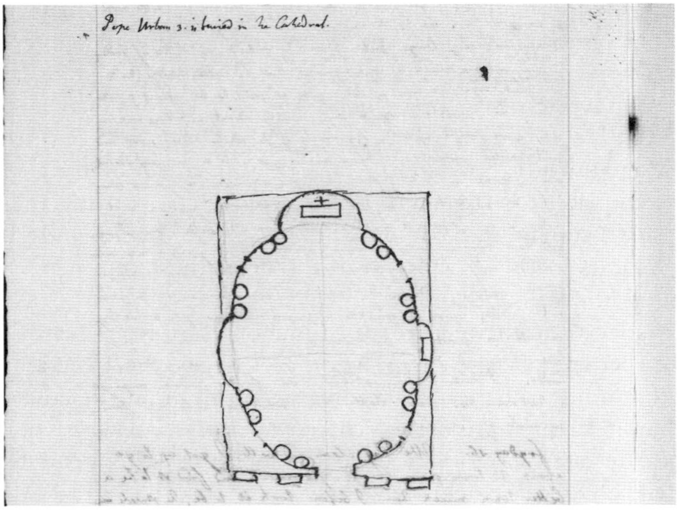

Figure 22. Folkes's sketch of the architectural plan of San Carlo Borromeo, Ferrara, on f. 55v of his diary. Courtesy of the Bodleian Library, University of Oxford.

[f. 56r] of St. Joseph, the Virgin and our Lord[2] assisting of him an uncommon subject. another Altar was adornd with a large brass Crucifix[3] a virgin and St John of the same metal as large as the life and well executed.[4] The Gothic front is very rich,[5] but I cannot

generale de Beni Culturali, https://catalogo.beniculturali.it/detail/HistoricOrArtisticProperty/0800214971 [Accessed 9 Jan 2023]. When Magalotti was appointed Bishop of Ferrara in March 1628, he took the work with him and set it up temporarily in the Archbishop's Palace. In 1636, Magalotti entrusted the restructuring of the apse area of Ferrara Cathedral to the architect Luca Danesi. 'Altare di San Lorenzo', Cattedrale di Ferrara https://cattedralediferrara.it/altare-di-san-lorenzo [Accessed 9 Jan 2023]. With Magalotti's death the following year, the painting, with its dynamism and dramatic chiaroscuro, was placed in the seventh chapel on the right of the cathedral nave. Here it remained until the beginning of the 18th century, when it was moved to its current location in the Cathedral. See also Salerno, *I dipinti del Guercino*, p. 215 (catalogue raisonné).

[1] Pope Urban III (1120–1186) is indeed buried in the Cathedral of San Giorgio Martire, as he had died suddenly in Ferrara on 20 Oct 1187.

[2] *Transito di San Giuseppe*, painted by Giacomo Parolini (1663–1733), part of a larger altarpiece designed by the Veronese sculptor, Angelo Ringheiri, during the rebuilding of this chapel in Ferrara Cathedral in 1767. The canvas was executed between 1720 and 1725, on behalf of Canon Giuseppe Fioravanti (d. 1725), Secretary of Cardinal Dal Verme. See 'Altare di San Giuseppe', Cattedrale di Ferrara https://cattedralediferrara.it/altare-di-san-giuseppe [Accessed 10 Jan 2023]; 'Morte di San Giuseppe', *Catalogo generale de Beni Culturali*, https://catalogo.beniculturali.it/detail/HistoricOrArtisticProperty/0800214986 [Accessed 10 Jan 2023].

[3] Here there was a cross inked in the right-hand margin.

[4] Niccolò Baroncelli (fl. 1434, d. 1453), Italian sculptor and bronze-caster, was commissioned in 1450 to make life-size statues in bronze for the high altar of Ferrara Cathedral of the Crucified Christ, the Virgin, St John the Evangelist, St George and St Maurelius, completing the first three with his son Giovanni. See 'Ferrara – Cattedrale. Il Crocifisso con la Vergine e S. Giovanni (Niccolò e Giov. Baroncelli', *Catalogo generale de Beni Culturali*, https://catalogo.beniculturali.it/detail/PhotographicHeritage/0800081939 [Accessed 10 Jan 2023].

[5] The facade was originally Romanesque, and the upper part of the Cathedral was built later in a Gothic style. There is an exceptional Last Judgment over the central loggia by an unknown sculptor. 'Cathedral', Ferrara terra e acqua, https://www.ferrarateraeacqua.it/en/ferrara/discover-the-area/art-and-culture/churches-baptisteries/cathedral [Accessed 6 Jan 2022].

so much like the Gothic of this country as our own for here is even some remains of the old Architecture spoilt whereas ours is a sort of taste by it self and consequently has its own beautys.[1] near the Cathedral is the Bishops Palace[2] now Cardinal [*blank*][3] a fine modern building and of large extent, the lower story are made into shops. Over against this is what is calld the Palace of Modena, now the place of the courts of justice and the court a sort of market; before the entrance are two brass figures upon sort of pedestals, the one Equestrian represents Nicholas Marquess of fferrara called [*blank*][4] the other sitting is of [*blank*][5] called Borso,[6] the figures are good enough and under the latter is *Hanc tibi viventi fferraria grata Columnam.*[7] a little way off is the castle or citadel a vast brick building with 4 towers and a large wet ditch, this is the residence of the Legate or Governor of the town[8] and here is a small guard kept at the Entrance.

[1] Folkes was correct about the cohesiveness of English Gothic. 'The Early English style combined such formal aspects of French Gothic as rib vaults with English pre-Gothic decorative and structural tendencies ... The English, apparently indifferent to the technology of French Gothic, were never able to accommodate its aesthetic.' There was also a uniquely English preoccupation with patterning, manifested for instance in the aisles of St Hugh's Choir in Lincoln Cathedral. See Woodman, 'Early English', *Grove Art Online*, https://doi.org/10.1093/gao/9781884446054.article.T024727 [Accessed 12 Jan 2022].

[2] Nothing remains of the original medieval episcopal palace in Ferrara. The present palace, known as the 'villa of the mensa', being part of the property belonging to the 'mensa' or revenues owed to the Church, was commissioned in 1482 by Bartolomeo della Rovere, then Bishop of Ferrara. The Bishop's Palace is not by the cathedral, as Folkes indicates, but rather was built in the Copparo area close to the town of Sabbioncello San Vittore. See Raimondi, 'The Bishop's Grand Palace', *Ferrara: Voci di una città*, https://rivista.fondazioneestense.it/en/2007/item/44-il-gran-palagio-del-vescovo [Accessed 12 Jan 2022].

[3] Folkes left a blank space here as he did not know the identity of the cardinal.

[4] Folkes left a blank space here. In 1451, Niccolò Baroncelli executed the horse for a bronze equestrian statue of Niccolò d'Este, Marquess of Ferrara (destroyed 1796 during the Napoleonic invasions). Commissioned by Niccolò's son, Leonello, it was the 'first Renaissance example of a form known only from Classical prototypes, such as the equestrian monument to *Marcus Aurelius* (Rome, Campidoglio)'. Antonio di Cristoforo (a pupil of Donatello) sculpted the figure, and the statue was placed on a column modelled after a Roman triumphal arch. See Carrington, 'Baroncelli, Niccolò', *Grove Art Online*, https://doi.org/10.1093/gao/9781884446054.article.T006454 [Accessed 10 Jan 2023]. The similarity to Marcus Aurelius 'highlighted the imperial associations clearly cultivated by Leonello and the humanists in his court'. Dandelet, *The Renaissance of Empire*, p. 33. There is still a copy of it in the central public square between the Castello and the cathedral in Ferrara.

[5] Folkes left a blank space here.

[6] A bronze effigy of Borso d'Este, the first Duke of Ferrara (r. 1450–71), was completed by Giovanni Baroncelli in 1453–54, and stood outside the Palazzo della Ragione. The statue was destroyed in 1796, but there is still a copy of it in the central public square between the castello and the cathedral in Ferrara, the inscription on its column that Folkes described still visible.

[7] The first four lines of an epigram by the humanist Tito Vespasiano Strozzi (1425–1505) are engraved on the base of Borso's statue:

Hanc tibi viventi Ferraria grata columnam | Ob merita in Patriam, Princeps iustissime Borsi, | Dedicat, Estensi qui Dux a sanguine primus | Excipis imperium, et placida regis omnia pace.

[This column, most righteous prince Borso, a grateful Ferrara dedicates to you while you are still alive in return for the services you have performed for your country, you who are the first duke of Estian blood to take the reins of power and rule all in untroubled peace].

Strozzi also wrote the *Borsiad* (finished 1505), a poetic tribute to Borso's role as a peacemaker amidst the war-torn states of northern Italy, its language and form based on the *Aeneid.* See Lippincott, 'The Neo-Latin Historical Epics', pp. 423–4.

[8] The Castello Estense di Ferrara, which is proximal to the cathedral. After Ferrara became part of the papal states in 1597, the Castello was the seat of the cardinal legate.

just by I took note of the Church of San Carlo[1] that I was extreamly pleasd with it is ~~extreamly~~ extraordinary little and within a perfect oval with little more besides a projection for the chief Altar and the Porch. there are 2 more altars in the middle spaces between the chief and entrance, and between the several Altars and entrance everywhere 4 Ionic columns round against the wall all which with their ornaments are only brick it is covered with an oval flatish dome and the whole tho small ~~is I d~~ and not expansive is as pretty as most things I have seen. between each altar is a small door and over it a Nich[e] with a statue.[2] I was then at the end almost of the town at the Church of St Mary in Vado,[3] famous for some pictures of Bonon di fferrara[4] a scholar of Caraccii[5] in the cieling, which are masterpieces in the manner of fore shortning, they are historys of the new testament and so painted as for the figures to appear upright tho on a flat cieling they are indeed very curious, and the ornaments between them for they are in compartments are done in a very masterly manner. A large round in the center of the cross is painted so as to repre=[f. 57r] sent a cupola standing upon pillars and has the effect very well from the middle underneath. there are also some large good pictures[6] particularly the birth of Christ, and the worshipping of the Shepherds, about the ~~uppr~~ upper end of the church, and some other good pictures of some of the old masters at the altars down the sides of the church, but these are ornamented only with wood painted like marble, and being too gay and not the best imagnd have but a tawdry effect [marring][7] of the rest of this pretty church. I was afterwards to the Ramparts which enclose a great quantity of ground besides the town, they are planted with trees and pleasant enough tho ill kept. the wall is very well faced

[1] The church of San Carlo Borromeo, Ferrara, a short distance from the Castello Estense. It was designed by architect, hydraulic engineer and stage designer Giovanni Battista Aleotti (1546–1636) who was appointed architect to Alfonso II d'Este, Duke of Ferrara and Modena, in 1575. Aleotti was profoundly influenced by Vitruvius and Leon Battista Alberti, as well as the writings of Serlio, Vignola, and Palladio 'which he united in an austere, ordered architecture'. Worsley, *European Classicist Tradition,* p. 29. The church's austere neoclassicism and central geometric planning would have appealed to Folkes, both as mathematician and as connoisseur. The church was dedicated to Cardinal Carlo Borromeo, Archbishop of Milan (1564–84), and a leading figure in the Counter-Reformation.

[2] The statues are of Saints Carlo Borromeo, Anthony of Padua, Ambrose and Augustine, attributed to Venetian sculptor Angelo Da Putti.

[3] Santa Maria in Vado, Ferrara, so called as you had to cross a 'vado' or ford to reach it. 'Its religious importance is tied to the miracle of the Eucharist that occurred on Easter Day 1171; at the moment of consecration blood spouted from the host staining the apsidal vault above the altar.' Afterward the church became a pilgrimage site. See 'Church of Santa Maria in Vado', Ferrara terra e acqua, https://www.ferrarainfo.com/en/ferrara/discover-the-area/art-and-culture/churches-baptisteries/church-of-santa-maria-in-vado?set_language=en [Accessed 11 Jan 2023].

[4] Begun in 1617, the five paintings in the nave, transept and apse ceilings are the work of Carlo Bononi (1569–1632), an artist from the School of Ferrara. As Amaturo indicated: 'For the nave vault he painted huge canvases, among which are the "Visitation" and the "Miracle of the Blood", and for the apse a vast fresco, the "Exaltation of God's Name". These works are indebted to Correggio and foreshadow the Baroque in their dramatic contrasts of light and shade and in the complex poses of the figures.' Amaturo, 'Bononi [Bonone], Carlo', Grove Art Online, https://doi.org/10.1093/gao/9781884446054.article.T009934 [Accessed 11 Jan 2023].

[5] Ludovico Carracci (1555–1619). Amaturo describes Bononi as amongst 'the last great painters of the Ferrarese school, his style uniting warm Venetian colour with the lyrical effects of light and elegant draughtsmanship of Ludovico Carracci'. Amaturo, 'Bononi [Bonone], Carlo', Grove Art Online, https://doi.org/10.1093/gao/9781884446054.article.T009934 [Accessed 11 Jan 2023].

[6] There is a cross inked in the right-hand margin here.

[7] This word is unclear.

and the bastions good and what amazed me some of them tho very old have orillons[1] not at all unlike the method of Vauban,[2] but I was indeed accidentally showd one old folio book of fortifications printed above 150 years since in Italy in which this is described.[3] the ditch is wide and good and there are lunettes before some of the gates. Mr Addison justly observes it would take too many men to defend this town, but he speaks too slightly of the fortification which is as good as in any of the old Italian towns I have seen.[4] the extent is indeed so great as to surprise. as I came back to my inn I passd what they call piazza nuova,[5] a sort of square unfinisht but designed to have had porticos round it like Covent garden, in the middle is a very fine marble pillar with foliages of vines and leaves curling about it, the Corinthian capital is adornd with the arms of P. Alexander the 7 and his figure in brass is on the top setting in his chair and blessing with his hand. I observd several more palaces and fine churches which I had not time to go into. coming in I wanted change for a gold Cequin and was surprized my landlord could not gives it me other than in brass pieces which he offerd but I did not have because of the weight, so can not tell just what they were yet. I had before heard of the scarcity of silver in the popes territorys, and which I conjectured from the exorbitant price I had heard at Venice they set on their

[1] An orillon is 'a projecting tower or shoulder at the corner of a bastion, between the face and a flank, used to defend the flank'. See *OED*.

[2] Sébastien Le Prestre de Vauban (1633–1707), a French military engineer who worked under Louis XIV, considered the greatest of his time. 'Vauban directed the building or renovation of more than 160 military sites, not counting ports and canals. … he literally covered France with his handiwork and gave it much of its present shape.' Langins, *Conserving the Enlightenment*, p. 49.

[3] Folkes may have been shown this book when he visited Marshal Schulenburg in Venice, as the Marshal had a collection of books and plaster reliefs of terrain to aid his military campaigns. Bastion forts, known as *trace italienne*, were built in response to the French invasion of the Italian peninsula in the 15th and 16th centuries. The original 'Italian' fortresses were simple structures, but with the advent of cannon became much more complicated, and 'first-class fortresses, such as those constructed by Vauban or Coehoorn around 1690, became ever larger'. Van Creveld, 'Technology and War', p. 212. In the specific case of Ferrara: 'the architectural model of the imposing structure echoed the form of the leading military fortresses of the late 16th century in Italy and the rest of Europe, including the pentagon-shaped fortifications of Parma, Piacenza, Turin and Antwerp.' See Figure 23. 'The Baluardo di Santa Maria and the Baluardo of San Paolo di Fortezza', Museo Ferrara, http://www.museoferrara.it/en/view/s/93766b37d047429a90543b95cf493f8c [Accessed 15 Jan 2022].

[4] Addison wrote: 'At Ferrara I met with nothing extraordinary. The town is very large, but extremely thin of people. It has a citadel, and something like a fortification running round it, but so large that it requires more soldiers to defend it than the pope has in his whole dominions.' Hurd, ed., 'Remarks on several parts of Italy', *Works of Joseph Addison*, I, p. 49. It is evident Folkes read Addison regularly to guide his travels, and his library auction catalogue shows he owned the 1705 edition of the work. See *A catalogue of the entire and valuable library*, p. 145, lot 4830.

[5] Today Piazza Ariostea, one of the main squares of the city in the northern part of the historical centre. It was known as the 'Piazza Nova', part of the so-called Erculean Addition, the urban expansion begun at the end of the 15th century by Ercole I d' Este (r. 1471–1505). 'Thanks to this expansion called "addizione Erculea" the city of Ferrara has been considered by scholars the first modern city of Europe, due to the social economic, and dimensional reason of the project.' Modena et al., 'Monument', p. 151. When Folkes was there, the piazza was dominated by a column designed by Ercole de' Roberti and carved by the workshop of Antonio di Gregorio, upon which was a statue of Pope Alexander VII (1675). Ibid., p. 153. Pope Alexander, originally Fabio Chigi (1599–1667), had been vice-legate in the papal service in Ferrara for five years from 1627. The statue was replaced by one of Napoleon in 1810, and finally by a figure of Ludovico Ariosto (1833), the Feronese poet. See Peterson, 'Pope Alexander VII', *Catholic Encyclopedia*, http://www.newadvent.org/cathen/C1294a.htm [Accessed 16 Jan 2022].

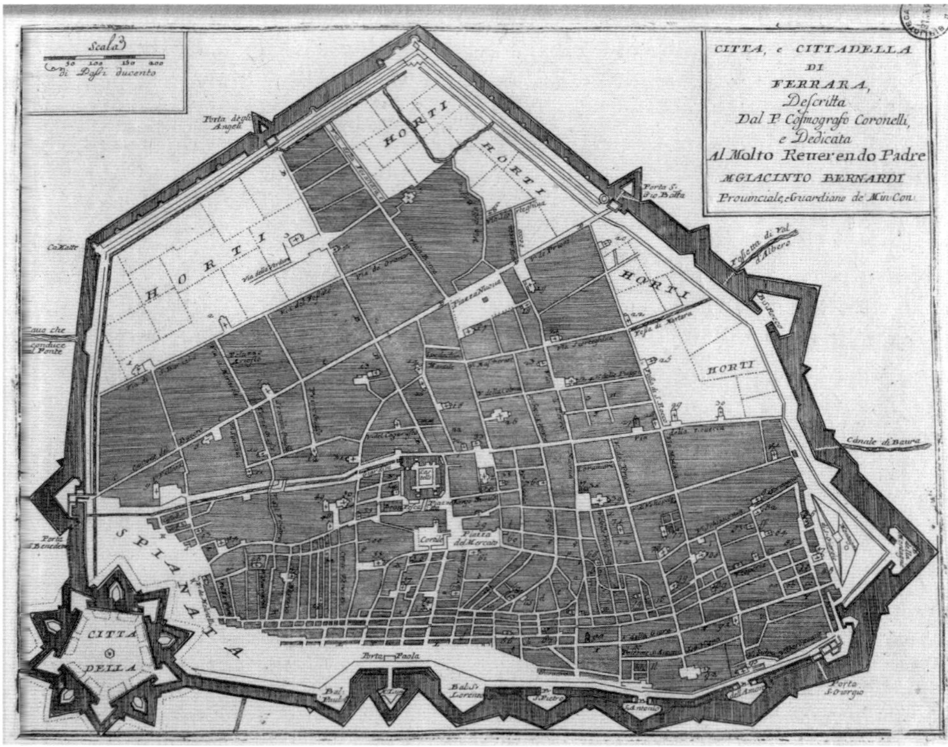

Figure 23. Vincenzo Maria Coronelli, *Città e Cittadella di Ferrara* (1696), engraving on paper, 29.2 × 23.4 cm. Biblioteca Nazionale Marciana, Venice. Folkes extensively described the city's fortifications. Courtesy of the Biblioteca Nazionale Marciana, Venice.

gold but did not think it had gone to this degree;[1] however it accountd for a little passage yesterday which I could not have believd. a \poor/ boy about 15 or 16 years old at least carrying along some grapes a few miles after I enterd these dominions my children had a mind to some, for which the boy askt 2 sols, and it being a poor creature I gave him a [f. 58r] half livre piece of silver (about 3 English pennys) or 20 sols[2] which the boy lookt at very simply and denied his grapes again, and could not be satisfied till I made a servant give him ten copper sols in the room of it[3] with which he seemd very well pleasd, this is a piece of ignorance would hardly be believd in England, and which I imputed to almost indolence till this passage this morning shewd me silver is so scarce that it is possible the boy might never have seen any silver coin at all. I left Ferrara a little after eleven a clock, and presently without the town passd a small branch of the Po calld here the River Vergine, which runs into the Lake of Comacchio.

[1] The Papal States had an underdeveloped economy with weak industrial production, largely controlled by oligarchic groups, 'amongst whom bankers were prominent'. The stagnant economy was underpinned by government bonds and taxation. Specie was also scarce as some of it was locked in the papal treasury as a security for government borrowing. See Korten, 'Financial Policies', p. 238.

[2] *Soldi.* See Appendix B on Venetian coins.

[3] 'In the room of it' is, 'in the place of', 'in lieu of', in this case, the silver half livre. See *OED.*

I then traveld the rest of the day through a very pleasant road the enclosures being like gardens almost all the way and several pretty villages but far fewer Gen[t][l]emens houses than in the territorys of Venice, about 4 a clock I came to another branch of the Po calld here the Po de Argenta about the size of the Thames at Henly[1] or hardly so large on the bank of which I went thro some little villages to Argenta[2] where I take up my quarters for this night in a very poor inn but where I hope to lie tolerably by the having my own sheets for our beds, but it is one of the most indifferent places we have yet been in. I came in just at the Sun set. My Veturinos have brought somewhat from fferrara for our supper so that I hope not to starve. Argenta is a little walld town on the Po, in which are several Churches[3] and a little marketplace it looks very thin and poor. our house is without the wall as a pretty many more houses are.

Saturday – 17. Our accommodation has been bad notwithstanding our own sheets we have been devoured with bugs and gnats, so are up early and setting out soon after five. we coasted the Po on its bank having had also for several miles the Valley of Comacchio all the greatest part overflowed like a lake, till about one a clock, when we baited at a lone house belonging to the village of St Alberto,[4] on some provisions we brought with us; the prospect is very dreary and melancholly the vale looking like a sea with here and there little islands what is not now overflowed is all rushes and sedges,[5] the houses on the way are very poor and there is a general face of poverty in the country. about 3 we set out again and crossing the Po immediately in a ferry, we presently after crossd on a wooden bridg another river not much less called the Sanerna,[6] here the country wended for a little way, but soon opend again into a vast [f. 59r] marshy pla[i]n full of rushes that held till within a mile and half of Ravenna where it is enclosd and seems rich, as a great deal more of it might be if cultivated, but as it is, it is the dullest flat I have seen. just before we enterd Ravenna[7] we passd between 2 old towers, and before the gate passd the River Montone;[8] the gate is very handsome adornd with Doric Pillars, and on one sides I observed large remains of the old ~~way~~ wall; there is a great deal of land within the gate[9] and the houses themselves are scattered so as to look but little of the appearance of a City, the streets are

[1] Henley-on-Thames, situated on one of the most famous stretches of the river.

[2] Argenta is situated in a flat agricultural region. It is proximal to the Valli de Comacchio lagoon wetlands, about 30 km south-east of Ferrara. Folkes was approximately midway between Ferrara and Ravenna.

[3] Folkes may be referring to these churches: San Nicolò in Borgo, destroyed in the Second World War and rebuilt, and/or San Domenico, which now serves as the local civic museum.

[4] Sant'Alberto, known for its Palazzone built by the Estensi (D'Este) family in the 16th century as a military outpost. It is in the neighbourhood of Ravenna, 14 km from the city.

[5] The Po Delta, Europe's largest wetland.

[6] The river Santerno, a major tributary of the river Reno.

[7] As the former centre of the Ostrogothic kingdom of Italy, Ravenna was an essential stop for antiquaries on the Grand Tour for its architecture and antiquities. Theodoric (r. 489–526) engaged in a major building programme in Ravenna, particularly his palace church, Sant'Apollinare Nuovo. Mullett et al., 'Early Christian and Byzantine Art', Grove Art Online, https://doi.org/10.1093/gao/9781884446054.article.T024432 [Accessed 23 Sept 2023]. See Figure 24.

[8] The river Montone is the most northerly river on the east-facing slopes of the Apennines to flow directly into the Adriatic rather than entering the river Po. South of Ravenna the Montone is joined by the Bidente-Ronco, the combined rivers called the Uniti.

[9] There is a cross in the margin here. The gate is the Porta Adriana, the western city gate. It was redesigned in 1583 by Cardinal Ferreri, on the 'supposed site of the Porta Aurea built by Claudian and ruined by Emperor Frederick II'. Murray, *Handbook for Travellers*, p. 92.

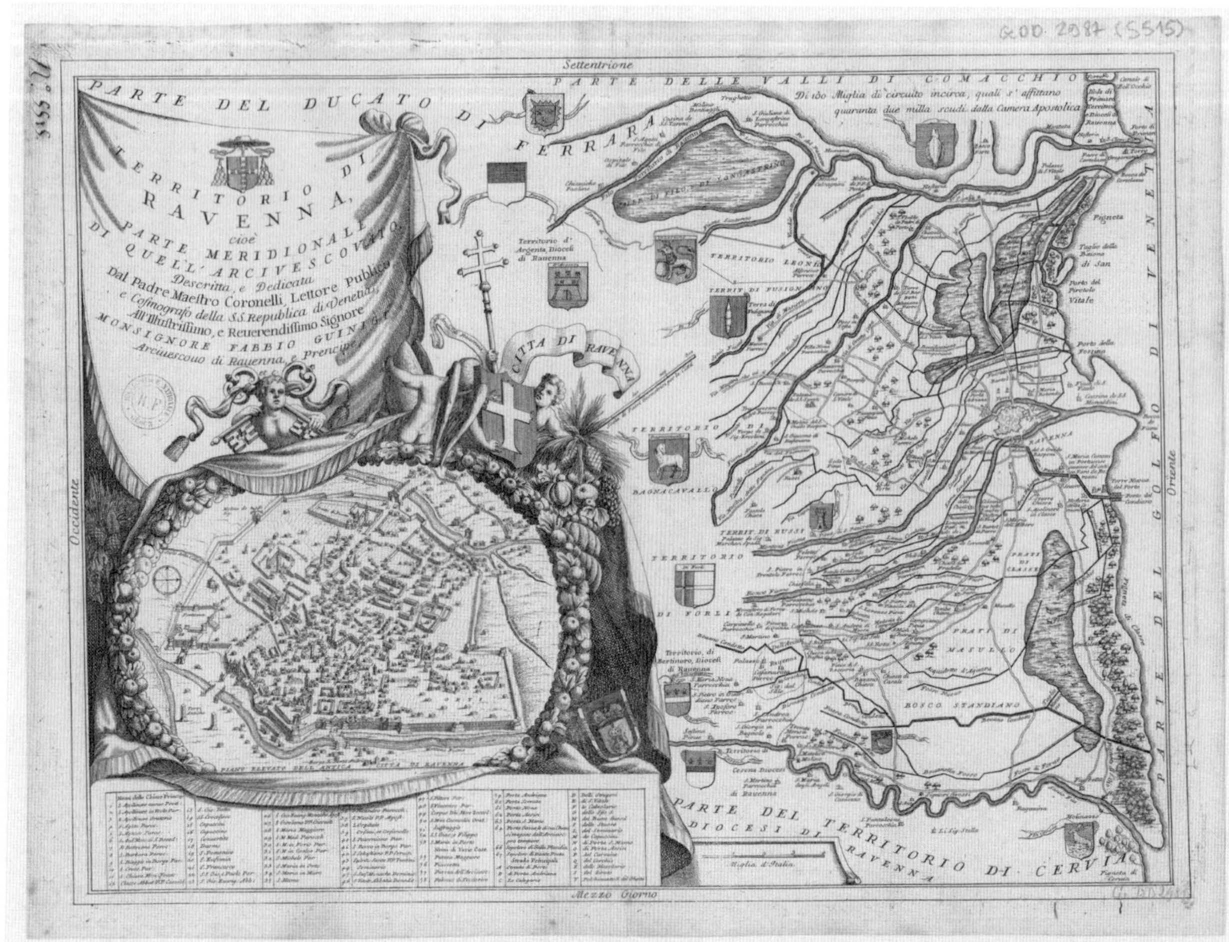

Figure 24. Vincenzo Maria Coronelli, *Territorio di Ravenna cioè parte meridionale di quell'archivescovato* (1690), engraving on paper, 45 × 59 cm. Bibliothèque nationale de France, Paris. It is possible to see the scattered nature of the town of Ravenna that Folkes described, as well as the portal over rivers through the gate and the marshy landscape. Courtesy of the Bibliothèque nationale de France, Paris.

however pretty good and well joind, I am at the Crown a middling inn, and propose staying tomorrow to rest and see what antiquitys may remain in town. I came in just at Sunset having gone to day about 35 miles, I think pretty good measure. about an hour after I got in a small accident happned the Kitchin chimney taking fire which put us in confusion not knowing at first what might be the matter, however it was soon extinguisht, and gave me occasion to take notice of a ready police in that regard, there came immediately some servants from the Cardinals who is the chief Magistrate here,[1] and directed what was fit to be done and kept the gate to keep others out that had nothing to do. afterwards when all was quiet we went to supper where we were the best treated we have been in Italy. we had a small accident the day before, the coach that carries my servants was overturnd going on to the bank of the Po, but there was no harm done. our windows here are only linen and shutters there having been little glass to be seen since we left the fferranese. the house where we dind at Alberto was a butchers, he had just killd half a dozen sheep, a Calf. and 2 hogs, a quantity of meat that surprised me in this country and in so thin a part of it but he told me there was a large village or 2 that he entirely servd and this was his market day. the house made a pretty appearance on the outside but was miserable within, there was no room but one within his shop or what ever one would call it where his meat hung and that without windows: it had been been a miserable place had we been to have lain here.

Sunday 18. It has been a very stormy tempestuous night, and the weather is turnd cold. I spent this day in taking a View of the City of Ravenna, which is but small and even for its bulk excessively thin of Inhabitants,[2] the streets are well pavd and there are many good houses one of the finest is off one Cavalier [Sputi?][3] a noble free stone palace near my inn. The Market place[4] is handsome at one end on a pedestal is a large brass statue of Pope Alexander the 7th sitting and blessing a fine figure at the other end and 2 granite pillars something like [f. 60r] those at Venice but less, with corinthian bases and capitals and the figures of St Vidal and another saint on the tops, the work of Lombard.[5] on one side of this place is the Palace of the Cardinal Legate[6] who governs here, and is at the time also ArchBp of Ancona; and joyning upon that are the courts of justice. just by one end is a handsome round or rather octangular Church with a good front and some good

[1] Ravenna was part of the Papal States until 1796. Cardinal Tommaso Ruffo (1663–1753) was Archbishop of Ferrara when Folkes visited; he served from 1717 to 1738. See Ritzler and Sefrin, *Hierarchia catholica*, V, p. 201.

[2] Folkes shared his impression with other travel writers. Philip Skippon in the 1660s recorded, 'It is true, the compass of the walls is large; but instead of houses it is all full of orchards, gardens, and farms, with some few remains of antient structures scatter'd among them'. Churchill, *A collection of voyages and travels*, VI, p. 46.

[3] This word is unclear.

[4] The Piazza Maggiore. The statue of Pope Alexander VII is now in Ravenna's Museo Nazionale.

[5] Pietro Lombardi.

[6] Now the Palazzo della Prefettura with its austere facade. It is on the south side of the present Piazza del Populo, rebuilt between 1694 and 1696 by the Legate Domenico Corsi and thought to have been designed by Carlo Fontana. 'Palazzo della Prefettura, già Palazzo del Legato Apostolico, 1694–1696', 'Notable Buildings in Ravenna, https://www.edificistoriciravenna.it/en/palazzo-della-prefettura-palazzo-del-legato-apostolico [Accessed 17 Jan 2022]. The Cardinal Legate in office during Folkes's visit was Cardinal Bartolomeo Massei (1731–45). See Ritzler and Sefrin, eds, *Hierarchia catholica*, V, p. 103.

[f. 59v]

Figure 25. Folkes's drawing of the Mausoleum of Theodoric (Church of Santa Maria della Rotonda) in Ravenna on f. 59v of his diary. Courtesy of the Bodleian Library, University of Oxford.

pictures within;[1] I also saw some in the Church of S[t] Francesco[2] over against my inn. I then went to the gate where we came in yesterday calld porta del Cibo[3] it is handsomely adornd. Not farr from this are the ruins of an old large brick castle with several large and

1 Santa Maria del Suffragio was constructed between 1701 and 1714, and designed by Francesco Fontana (1668–1708), a restrained Baroque church with an octagonal plan and side chapels. The most noteworthy paintings are by Andrea Barbiani (1708–79), such as the *Nativity of Mary* behind the high altar. See 'Notable Buildings in Ravenna', https://www.edificistoriciravenna.it/en/santa-maria-del-suffragio/ [Accessed 17 Jan 2022].

2 The Basilica of San Francesco was erected between the 9th and 10th centuries over the site of the 5th-century Basilica Apostolorum of Bishop Neone, originally dedicated to the Ten Apostles. In 875, the original church was demolished and replaced by a larger structure, and in 1261 it was given to the Franciscans and dedicated to St Francis of Assisi. The church is most famous as being the site of the tomb of Dante Alighieri, who spent his final years of exile from Florence in Ravenna. Folkes may have seen Francesco Longhi's *Our Lady of the Assumption with the Saints Naborre and Felice and a Child* or possibly the *Madonna and Child with Saints Sebastian, Roch, Francesca and Camilla Dal Corno* (1521) by Gasparo Sacchi (active *c.*1517–36) in the Chapel Dal Corno. See Ribuffi, *Guida di Ravenna,* p. 115; Raffa, 'Bones of Contention', p. 380.

3 Porta Serrata, a gate at the northern extremity of Ravenna, called so because it was closed during the Venetian possession of Ravenna to strengthen its defences, and reopened by Pope Julius II who called it, immodestly, the Porto Giulia.

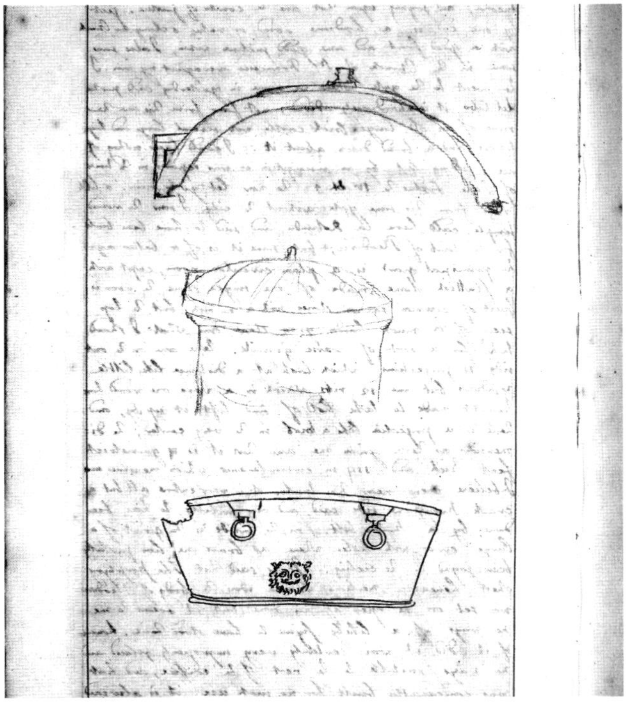

Figure 26. Folkes's sketches of the dome of The Mausoleum of Theodoric in Ravenna, as well as the porphyry sarcophagus of Theodoric on f. 60v of his diary. Courtesy of the Bodleian Library, University of Oxford.

high towers and a broad ditch about it:[1] I could hear nothing of this building but by an inscription it was repaired in the time of Pope Sixtus the IV and 5 tho now let go to ruin. a little way from this same gate without the city I saw the round temple calld here the Rotunda and said to have been built for the tomb of Theodoric;[2] but sure it is of a better age. the principal part is a plain circular room, capt with a flattish dome made of one single stone[3] the room is built of common squard stones with a cornish,[4] but the top

[1] The Fortress of Ravenna, the Rocca Brancaleone, which was erected by the Venetians in 1457, taking 10 years to complete. In 1735, shortly after Folkes's visit, it was demolished to provide materials for the Ponto Nuovo over the rivers Ronco and Montone. Little remains except foundations. See Murray, *Handbook for Travellers*, p. 541.

[2] The Mausoleum of Theodoric, now known as Santa Maria della Rotunda, created by Theodoric, King of the Goths *c.* AD 526, as his future tomb. A late example of Roman imperial mausolea, the unfinished tomb is constructed of large ashlar blocks from Istria, without the use of mortar, an ancient Roman masonry technique called *opus quadratum*. The Mausoleum is decagonal in plan, and two storeys high. As Theodoric was an Arian, his ashes were ejected from the mausoleum and the sepulchre was stripped of its ornaments. The facade and portico were rebuilt in the 16th century. The mausoleum was a mandatory stop on the Grand Tour, appearing in engravings by Giovanni Battista Piranesi. See Secondo, 'Mausoleum of Theodoric', pp. 32–6.

[3] The roof is flat domed and monolithic, with a diameter of 11 metres, weighing *c.*300 tonnes. See Nordhagen, 'Mausoleum of Theodoric (Ravenna)', Grove Art Online. https://doi.org/10.1093/oao/9781884446054.013.90000369940 [Accessed 30 Jan 2023].

[4] The cornice is, 'carved with a "tongue frieze" or pincer motif, derived from either debased classical ornament or Ostrogothic ornament'. Frazer, 'Architecture', p. 122.

seems of a much harder grey stone and what I should take for a sort of coarse granite. there are on the out side 12 projections which look at a distance like little windows but are 12 ribs[1] which in a vase one would have thought made to take hold of and lift it up by, and there is a projection like a knob in the very center; the dimensions on here given me are that it is 4 geometrick feet thick and 114 in circumference which measures I believe very near to the truth. it is very entire all but a crack throug[h] one side said, and very probably, to have been done by a thunder bolt. on the inside is the print of a large cross with holes where a brass one has probably been joynd to the cieling. It is said The porphyry chest[2] hereafter mentioned wherein was the body of Theodoric was set on the apex of this dome but it seems to me no ways of a likely figure to have stood there.[3] however if it did it was certainly very improperly placed and no ways suitable to the rest of the edifice, and that was consequently built for no such use. it is also said that the twelve Apostles in Porphyry stood on the 12 projections above spoken of but were beat down in the time of Lewis 12 by his canon;[4] but besides the im- [f. 61r] probability that just these should have beaten down and none of the pedestals hurt; it seems to me impossible that statues should ever had stood here much more than those pedestals should have been designed for that use being ridgd like the roof of a house. the names of the Apostles are certainly here but those may easily have been added when on, I believe this temple was turnd from a heathen to a Christian Church,[5] and that I am willing enough to believe might have been about the time of Theodorick, but surely such a work could not have been made in this time.[6] under his temple was another Archd room of a larger circumference so as for there to be a wall on the top of it round the first but this is now 2 kinds filld up with rubish and there has been joyned an ugly long building to it to make the nave of a church of which the Rotunda may be lookt on as a chancel, but so gothic and bad as only to serve to hide the great beauty of the rotunda it self which has really given me as much pleasure as any thing I have seen since I left England, and I am surprised it is not more talkt of then it is, appearing to be the roof as great a work I think as I have heard of, and which I should be perswaded came from Egypt where it was made for one of those temples of one stone mentioned in the Antients. I would suspect it to have been designed for a temple

[1] The roof is indeed rimmed with twelve pierced, projecting spurs, each inscribed in Latin with the name of an apostle. Ibid.

[2] The Porphyry sarcophagus of Theodoric.

[3] Folkes was correct. The room in which the sarcophagus (probably a Roman bath) is located was likely a chapel for funeral liturgies.

[4] A reference to King Louis XII of France's sacking of Ravenna (1512) during a major battle of the War of the League of Cambrai. The French brutally plundered the city for 3 days. Folkes probably mentioned the French cannon, as it was the first time in history that artillery such as cannon had played a key role in such a pitched battle. Donvito, 'Storm of Steel', p. 21.

[5] Folkes was correct in his supposition. Until 1719, the building served as the church of the monastery of Santa Maria al Faro. See Nordhagen, 'Mausoleum of Theodoric', *Grove Art Online*, https://doi.org/10.1093/gao/9781884446054.013.90000369940 [Accessed 26 Sept 2023].

[6] Folkes's puzzlement about the time of the construction of the mausoleum was probably because 'the tradition of stone building and the use of double-storey tombs had survived in the 6th century only in Syria and Asia Minor'. Theodoric probably thus 'employed a group of Eastern masons to construct his mausoleum'. Nordhagen, 'Mausoleum of Theodoric', *Oxford Art Online*, https://doi.org/10.1093/gao/9781884446054.013.90000369940 [Accessed 17 Dec 2023].

of the Sun and the 12 projections to allude to the 12 signs or months of the year.[1] they have also joined to it another house which yet more clogs up the view. the room within tho plain and unadorned is one of the prettyest I ever saw. I went from hence to the convent of St Apollinarius[2] and there is built into the wall the broken Porphyry chest just mentiond as having stood on the apex of the rotunda, it is about 8 foot and half long on the top, is adornd with a lyons head and the representation of 2 rings. not farr from hence is a fine church of the Canons regular[3] the front is unfinisht but designd to be with Ionick pillars. there are within 12 Altars on the sides besides the great Altar and those in the two ends of the cross, here are all very curiously inlaid with marbles of several colours they have good pictures over them tho some copys and are very handsomely adornd, the Church tho less seems somewhat in the taste of St Justinas at Padua.[4] I was also in another very good old Church flat rooft with pillars and a lower Apse [f. 62r] on each side, and in 2 or three more in all which were things worth seeing and some good paintings. as I cam home I went into the little Church of the Theatins[5] remarkable by the Miracle much talkt on here of the 2 Holy ghosts coming in a window when they chose an ArchBishop and nominating the person by perching upon him; the hole in the roof is still open[6] and the representation of the Dove made in it. this story was very solemnly told me in the Church and that the last time the Dove came in and instead of seating it self on any of the priests, came on the shoulder of one Severin a poor cobbler that livd over the way, upon which they saying he was nominated replyd it could not be for he was a married man, but upon going home to his stall they found his wife that instant dead of an Apoplexy. so he was enthroned and immediately tho quite illiterate before he was inspird

[1] Folkes was patently incorrect here, but he may have been influenced in his ideas by the work on Stonehenge by William Stukeley, or by his interest in Egyptian antiquities, manifested later in his membership of the Egyptian Society in London. See Roos, *Folkes*, pp. 208–36.

[2] Basilica of Sant'Apollinare Nuovo, the palace church of the Gothic King Theodoric. After Theodoric's conquest of Ravenna, this church was planned by Theodoric as a religious building for the Arians, the work still in progress after his death in AD 526. After that, in AD 561–9, the basilica was re-consecrated by Bishop Agnellus, under the supervision of Justinian, who converted it into a Catholic church. Ugolini, 'A New Insight', p. 289.

[3] Santa Maria in Porto. The church was built by Bernardino Tavella between 1554 and 1557, based on his design, as a new location of the church of Santa Maria in Porto Fuori for the Canons Regular. Its neoclassical facade was not completed until 1780. As Folkes stated, there are twelve symmetrical chapels, six on each side. They are decorated with marbles and several paintings dating between the end of the 16th and the 17th century, including works by Ippolito Scarsella, Cesare Corte, Palma il Giovane, Francesco Longhi, Giovanni Barbiani, Giovanni Battista and Andrea Barbiani. As Folkes indicated, the altar has a ciborum made of polychrome marbles, whose construction was started by Giuseppe Vivoli in 1633 and ended by Ascanio Mula in the following years. The church is most famous for its possession of the Greek Virgin, a 9th-century marble bas-relief. See 'Sant Maria in Porto', Ravenna Turismo, https://www.turismo.ra.it/en/culture-and-history/religious-buildings/basilica-of-santa-maria-in-porto [Accessed 26 Jan 2023].

[4] Abbey of Santa Giustina Padua, a 10th-century Benedictine abbey complex, in the layout of a Latin cross. Its altars and floors also feature polychrome stone.

[5] Basilica Spirito Santo, intended by Theodoric to be an Arian cathedral. Consecrated as Hagia Anastasis in the 6th century AD, it has a standard three-aisled Baptistry. After the conquest of Ravenna by the Byzantines, the church was converted to the Catholic religion in the second half of the 6th century, and was dedicated to San Teodoro.

[6] Folkes picked up the story of the miracle from Addison, who noted: 'There is a convent of Theatins [Theatines], where they show a little window in the church, through which the Holy ghost is said to have entered in the shape of a dove, and to have settled on one of the candidates for the bishopric .... I should not indeed, think it impossible for a pigeon to fly in accidentally through the roof, where they still keep the hole open ...' See Hurd, ed., 'Remarks on several parts of Italy', *Works of Joseph Addison*, II, p. 49.

and preachd an excellent sermon full of prophecy one part of which was that the Holy Ghost would come no more personally to their elections of their bishop and he accordingly has not since. St Severins \house/ for he was sainted after has been turnd into a little church with an altar of marble in it. and the dove is represented in many places as a particular protector of Ravenna. there are very many granite and other columns lying about in the Church yards and other places in the streets of Ravenna. the town is not walled but has a sort of ditch and all the gates are handsomely adornd with pillars etc and the Arms of Popes ~~without~~ \upon/ that calld porta Pamphilia[1] is an inscription imparting that as Ceres had in that place succeeded to Neptune, now again Neptune succeeds to Ceres alluding to the River Ronco lately made navigable just by that gate, the sea it is agreed came up formerly to the town but now there is no appearance of it the whole bed being so raised as not to leave any resemblance of what it formerly was, the light house[2] which antiently stood into the sea is now 3 miles within land but this I had not time to see being 2 miles out of town and the coming season hastening my away towards Rome. just before dinner as I was walking with my ffamily to the market place one Sig Petro Affelli[3] who is the Popes receiver and an officer under the Cardinal Legate came to us and as civilitys offrd us very kindly his coach which I accepted for my wife to go and see the Rotunda out of town and he himself came to us in the afternoon and carried us to the Church of the Benedictines[4] who have the finest convent in this place the Church is circular [f. 63r] with a dome richly painted and the whole inside of the finest marbles, the paving is also magnificent, and 2 of the Altars which are in projections added to the Area, are adorned with groups of marble figures the one a Pieta the same as that of St Justina,[5] the other St Vidal and the Patron Saint of this church the great Altar is in a lesser

1 The city gate, Porta Nuova, previously known as Porta Pamphilia. In 1653, Cardinal Giovanni Stefano Donghi (1608–69) wished to revive Ravenna by reconnecting the city to the sea via a new navigable canal. At the inauguration of the canal, called Pamphilio, the gate was also renamed, assuming the name of Pamphilia in honour of Pope Innocent X, born Giovanni Battista Pamphili (1574–1655). In memory of the event, the following year, inscriptions were placed on the sides of the external facade. The inscription to which Folkes referred read:

> COLUMBA NOVUM MARE | COLUMBUS NOVAM TERRAM | INVENIT | NEPTUNUS CERERI NAVES SUCCEDUNT ARATRO | UBI OLIM | NEPTUNUS CERERI | NAVES CESSERUNT ARATRO
>
> [A dove discovered a new sea, [*when Noah's ark came to land*] | Columbus discovered a new land | Neptune gives way to Ceres | Ships give way to the plough | Where once | Neptune gave way to Ceres | Ships have given way to the plough].

Pope Innocent X's heraldic symbol was the dove. The reference to Neptune and Ceres symbolizes Ravenna's new role as a port, instead of its older agricultural purpose. See Mauro, *Mura porte*, p. 256.

2 A port serving Ravenna grew up to the south-east of the city, the Emperor Augustus dedicating this port – Classe – as home port for his eastern navy. Pliny 'described the tower at Classe as having a Pharos-like structure [a lighthouse] ... it may be no coincidence that a similar building attached to the church of Santa Maria in Porto Fuori was used as a bell tower ... such a tall tower would have required better foundations than were available in Classe, and its great height would have made a light available over a great distance'. The tower existed until 1944 when it was completely destroyed. See Trethewey, 'Ancient Lighthouses', p. 29. See also Hutton, *Ravenna*, pp. 276–7.

3 This man has not been identified.

4 The late antique Benedictine monastery of San Vitale, now the Museo Nazionale of Ravenna, renowned for its Byzantine mosaics and architecture. It is the only major complex to survive virtually intact from the period of the Emperor Justinian I.

5 The Pietà Chapel in the Basilica of San Vitale was demolished during the restoration works executed in the early 20th century. It contained a marble group depicting the Pietà by Giovanni Toschini, today conserved in the church of Suffragio in Ravenna. Europeana, https://www.europeana.eu/en/item/22/_100092 [Accessed 27 Jan 2023].

dome joyning to the great one and is adornd with a fine picture of Baroccis.[1] there are 2 other very fine chappels on either hand the chief entrance one of which is but just finisht and with great expence it was towards evening when I saw this Church yet it made as good an appearance as most things I have seen, and is adorned in as rich a manner. one of the fathers carried me into their garden in which is an old little Gothick Chappel, in which is the sarcophagus of the upper end of the Emperor Placidia[2] and at the ends of the little cross here of Honorius on one side and Constantine on the other the Altar stands before that of Placidia, and by the entrance are 2 marble chests that on the right hand with a cross and that on the left without, these are said to have been for the urns of her domesticks the Christians on one side and the pagans on the other. the chapel is very pretty and adornd with mosaick. I saw about this convent[3] several pieces of old inscriptions and some ancient sarcophagi and bass reliefs but mostly of the lower Empire. if my time would allow I find I could have been longer entertained at Ravenna; and I ~~receд~~ receivd really so much civility from Sig[r]: Arfelli as I could noways have lookt for from a quite stranger: he offerd to have introduced me to the Cardinal Legate if I could have staid, and that Prelate is here spoken of in a very amiable light. Money at Ravenna is excessive scarce and no where to be met with but in the ornaments of the Churches and these are truly magnificent. ~~The poverty I fear of~~ provisions are therefore very cheap so that the treatment is surprizingly good. The poverty I doubt of the people is excessive, I judg from the difficulty of getting small money of any sort for a cequin, ~~I~~ which my Landlord could not do neither by himself nor his ffriends. M[r] Arfelli changd me 3, and I would have had five but that he could not do other than in copper, but even that could not be done at my inn. what sort of trade is in the town I know not being here only of a sunday, but by the river being made navigable and the ground lying well about

[f. 63v] cut as at Cervia[4] to the Salt works of which this town is also full. They cut here narrow and shallow trenches for the sea under which the sun evaporates and leaves the salt at the bottom

[f. 64r. the town I suppose the people do not want industry. so that it is I believe the riches of the religious that occasions the excessive want of money with every body else.

[1] Federico Barocci (*c.*1532–1612, originally Federico Fiori), 'the most famous and well-paid Italian artist of the later sixteenth and early seventeenth centuries', known for his skilful Mannerist style and his draughtsmanship. See Ian Verstegen, 'Federico Barocci', *Renaissance and Reformation* https://www.oxfordbibliographies.com/view/document/obo-9780195399301/obo-9780195399301-0442.xml [Accessed 27 Jan 2023].

[2] Mausoleum of Galla Placidia, a late Antique building built between AD 425 and AD 450. As Schoolman has indicated, the Empress was not buried here, but rather died in Rome in 450 and was likely buried in the imperial mausoleum at St Peter's; it was a persistent misconception until the 19th century. Schoolman, *Rediscovering Sainthood in Italy*, p. 12. See also Deliyannis, 'Bury Me in Ravenna'. The sarcophagus on the right is reputed to hold the remains of her half-brother, Emperor Honorius (395–423), and the one on the left the remains of her husband, Emperor Constantius III, and her son, Valentinian III.

[3] There is a cross in the margin here.

[4] Cervia, now a seaside resort, still has its salt pans. Most have been subsumed into a single collection system that is mechanized. The salt pans used to be in small lots that were harvested individually by salt workers and their families. See 'The Salt Pan', Cervia, https://www.turismo.comunecervia.it/en/discover-the-area/nature-and-environment/parks-natural-reserves/the-salt-pan [Accessed 4 Feb 2023].

Monday. 19. I left Ravenna by 4 a clock, and passd Classe a convent[1] about two miles off before it was light which prevented my seeing some antiquitys that I was told were there particularly some very large pillars of black and white antient marble.[2] Further on I passd the River Savio on a bridge at the foot of which stands a large granite pillar, we then traveld some miles through a wood of Pignolis[3] a tree that bears a fruit very like the cones of the common fir but very different colour from those I observd in Germany. here is a village called Sta Maria del Pino. a little farther is a little town called Cervia consisting altogether of saltworks and a channel to sea is cut and brickt for the vessels to come up to the offices[4] which are very handsome and look like a palace. a few miles further is a little town calld porto Cesennatico[5] where we baited at a poor inn and eat what we had brought. Here is a handsome bridge and \over the river or channel +/[6] and at the end 2 granate pillars, with modern capitals etc of the Corinthian order and the Popes keys etc on them. there are some pretty convents in this little town, I was in that of the ffranciscans,[7] and there are some ruins of walls a tower and a gate which seem to be Roman and several fragments of ancient marble in the street.[8] A few miles from hence we passd the Rubicon[9] fording it in 2 channels that immeditately joynd, it is now very inconsiderable but somewhat rapid and may be a good deal larger when the snows melt.

1 The former monastery of Classe, one of the largest Camaldolese monasteries in Italy, active from 1515 until the Napoleonic suppression. The complex today is the headquarters of the Classense Library, which has a world-renowned collection of 800,000 printed books as well as 750 manuscripts, including the only extant manuscript with the eleven comedies by Aristophanes. It is a pity that Folkes did not have time to visit, as his commentary upon the library's holdings may have been illuminating. See 'Former monastery of Classe', Ravenna Turismo, https://www.turismo.ra.it/en/myravenna/city-of-art/monasteries-of-ravenna/ [Accessed 4 Feb 2023].

2 The 24 columns of black and white Greek marble, topped with Corinthian capitals, in the Basilica of Sant'Apollinare in Classe, built in the 6th century. The Basilica was dedicated to Sant'Apollinare, the first Bishop of Ravenna and is internationally known for its Byzantine mosaics.

3 Domestic pine (*Pinus pinea*), also known as the Italian stone pine, source of the pine nut.

4 The brick Torre storehouse was built in 1691 and is next to the San Michele Tower. Romagna Treasurer Michelangelo Maffei had the storehouse constructed to store produce from the salt pans. On the opposite side of the Canal, the Darsena storehouse, built in 1712, served to keep salt after it had been washed and dried in the Torre storehouse, so it could be weighed and divided into bags for sale. The Torre storehouse is now the site of MUSU, the Salt Museum, featuring the burchiella, the iron craft in this area used for salt transport. See 'Torre Salt Storehouse', Cervia, https://www.turismo.comunecervia.it/en/discover-the-area/art-and-culture/industrial-archaeology/salt-storehouses?set_language=en [Accessed 4 Feb 2023].

5 Cesenatico, a port town on the Adriatic coast, its port and canal built in 1314. In the 1660s, Philip Skippon described it as, 'a small village, with a port for little barks'. Churchill, *A collection of voyages and travels*, VI, p. 704. Most of its buildings were razed in the Second World War, and it was rebuilt as a tourist resort.

6 Folkes's writing of a plus mark referred to the text he wrote on f. 63v.

7 Probably the church of San Nicola, a Franciscan friary in Cesenatico, 'built in 1611 and dedicated to the saints Nicola di Mira and Francesco di Paola, patron saints of seamen'. See 'San Nicola's Church,' Visit Cesenatico, https://visitcesenatico.it/en/pointofinterests/san-nicolas-church/ [Accessed 7 Feb 2023].

8 The *Ad Novas*-Cesenatico Project (2008–13) at the University of Leicester, organized by Denis Sami and Neil Christie, retrieved evidence of late antique wooden structures and a substantial paved Roman road here. As they noted: 'the Tabula Peutingeriana records a settlement named *Ad Novas* depicted immediate north of the Rubicon River and south of *Sapis* (modern Lido di Savio) whose location roughly matches the (southern) area of modern Cesenatico'. *Ad Novas* saw 'important traffic from Ravenna across the late Roman and Byzantine periods'. Sami et al., '*Ad Novas*-Cesenatico', pp. 1–2.

9 Rubico (commonly called Rubicon), reddish stream flowing into the Adriatic and marking the boundary between Italy and Gaul (Cisalpine): possibly the modern Pisciatello. Edward Salmon, 'Rubicon', *Oxford Classical Dictionary*, https://doi.org/10.1093/acrefore/9780199381135.013.5623 [Accessed 7 Feb 2023].

it is now calld pisatello[1] and some poor people forded it on foot as we went over. we a little after passd a larger river on a bridg calld in the map Luso[2] and got to Rimini between four and five in the afternoon, our way having been very bad most part of the day. the cottages we have seen have all a wretched air of poverty and the people look thoroughly miserable. they call it 36 miles from Ravenna to Rimini but I thought em long. as I had but little time I went out immediately to have a view of the town, it is I think near as big as Ravenna but not so handsome near as the streets are not so strait nor well pavd yet is here some good houses and convents, the churches I had not time to go into. in the chief Piazza or market place[3] is a pretty marble fountain built by P. Paul the 3^{d}.[4] as also an admirable statue of him setting in his chair and blessing, of brass much bigger [f. 65r] than the life. one side and end of this piazza is filled with a new and very fine Palace for the Magistrates the old one was destroyd some years since by an Earthquaque. a little way from this market place is a fine old castle with a good ditch and walls of the old gothic make in which somewhat like a guard is kept. it is one of the most entire of those sort of things I have seen. There is also another pretty good market place in which are some very handsome houses. And here stands the suggestum[5] on which Caesar is said to have harangued his Army after passing the Rubicon when he began the Civil war.[6] it was probably the pedestal of a pillar or statue, the inscription is certainly more modern much so that what foundation there is for the fact I cannot tell, if it is true he must have used it only as somewhat that was accidentally there.[7] for I agree entirely with Mr Addison that those on the trajan pillar etc are quite of another sort.[8] but the 2 most remarkable

[1] River Pisciatello. Addison also called it Pisatello. See Hurd, ed., 'Remarks on several parts of Italy', *Works of Joseph Addison*, I, p. 95. There was a Renaissance humanist debate over which waterway corresponded to the ancient Rubicon with the Uso or Fiumicino rivers considered probable candidates.

[2] The river Luso, which crosses the high road from Ravenna to Rimini, about 16 km from Rimini.

[3] Campo del Comune, now the Piazza Cavour. Philip Skippon noted in the 1660s: 'the piazza is fair, with some portici about it; and here is preserved the stone Julius Caesar stood on when he spoke to his followers after he passed the Rubicon; these modern inscriptions on it: C. Caesar Dict. Rubicon Superato civili Bel. commilit. suos hic in Foro R. adlocut. Suggestum hunc Vetustate collapsum Coss. Ariminensium Novembris et Decembr. MDCV. Restit'. Churchill, *A collection of voyages and travels*, VI, p. 703.

[4] It is a statue of Pope Paul V (r. 1605–21), designed by Nicola Cordier and cast by Sebastiano Sebastiani in 1613. Rimini was in the Papal States at that time.

[5] A *suggestum* is a raised platform, stage, mound or tribune for a speaker.

[6] 'In 49 BC Caesar, after some hesitation, precipitated civil war by crossing it'. Salmon, 'Rubicon', *Oxford Classical* Dictionary, https://doi.org/10.1093/acrefore/9780199381135.013.5623 [Accessed 7 Feb 2023].

[7] Again, Folkes used Addison as the basis of his remarks in his diary. Addison noted: 'The Suggestum, on which it is said that Julius Caesar harangued his Army after having passed the Rubicon. I must confess I can by no means look on this last as Authentick: It is built of hewn Stone, like the Pedestal of a Pillar, but something higher than ordinary, and is but just broad enough for one Man to stand upon it. On the contrary the ancient Suggestums, as I have often observed on Medals, as well as on Constantine's Arch, were made of Wood like a little kind of Stage, for the Heads of the Nails are sometimes represented, that are supposed to have fastened the Boards together. We often see on them the Emperor, and Two or Three General Officers, sometimes sitting and sometimes standing, as they made Speeches, or distributed a Congiary to the Soldiers or People. They were probably always in readiness, and carried among the Baggage of the Army, whereas this at Rimini must have been built on the Place, and required some time before it could be finished.' Hurd, ed., 'Remarks on several parts of Italy', *Works of Joseph Addison*, I, pp. 95–6.

[8] Addison continued: 'If the Observation I have here made is just, it may serve as a Confirmation to the Learned Fabretti's Conjecture on Trajan's Pillar; who supposes, I think, with a great deal of Reason, that the Camps, Intrenchments, and other Works of the same Nature, which are cut out as if they had been made of Brick or hewn Stone, were in reality only of Earth, Turf, or the like Materials; for there are on the Pillar some

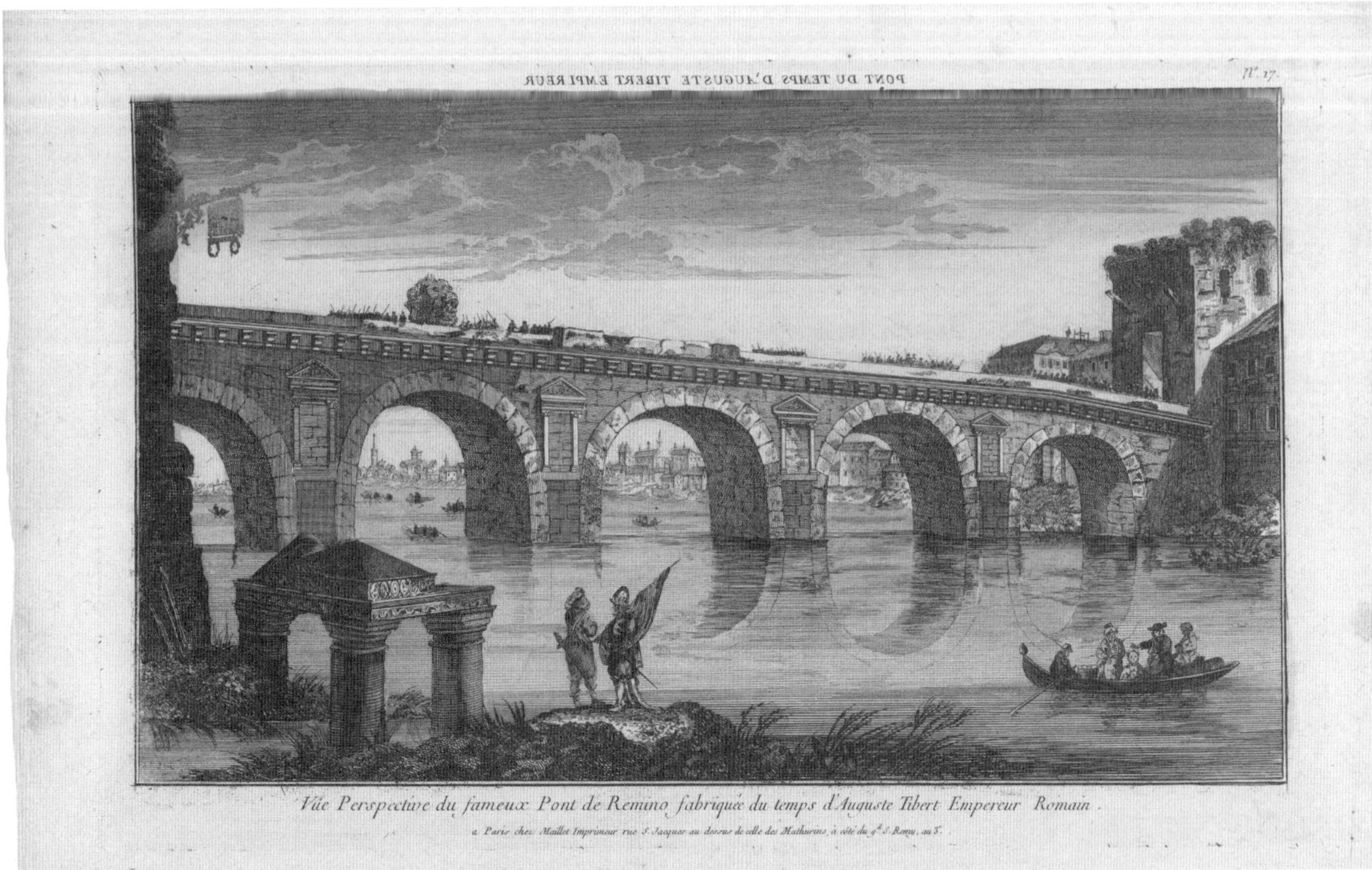

Figure 27. Unknown maker, *Gezicht op de Brug van Tiberius over de rivier de Marecchia te Rimini, Pont du temps d'Auguste Tibert empireur* (View of the Bridge of Tiberius over the Marecchia River in Rimini, Bridge from the time of the August Emperor Tiberius) (*c.*1744–60), etching with watercolour and ink on paper, 27.5 × 43.8 cm. Rijksmuseum, Amsterdam, courtesy of C. J. van der Feen. Folkes greatly admired this bridge in Rimini, calling it remarkable and 'very fine'.

antiquitys[1] here and what are really very fine are the bridg going into the town over the river Marecchia, of 5 arches of marble built by Augustus and Tiberius, the inscriptions on the parapet walls tho broken to pieces are still very legible the letters near a foot long.[2] but as they are shattered they have been copied on 2 marble tables on the inside the gate of the town that is just at the end of the bridg. the other is the Triumphal Arch of Augustus which is at the gate next Pesano, it is a noble arch about 30 foot wide and adornd with corinthian Columns those on the inside are about half gone on the outside they are pretty entire, and so is a good deal of the inscription,[3] part of the top is tumbld down but it has been repaird with brick to keep it together it is a fine piece of Antiquity. I had not time to copy either this inscription or that on the bridge, but they are both in print, and I think I have a stamp of the Arch itself that will refresh my memory of it, if I live and it please God I get home. they talk also of another piece of Antiquity the remains of an Amphitheater[4] near the Capucin Convent,[5] but it is so imperfect and so little

---

of those Suggestums which are figured like those on Medals, with only this difference, that they seem built of Brick or Free-Stone.' Hurd, ed., 'Remarks on several parts of Italy', *Works of Joseph Addison*, I, pp. 97–8. Folkes himself had a keen interest in Trajan's Column in Rome, analysing it metrologically, but also its use of optical illusion on its friezes. See Folkes, 'On the Trajan and Antonine Pillars', pp. 117–21; Roos, *Folkes*, pp. 153–6.

[1] A cross is inked in the right-hand margin here.

[2] Bridge of Tiberius (Ponte di Tiberio) or Bridge of Augustus (Pons Augustus), a Roman bridge with five semicircular arches of Istrian stone, constructed where the Via Aemilia entered the city from the north. This major road 'had been completed by Marcus Aemilius Lepidus in 187 BC, to abet Rome's control of what was then called Gallia Cisalpina (literally, "Gaul on this side of the Alps")'. See Cassibry, *Destinations in Mind*, p. 35.

[3] During the reign of Augustus, the renovation and expansion of the bridge and road network took place on a grand scale. This is seen on the inscription on his Triumphal Arch at Rimini (27 BC), commemorating his achievements:

> Senatus populusq[ue Romanus] | [Imp(eratori) Caesari Divi Iuli f(ilio) Augusto imp(eratori) sept(imum)], | co(n)s(uli) sept(imum), designat(o) octavom, v[ia Flamin]ia [et reliquei]s | celeberrimeis Italiae vieis consilio [et auctoritate ei]us m[un]iteis
>
> [The senate and people of Rome dedicated this to Emperor Caesar Augustus, the son of a God, [*Augustus was the adopted son of Julius Caesar*], emperor for the 7th time, appointed consul for the 7th time and designated for the 8th time, because on his initiative and at his expense the Flaminian Way and the other most well-known roads of Italy were restored].

See EAGLE (Electronic Archive of Greek and Latin Epigraphy), inscription EDR12890, http://www.edr-edr.it/edr_programmi/res_complex_comune.php?id_nr=EDR128590 [Accessed 30 Dec 2023]. 'The emperor [Augustus] was in many ways synonymous with the means of land travel and transport, the network of roads that bound many parts of the empire together'. Quilici, 'Land Transport', p. 560.

[4] The amphitheatre, built during Hadrian's reign in the 2nd century AD, was nearly the size of the Colosseum in Rome. Due to being systematically dismantled to provide material for the city's defensive walls and buildings, and then intensive Allied bombing during the Second World War, the amphitheatre is now in an advanced state of ruin. In 1701, De Blainville noted: 'Towards the Sea, and behind the Capuchine Garden, we see the Remains of an Amphitheatre, which are next to nothing. There is only an Inscription on the Wall with a Hand pointing to them; and this Inscription is as laconic as the Ruins of the Amphitheatre are inconsiderable. It is as follows: *Amphitheatri olim P. Sempronio Cos. exercitati Reliquias indigitat Sen. Ar.* That is "The Senate of Rimini points out the Remains of the Amphitheatre formerly built by the Consul Pub. Sempronius." But with all the Pains which the Senate of Rimini has been to point them out, one may be long enough in Search of them before he can find them.' De Blainville, *Travels*, II, p. 226.

[5] Santarcangelo di Romagna. The Franciscan Friars were present in Santarcangelo from the 14th century. Originally, the Capuchin friars lived in St Mary's church, close to the Uso River bridge, on the 'Via Emilia', but had to move to the top of Mount Giove in 1656 because of the humidity. The construction was led by Cardinal Frederick Sforza, Bishop of Rimini. See 'Santarcangelo di Romagna', https://www.santarcangelodiromagna.info/en/chiesa-e-convento-dei-frati-cappuccini-25 [Accessed 7 Feb 2023].

Figure 28. Thomas Salmon, *L'Arco di Rimini vicino alla Porta Orientale fatto per ordine dell'Imp. Augusto*, or *View of the Arch of Augustus in Rimini* (*c.*1757), etching on paper, 17.5 × 11 cm, in *Lo stato presente di tutti i paesi,* Venice, 1757, XXI, p. 452. Biblioteca Nacional de España, Madrid. This prospect of the arch of Augustus that Folkes described is roughly contemporaneous with Folkes's journey. Courtesy of the Biblioteca Nacional de España, Madrid.

remaining that I could make scarce any thing of it. they say there are some pillars and statues that were part of it in the Convent but those I had not time to see. I have traveld to day with the sea all along pretty[1] near and mostly in sight to the left, and some hills and mountains at some miles to the right which we had not before seen since we left Montefelice, we have never been out of a most perfect plain.[2] this town has a sort of port.

[f. 66r] Ravenna, Cervia, and Sesennatica have also, and there are a pretty many sailors about the town, the people look as employd in business yet is there the same air of poverty I before took notice of superlatively since I came into these dominions; there is no body seems to have any money or be able to change the least peice. I am at the Cross keys a tolerable inn for this country and where I am not ill treated. My Voiturin seems to pique himself on giving us satisfaction on to what ever is possible in the places we pass through.

Tuesday 20. We left Rimini between 5 and 6 a clock, \crossing the Ausa and/[3] travelling all the morning pretty near the sea ~~and~~ on our left and the hills we saw

[1] There is a cross in the right-hand margin here.

[2] De Blainville made much the same remarks about the landscape as he travelled down the old route of the Via Flaminia, writing 'we are almost all along in View of the Adriatic Sea'. De Blainville, *Travels*, II, p. 226.

[3] The river Ausa is 27 km long, its source is Monte Titano. It flows north-east past Serravalle, crossing the border into the province of Rimini.

yesterday approaching on our right to a Village calld Catholica[1] having passd several little rivers particularly the Conca[2] which went to the City of that name now swallowed by the sea. Catholica is a little town with gates that seem to have been ancient and some other fragments lying about the country here grows hilly, and has sufferd some years since very much from an Earthquake, an inscription concerning which is set up on the Church at Catholica.[3] we immediately after passed the tavolo[4] and leaving the Sea went through some very high hills or rather mountains to Pesaro,[5] these mountains have a pavd way quite through them and are very pleasant the sunny side finely coverd with vines and plantations, just before the gate of Pesaro we crossd the Foglio formerly the Isaurus.[6] the town is walld and a Rampart facd with stone and bastions very good with a dry ditch, the town it self is very pretty the best part of it pavd with brick, and several good houses, the people look more busie and substantial and cleaner than in any place I have seen hereabouts and Imagin therefore there is some trade stirring in the place.[7] the Piazza is very handsome and the Palace of the magistrate in it,[8] there stands a large Marble statue of Pope Urban the 8th very finely done, and there is also a noble fountain of Marble and brass in a good taste and which makes the palace look exceeding pleasant. the Churches at least those I saw tho fine do not equal what I have seen in other towns

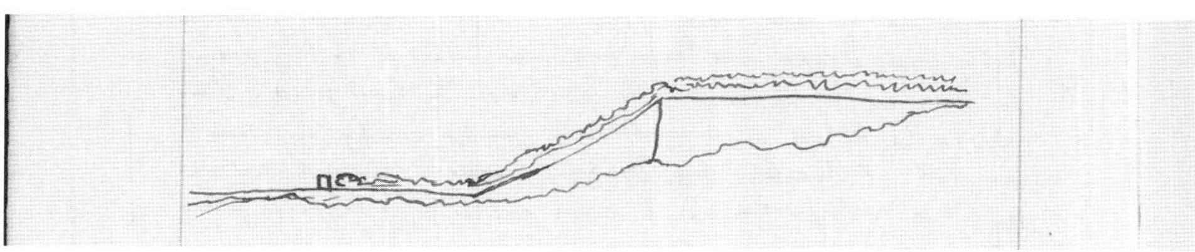

Figure 29. Folkes's drawing of the waterworks of Fano on f. 66v of the diary. Courtesy of the Bodleian Library, University of Oxford.

[1] Cattolica was a Roman settlement on the Via Flaminia, situated on a cape called the Promontorium Focariae, and it was a stopping point with taverns. It is now a seaside resort on the Adriatic Riviera, approximately 20 km from Rimini.

[2] The river Conca originates in the Marche (Mount Carpegna) and ends in the Adriatic Sea between Cattolica and Misano.

[3] Chiesa di Sant'Apollinare.

[4] Flat areas.

[5] Pesaro, part of the Papal States governed by cardinal legates until the Napoleonic occupation. It is now the capital of the Province of Pesaro e Urbino in the Marche region. Eiche and Guidotti, 'Pesaro', *Grove Art Online*, https://doi.org/10.1093/gao/9781884446054.article.T066661 [Accessed 17 Feb 2023].

[6] The river Foglia. Pesaro is located at the river's mouth. The Roman poet, Lucan, referred to the river as Isaurus. See *M. Annæus Lucanus, Pharsalia 1831,* II, 406.

[7] Folkes's view of Pesaro is shared by De Blainville, who in 1701 noted: 'It is, next to the Capital, the finest and most agreeable City in all the Duchy of Urbino'. De Blainville, *Travels*, II, p. 231.

[8] The Old Ducal Palaces where the Dukes of Urbino used to winter and there administered their domains. The Dukes came from the families (successively) of Montefeltro, Sforza and Della Rovere. The facade was built by Alessandro Sforza in the mid-15th century. With the transfer of the court from Urbino to Pesaro, from 1564, the palace became the administrative centre of the Duchy. In 1631, with the death of the last Della Rovere Duke, Francesco Maria II, the Duchy of Urbino was annexed to the Holy See. The Ducal Palace then became the seat of the cardinals legates from 1631 to 1797. See Patrignani, *Pesaro*, pp. 36–7; Tesini, *Pesaro*, pp. 52–5; 'Palazzo Ducale', Commune di Pesaro, http://www.comune.pesaro.pu.it/cultura/da-vedere/luoghi-da-vedere/edifici-storici-e-monumenti/edifici/palazzo-ducale [Accessed 7 Feb 2023].

hereabouts, however they abound in very fine pictures of Guido,[1] Baroccio,[2] and Cantarini[3]; I took particular notice of 3 very capital ones of Baroccio viz, the Christ on the water, ~~and~~ at St Andreas[4] a little ~~round~~ church with a ~~dome~~ cupola, the Circumcision at the Schola of the Name of Jesus,[5] and the St Michaelina[6] at the Church of St Francis; this last is one of the sweetest pictures I have ever seen; here is her body which does miracles,[7] and a stone crucifix that one of the ffathers told [f. 67r] me spoke to her. I dind at the Port house and set out again[8] a little ~~before~~ after 3, and turning immediately to the left came on the sand under a small cliff to Fano[9] in about 2 hours, they \count/ ~~count~~ the whole journey today 36 miles. Fano is a middle sizd town with an old wall gates and towers most of which seem very antient at coming in to the town we passd the River Arqila,[10] which running down pretty eagerly they have leveld the lower part from the sea for the conviency of the navigation of the boats and [[*one word*]] to the town, and have thrown the whole fall into a short compass so that it falls with vehemence like a sort of a cascade and after meets a stop to moderate its course; it is brickt along the side and the Cascade part along the bottom and makes a very pleasant appearance; the streets here are not by a great[11] deal so good as at Pesaro, but there is a pretty market place adornd with a marble fountain, and a [[~~xxx~~]] hall has been turnd into a Theater[12] that they talk much of but I could not see the inside it being just dark

[1] Guido Reni (1575–1642), Baroque painter from Bologna. It is unknown which is the painting to which Folkes refers, but there is a *Fall of the Titans* (1636/7), that is now in the Museo Civico, Pesaro. Reni painted the *Virgin and Child with Sts Thomas and Jerome* (1634) for the Olivieri Chapel of the Cathedral of Pesaro (now in the Pinacoteca Vaticana). See Pepper, *Guido Reni,* p. 271.

[2] Federico Barocci.

[3] Simone Cantarini or Simone da Pesaro (1612–48), an Italian painter and engraver known for history paintings and portraits. He was an 'eminent pupil' and then rival of Guido Reni, with a 'highly original style, which united aspects of Bolognese classicism with a bold naturalism'. Garofoli, 'Cantarini, Simone', *Grove Art Online*, https://doi.org/10.1093/gao/9781884446054.article.T013753 [Accessed 13 Feb 2023].

[4] Chiesa di St Andrea Apostolo, located just beyond the Arco della Ginevra, the former church of the Confraternita della Vecchia Misericordia, built in the 14th century.

[5] Federico Barocci's *Circumcision* (1590), now in the Louvre, commissioned by the confraternity the Compagnia del Nome de Gesù at Pesaro.

[6] St Michelina of Pesaro (1300–1356), a beatified Franciscan penitent. In 1606, Alessandro Barignani commissioned Barocci to paint the Beata Michelina for the Capela della Beata Michelina for San Francesco in Pesaro, for 150 *scudi.* Folkes was correct that it was a painting characterized by sweetness, as the saint is represented in *dolci estasi.* Gillgren, '*Una dolcissima Estasi*', p. 158. The painting is now in the Pinacoteca Vaticana. See Turner, *Federico Barocci,* p. 197; Mvsei Vaticani, https://catalogo.museivaticani.va/index.php/Detail/objects/MV.40378.0.0 [Accessed 13 Feb 2023].

[7] St Michelina's relics are in the reliquary chapel in the Basilica in Pesaro.

[8] There is a cross in the margin here.

[9] Fano is 12 km south-east of Pesaro, where the Via Flaminia reaches the Adriatic Sea. Originally called *Fanum Fortunae*, the settlement took its name from its 'Temple of Fortune', founded in the 3rd or 2nd century BC, and occupied by Julius Caesar in 49 BC. Between 31 BC and 27 BC, 'Augustus planted a colony of veterans there, and built wall fortifcations ... a triple arch erected in his honour still stands'. 'Fano', *Encyclopaedia Britannica*, https://www.britannica.com/place/Fano-Italy [Accessed 17 Feb 2023].

[10] Torrenta Arzilla.

[11] Here there is a cross in the margin.

[12] Teatro della Fortuna. The current theatre, whose facade is the ancient Palazzo del Podestà, founded in 1299, was built to a design by the Modenese architect Luigi Poletti between 1845 and 1863. Poletti's design replaced the famous theatre of the same name erected by the renowned Fano scenographer and stage technician, Giacomo Torelli, between 1665 and 1677. It is unfortunate Folkes could not have visited, as its sets may have

when I was here; it stands on a fine collonade so is above stairs and the Edifice makes a good shew, near it is the palace pretty huge for the Governour and magistrates. I was showd the Arch of Augustus it is less that that at Rimini and more plain but very handsome and most of the Inscription pretty legible[2] they have joined a church to it[3] which hides some part of it and other part is fallen down there is however enough left to give a good Idea of it; they have cut the representation of it as entire on the wall of the church next to it with the inscriptions;[4] there they represent a sort of 2^{d} story above the Architrave and cornish of the building I know not from what Autority for nothing of it is now to been seen, and I should doubt my self if any such had ever been. I enquird of several people for the ruins of the Temple of Fortune[5] whence this town takes its name, but could hear nothing of it, at last an Abbè told me he had heard it had stood where a new church now is but there are no remains at all of it. I saw some ruins about half a mile out of town but what they are could not learn. they have set fortune in the fountain of the Market place to preserve her memory. I am at the Port house a very indifferent inn seemingly.

---

influenced current cosmological portrayals. As Ait-Touati has noted, 'Torelli's sets and engravings were widely known, distributed, quoted, and used throughout the second half of the [17th] century', and may have influenced visualizations of the universe by Bernard le Bovier de Fontenelle in his, *Conversations on the Plurality of Worlds [Entretiens sur la pluralité des mondes]* (1686). See Ait-Touati, 'Clouds and Meteors', p. 191. See also 'Storia del Teatro della Fortuna', Fondazione Teatro della Fortuna, http://www.teatrodellafortuna.it/index.php/teatrodellafortuna [Accessed 17 Feb 2023].

[1] The arch was erected in 9 BC to commemorate the victory over Hasdrubal Barca in the Second Punic War, and was built over the Via Flaminia at the western entrance into Fano. The upper part of the arch, a loggia of the Corinthian order, was destroyed in 1463, when Count Federico da Montefeltro of Urbino, mercenary captain of Pope Pius II, laid siege to the city of Fano to recover it for the pontiff from Sigismondo Malatesta. See Fritsen, *Antiquarian Voices*, p. 181; See also Poggi, *Origine e antichità di Fano*, pp. 41–2.

[2] The Epigraphic Database Heidelberg records the inscriptions in this manner:

Divo Augusto Pio Constantino patri dominorum Imp(erator) Caesar Divi f(ilius) Augustus pontifex maximus co(n)s(ul) XIII tribunicia potest(ate) XXX<V>II | imp(erator) XX{V}I pater patriae murum dedit Curante L(ucio) Turcio Secundo, Aproniani praef(ecti) urb(i) fil(io) Asterio v(iro) c(larissimo) corr(ectore) Flam(iniae) et Piceni.

[The Emperor Caesar, son of a God, Augustus, Chief Priest, Consul 13 times, Tribune 37 times, Emperor 26 times, Father of the Fatherland, dedicated this wall to the late Augustus Pius Constantinus, Father of Rulers. The work was supervised by Lucius Turcius Secundus Asterius, son of Apronianus, prefect of the city, a most distinguished gentleman and governor of Flaminia and Picenum].

See Epigraphic Database Heidelberg, inscription HD032644, https://edh.ub.uni-heidelberg.de/edh/inschrift/HD032644 [Accessed 20 Dec 2023].

Part of this inscription was added after Constantine's death (AD 337) to one of the monumental city gates of Augustan date in Fano (*Fanum Fortunae*), associating the two reigns to generate prestige for Constantine's memory. In addition, 'L. Turcius Secundus *signo* Asterius (son of a *praefectus urbi*), also added his own name to the monuments, showing that not only Emperors could profit from earlier imperial inscriptions'. Machado, 'Dedicated to Eternity?', p. 344.

[3] Folkes is referring to St Michele, which was partially built with marble taken from the triumphal arch.

[4] A representation of the arch as it was in antiquity was sculpted in bas-relief by Antonio Constanzi, a citizen of Fano. It is located in the marble wall of the church of St Michele. Poggi, *Origine e antichità di Fano*, pp. 41–2.

[5] This is probably a reference to the amphitheatre of Fano. Fano was the largest Roman settlement on the Adriatic coast and featured an amphitheatre, which could seat 10,000 people. one of the biggest Roman theatres in the region. 'Roman theatre', *Archaeology News Network*, https://archaeonewsnet.com/2016/03/roman-theatre-of-fano-forgotten-under.html [Accessed 17 Feb 2023].

Wednesd. 21. Setting out from ffano about 6 a clock[1] at about 2 miles distance through a plain quite flat and open I passd the River Metaurus on a wooden bridg somewhere hereabouts Asdrubul was killd in the second Punic war.[2] The [f. 68r] bed of the River is very large where I crost in [*sic*] runs now in 3 channels, but when the snows melt those all unite and it becomes a very formidable river I believe the lenght [*sic*] of the bridg a good deal exceeds London bridg, but none of the streams at this time were so wide as the Thames at Maidenhead, it was not much more than half a mile from the mouth and it runs very rapid, the same plain continues some miles on the other side about 2 from the river we came again on to the sand of the sea and kept on that or very near it till we came to Singaglia[3] passing by the way the river cesano[4] seems very near its mouth on a wooden bridge of a great length, the river running in several channels and being at this time of the year like a wash. at Singaglia, it is handsomely ~~bred~~ brickt on the side for a good way and there are at its mouth 2 moles[5] that run above ¼ of a mile into the Sea on both sides I walkt on one of them which is very handsome and by an inscription was made by P. Alexander the 8th[6] they are both set with stone parts to moor the vessels to. the principal part of the town is on the South Side of the River, and makes a pretty regular square with 4 bastions of brick wall built but without a ditch except on one side. The rest of the town is on the other side the River and walld like the other on 3 sides. the town within is dirty and the streets narrow and ill pavd otherwise here is the most appearance of business and trade I have seen any where since Venice tho the place is but small there were several Vessels lying at the key and some building and I was told the place is one of the richest in this country. I was in 2 churches in both which were some very good paintings, I saw there the taking down from the Cross[7] and the rest in Egypt that C. Cort[8] has graved being two of the Capital pictures of Baroccio, the Priests that were singing in the Choir of one of the Churches were habited in purple with linnen and lace underneath like Cardinals. I took notice here of a manner of fishing I had not seen elsewhere, which is that a very large Net is set out with a frame made of long poles and so suspended and let

1 There is a cross in the margin here.

2 Hasdrubal Barca (brother of Hannibal), who was defeated and killed in the Second Punic War in 207 BC at the Battle of the Metaurus, near the river Metauro in Italy. It is considered one of the most significant battles in history, removing the threat of Carthage to Rome's ascendancy. See Davis, *100 Decisive Battles*, p. 39.

3 Senigallia, a port town on the Adriatic coast, named after the Senones, the Gallic tribe that settled there in the 4th century BC. At the time of Folkes's visit, it was annexed to the Papal States legation of Urbino. The river Misa is joined by the river Nevola before it enters the Adriatic Sea near Senigallia.

4 River Cesano. Three rivers are near the port, the river Misa, which runs through Senigallia, the river Cesano and the river Esino.

5 A mole is 'A massive structure, esp. of stone, serving as a pier, breakwater, or causeway'. See *OED*.

6 Pope Alexander VIII (1610–91), who was Pope from 1689 to 1691.

7 *The Entombment of Christ* (1592) by Federico Barocci, for the altarpiece in the Chiesa della Croce in Sanigallia. Information about the iconography is here: Barocci, *The Entombment of Christ*, Art Institute of Chicago, https://www.artic.edu/artworks/100036/the-entombment-of-christ [Accessed 17 Feb 2023].

8 Cornelius Cort (1533–78), engraver and draughtsman born in Hoorn or Edam, who worked in Venice and Rome, engraving at least a dozen prints for Titian. 'Cornelis Cort', *British Museum*, https://www.britishmuseum.org/collection/term/BIOG23700 [Accessed 17 Feb 2023]. Cort engraved Barocci's, *The Rest on the Return from Egypt*, in 1575. The auction catalogue of Folkes's collection showed he owned prints of Barocci's works. See *A Catalogue of the Prints and Drawings,* p. 10, item 54.

down by a pully and taken up again.[1] I dind very badly at the Post house which is out of town, and from whence the town with its walls looks very pretty. we went but a very little way from hence before we came on the sand again[2]

[f. 68v] +at a place calld l'Ulmo[3]

[f. 69r] continued either on them or at a very little distance for about a dozen miles, passing by the way the River Fiumesino[4] formerly the Aesis where it runs very broad, and on a like bridg with those over the Metaurus and Cesano. just before we passd this river we passd a very remarkable and large old Gothic Castle,[5] ~~at a Place calld Jesisino~~ it is entire tho not kept in any order, and over the gate and in other places about it was carvd a very barbarous representation of a Knight on horseback, over the gate was writ under the horse Ancona. a few miles further we began to come into the mountainous country, and went up a pretty high hill the sea always on our left, at the foot of this hill was a little village the only thing I can call such all day, of excessive poor houses, and where all the children in general go a begging. I believe I had 40 after my coach, they have a sort of begging song hereabouts relating to the Madonna the Casa Santa etc, and as we approach Loretto the road was full of Pilgrims[6] coming and going; we lie to night in a very poor lodging on the top of the just mentioned hill \at a single house +[7] ~~belonging to a village calld~~/ about 3 miles from Ancona. we have seen Ancona all day it is a remarkable scituation being a hill or little mountain advancing into the Sea we at first saw it only as a hill, but for several miles we have seen the buildings plain over the bay that the sea here makes, and which we have coasted, they seem to stand one above another so as to cover the side of the hill next us, and their seems a good deal of shipping lying before the town.[8] The Pope has lately made it a free port and granted some

[1] Folkes's description probably refers to a *trabucco* (also *trabocco* or *travocco*), a fishing shack with decks attached that are built on stilts. Nets (*trabocchetto*) are rigged on long Aleppo pine poles (antennae) extending out over the water and dropped in the paths of passing schools of fish, with winches. This allowed fishing on the Adriatic when the weather was too inclement for sea fishing. See Barone et al., *I trabocchi*, pp. 1–48.

[2] There is a cross in the margin here.

[3] re l'Ulmo, cf. p. 144, n. 5; p. 205, n. 15, is unknown other than its proximity to Ancona.

[4] River Esino. The river Esino originates from Monte Cafaggio, in the Province of Macerata, and flows north for a length of about 90 km, flowing into the Adriatic Sea near the Municipality of Falconara Marittima. Grossi and Polsinelli, 'A Sustainable Management', p. 44.

[5] Castello di Rocca Priora, built in the 12th century. The people of Jesi erected this 'Rocca di Fiumesino' to mark the territory and protect it from Anconetani expansionist threats. The disputes between the Jesi and the Anconetani lasted for centuries, but around 1356, the Rocca became part of the system of defences for Ancona. In the mid-18th century, the castle became a stately home. See 'Castello de Rocca Priora', *Museoitalia*, https://www.museionline.info/tipologie-museo/castello-di-rocca-priora [Accessed 17 Feb 2023].

[6] Reflective of the cult of Our Lady of Loreto, the pilgrims going to the Basilica della Santa Casa or Basilica of the Holy House, thought by the pious to be where the Virgin Mary had lived. For more about the history of the Holy House, see Vélez, *Miraculous Flying House*.

[7] This cross refers to the note Folkes made on f. 68v.

[8] Ancona hugs the Adriatic coast and is elbow shaped, allowing one to see the sun both rise and set over the sea. It was originally settled by the Piceni, followed in the 4th century BC by the Dori, the founders of Ancona, a natural port and important defence of the Adriatic. Ancona became a Roman colony during the Republican Era. An international port, it is now capital of the Le Marche region. See Salucci, 'Development, Recovery, and Reuse', p. 1296.

priviledges,[1] that gave a good deal of uneasiness to the venetians when I was at Venice. Our view from the mountain where I am is very pleasant being now higher than Ancona and looking down upon it.

Thursday 22. I got up early and went with Martin[2] to Ancona before the rest of the family was up. It is an exceeding odd scituation, the town stands on a Mountain projecting into the sea high and rocky, and this mountain may be lookt on as made up of 3 steep hills on the first I mean that next the country is a large Castle or Citadel,[3] with very high and good walls, facd with brick and flanking in the modern manner,[4] tho without a ditch the scituation not being well capable of it but the precipices and steepness of the rock give a notional strength, the way to the town was between this [f. 70r] castle and the sea upon which one looks down from a great height, the gate of the town by which I went in and which I take to be the only one is new and very handsome with an inscription in honour of the present Pope Clement XII.[5] as there are several others upon the priviledges and encouragements he has given to bring commerce hither. The streets of the town lie very[6] narrow the town being built all on the steep side of the hill, and I observed from one place I could see ten line of houses one above an other, from whence I enterd I took notice of a steeple in the lower part over the top of which has of some height I ~~saw~~ saw some rows of houses, the descent was so steep a coach is forced to take a compass about to get into the middle and even so very difficultly, what ever is not surrounded by sea has the same sort of defense as the castle just mentiond but much the greater part is sea and that a steep rocky coast except to the port which is defended by the Mole.[7] on the top of the 2[d] hill is another Castle, and on the very top of the 3[d] is the Cathedral Church[8] and Bishops

[1] Ancona lost its independence as an oligarchic republic in 1532, and became part of the Papal States under Pope Clement VII. Ancona was made duty free (a *porto franco*) in 1732 and, as Nardone has noted, the French saw this as an opportunity to create a commercial link between the Mediterranean provinces of southern France and the Papal State's Adriatic provinces. The French consulate also saw it as a means of damaging British maritime trade, and the creation of the Ancona free port in 1732 dealt Venetian trade a harsh blow. The Venetians, mindful of the alliance between the papacy in Ancona and France, for their part tried in every way to hamper trade in Ancona, including a maritime blockade, confiscation of goods from French ships, and indiscriminate use of quarantine legislation. The 'uneasiness' that Folkes recorded was thus probably a reflection of the commercial rivalry between the Venetians and Ancona, the Papacy and France. See Nardone, 'Diplomacy and Internal Commerce,' pp. 165, 183–4.

[2] Folkes's only son, also named Martin (1720–40). See above, p. 81, n. 4.

[3] Castello di Rocca Priora.

[4] During the era of the Roman Republic: 'the defensive system in Ancona consisted of a 2300-m long wall, and a total area of 25 ha. After the year 1000, flourishing mercantile activities brought a notable impulse to urban development and the walls grew to 4200 m in length.' Salucci, 'Development, Recovery, and Reuse', p. 1297.

[5] Cardinal Lorenzo Corsini (1652–1740) was elected Pope Clement XII. He had held the post of *Commissario del Mare* (an administrative role overseeing port trade and shipping) during his career in the papacy, and in his professional capacity determined that Ancona needed a new port facility. Salucci, 'Development, Recovery, and Reuse', p. 1297. In 1733, the Pope commissioned Luigi Vanvitelli (born Lodewijk van Wittel, 1700–73) to make a series of improvements. These included a *lazaretto* (quarantine station), an 18th-century pentagonal structure built on an artificial island that was constructed between 1733 and 1738. Vanvitelli also built the Nuovo Molo and a triumphal arch (Clementine Arch) by the new quay.

[6] There is a cross in the margin here.

[7] For 'mole', see above, p. 140, n. 5.

[8] Cathedral of San Ciriaco, built in 995 and 1015, with enlargement works between the late 12th and early 13th centuries to a Greek Cross plan. The Cathedral incorporates the remains of a 6th century AD basilica and was built upon an Italic temple to Aphrodite. The church was dedicated to St Cyriacus the Martyr, and the

pallace; I climd up to this as I may call it mostly by bad irregular stairs, from whence I had a Noble view of the Sea and the coast by which I traveld yesterday with the hills beyond. the Cathedral is a good old Gothick church tho with a sort of Cupola, but has nothing in it very remarkable, what is best is the \high/ Altar and pulpit with its support made up of several sorts of rich marble. there are in the town some good houses but[1] the buildings in general are bad and the streets very irregular narrow and ugly; the building of most note is the Sala de Mercatanti,[2] an old Gothic outside but good in its way and the inside a fine room painted and adornd by Pelegrino Tibaldi, here are shops in it as in Westminster hall; there is a fountain in the town which is their only fresh water, and I met abundance of women carrying about great pitchers of water on their heads; these pitchers were all alike and of a very elegant shape, the same must have continued for they are true antique like vases. there are here a great many jews[3] I walkt throug their quarter a long very nasty street where they are lock up by gates at each end. they are all obligd to wear a red crown to their hats.[4] The port which seems good is made partly by a natural bay, and partly by the mole a noble work of the Emperor Trajan, which runs out in the form of [f. 71r] an Arch from almost the Extremity of the town,[5] near one end of it is the

---

church houses his relics. The dome was built over the crossing of the apse and transept in the 13th century, and it is one of the few contemporary examples of a dome in Italy built in a church, rather than in a baptistery. For a complete history of the Cathedral, see Polichetti, *San Ciriaco.*

[1] There is a cross in the margin here.

[2] Merchants' Loggia, designed by architect Giovanni Pace in 1442, with a Venetian-Gothic facade by architect Giorgio da Sebenico. After a fire, the building was restored in 1558–61 by Pelligrino Tibaldi who, as Folkes indicated, also frescoed the central hall

[3] As Dorfman and Dorfman indicated, in 1555, 'Pope Paul IV began to institute anti-Jewish measures in the Papal States ... the Ancona community was reduced to a state of debility that lasted through two centuries. Any temporary improvement that occurred was prompted by economic considerations.' Dorfman and Dorfman, 'The Jewish Community of Ancona, Italy', *ANU Museum of the Jewish People*, https://www.anumuseum.org.il/jewish-community-ancona-italy [Accessed 17 Feb 2023].

[4] The requirement for most Jews to wear red hats was common in Venice and apparently in Ancona. In Venice, only Levantine Jews wore yellow head coverings. All other Jews wore red ones. The practice stemmed from the Fourth Lateran Council, convened by Pope Innocent III in Rome in 1215 in which Jews were to 'be distinguishable from the rest of the populations by the nature of their clothes'. See Ravid, 'From Yellow to Red', p. 180. The purpose was to eliminate forbidden sexual relations between Christians and Jews and regulate the status of Jewish people. My thanks to Janet Hartley for alerting me to this article.

[5] A Roman marble triumphal arch of Emperor Trajan on the old quay, erected in 114 AD. 'It commemorates the completion of the harbour at Ancona, and, besides the inscription in honour of the emperor, bears the names of his wife and his sister. The letters of the inscription were made of bronze; the letters have been lost, it is only by means of the holes left by the rivets, that the words have been read.' See Sandys, *Latin Epigraphy,* pp. 125–6. The inscription reads:

> Imp. Caesari divi Nervae f(ilio) Nervae Traiano | Optimo Aug(usto) Germanic(o) Dacico, pont(ifici) | max(imo), tribuniciae pot(estatis) XVIIII, imp(eratori) | IX, cos. VI, p(atri) p(atriae), providentissimo | principi, senatus p(opulus)q(ue) R(omanus), quod | accessum Italiae, hoc etiam addito ex pecunia sua | portu, tutiorem navigantibus reddiderit
>
> [The Senate and the People of Rome have dedicated this monument to the Emperor Caesar Nerva Traianus Optimus Augustus Germanicus Dacicus, the son of the late Nerva, Pontifex Maximus, Tribune for the 19th time, acclaimed Emperor 9 times, consul 6 times, Father of the Fatherland, and most far-sighted ruler, because he rendered the approach to Italy safer for those at sea through the construction of this harbour, which he personally paid for].
>
> On the left of the main inscription is the name of Emperor Trajan's wife:
> Plotinae | Aug(ustae) | Coniugi | Aug(usti) [To Plotina Augusta Wife of the Emperor]

Arch of Trajan all of White Marble which has kept its colour and is the most entire piece of Antiquity I have seen, it is not very wide being equal to the breadth of the Mole,[1] the arch it self is not above 20 foot on each side of which are 2 cornithian pillars, above the Cornish is a sort of attick[2] in which is the inscription very plain and legible the letters cut extremely deep but the new wall of the Mole comes so in the way there is now no good coming to this arch so as to have a direct prospect of it. The Inscription is on that side next to town but next to sea has no inscription but 4 marble columns on pedestals like the other side. It looks so white and new I could not persuade my self at a distance but it was new built till coming near I perceivd the evident flawings of time the Mole is all coverd with new walls and at the Extremity here is a circular fort or castle in which a sort of guard is kept. there is a key around the inside of the Mole. the form of this Mole and Arch is not very ill exprest on a Medal of Trajans,[3] by which it also should seem there was formerly an old building at the other end where the fort is now. over against the first Castle they have just now laid the foundation with piles of a pentagon in the water which is designd for a Lazaret and at the same time a further defense of the port.[4] the town has some appearance of business, and tho it looks poor in many respects one cant help observing it has an air of recovering it self. I enquird what I could in the little time I was there of the mischief done by the hurricane but found it much less than the report I had heard of it at venice, there being no building of any consequence blown down or even much damaged, and all that was of the shipping was only the loss of 3 vessels and about 36 persons. As I returnd to my inn I went into a little circular church which seems to me to have been an old temple at least to be built on the foundation of one it is meanly fitted up yet has a pretty effect. It is not half a mile from l'Ulmo where I lay. It has the kings a horseback over the door which is the arms of Ancona.[5] being returnd I set out from my inn about noon taking some provisions in the coach, so went on without stopping to Loretto where I came just after the Sun Set, the way mountainous all along but the road pretty good, and the mountains [f. 72r] not so steep but to be cultivated finely with vines

---

On the right of the main inscription is the name of the Emperor's sister:

Divae | Marcianae | Aug(ustae) | Sorori | Aug(usti) [To the late Marciana Augusta Sister of the Emperor]

See Sandys, *Latin Epigraphy*, pp. 125–6. (The transcription is by Sandys, the translation of the inscriptions by the editor of this volume.) Benelli indicated that the arch was a: 'civic and political symbol of the city that could provide Ancona with an imperial lineage. This new identity emerged out of Ancona's rivalry with Venice, the city's most important commercial competitor'. Benelli, 'The Arch of Trajan', p. 37. Folkes himself measured Trajan's column in Rome and analysed its carvings from a standpoint of how they utilized optical illusion, which may explain his pronounced interest in Trajan's Arch in Ancona.

[1] For 'mole', see above, p. 140, n. 5.

[2] The arch is a single bay with engaged Corinthian columns, devoid of figural decoration, save for a 'lost statuary group once on the attic'. De Grummond, 'Arch of Trajan', in *Encyclopedia of the History of Classical Archaeology*, p. 76.

[3] Folkes was a collector of medals, the auction catalogue of his coin and medal collection, showing that he owned this particular example. See *A catalogue of the ... collection of coins*, p. 23, lot 31. Folkes may have included this remark due to his reading of Addison. In his *Remarks on several parts of Italy*, Addison noted: 'The Medal which was made for Trajan in Remembrance of his Benificence [*sic*] to Ancona is very common. The Reverse has on it a Port with a Chain running across it, and betwixt them both a Boat with this Inscription, S. P. Q. R. OPTIMO PRINCIPI. S. C.' Hurd, ed., 'Remarks on several parts of Italy', *Works of Joseph Addison*, I, p. 407.

[4] A reference to Pope Clement XII's projects for the port, which included a quarantining facility or Lazaretto. See above, p. 93, n. 1.

[5] There is a cross in the margin here. The location of l'Ulmo is unknown, other than its proximity to Ancona.

Figure 30. Giovanni Battista Piranesi, Plate 28: Arch of Trajan in Ancona (Arco di Trajano in Ancona), from *Alcune vedute di Archi Trionfali ed altri monumenti inalzati da Romani parte di quali se veggono in Roma, e parte per l'Italia* (Some Views of Triumphal Arches and other monuments erected by the Romans, some of which are in Rome and some elsewhere in Italy) (*c.*1748), etching, 14.1 × 27.6 cm. Metropolitan Museum of Art, New York. Courtesy of Harris Brisbane Dick Fund, 1937. Folkes described the Arch of Trajan in Ancona extensively, noting its remarkable preservation. He was not convinced there was a good prospect of it, though Piranesi managed to create a compelling image for Grand Tourists to collect.

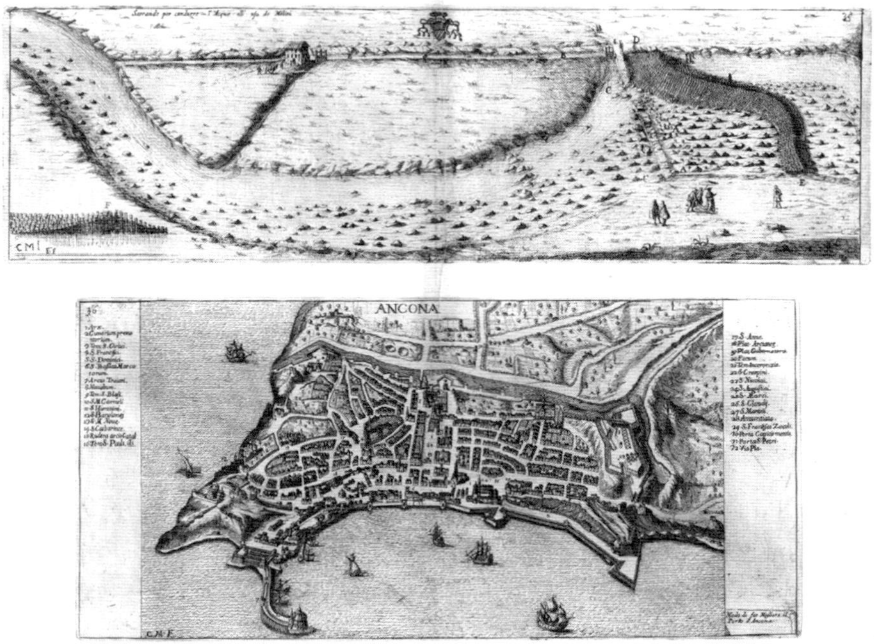

Figure 31. Cornelis Meijer, *Twee voorstellingen van een rivier en een kaart van Ancona* (Two Representations of a River and a Map of Ancona) (1683 and/or 1696), etching on paper, 14.9 × 50 cm; 41.9 × 20 cm; 39.8 × 52.5 cm. Rijksmuseum, Amsterdam. Trajan's arch is labelled No. 7, and it is possible to see the topography of the port that Folkes described from this representation. Courtesy of the Rijksmuseum, Amsterdam.

plowd tilld etc all the way. I saw no villages, but there are little houses scatterd all about at the distance of about half a mile from each other. we saw at a distance some towns looking pretty large on the tops of high hills 2 of which I was told were Recanati[1] and Fidardo.[2] we came through one town calld Camarino[3] on the top of a mountain which made a pretty show at a distance and in which have been some handsome houses on the outside but now it is very desolate and the whole town goes a begging, such quantity of beggars as I saw here and all the way to day on the road is not to be imagind any more than their importunity without seeing it. it is now no longer S^{t}Antonio but for the sake of the Santa Casa. at the foot of the hill of Loretto we[4] passd over a bridg the River

1 Recanati is a hilltop town and commune in the Marche province, 10 km from the Adriatic coast, that was well known for its market and Benedictine convent.

2 This may be Castelfidardo which sits on a rocky spur at the foot of the Conero Promontory.

3 Camerino is in the province of Macerata in the Marche, situated in the Apennines bordering Umbria. It is positioned between the valleys of the rivers Potenza and Chienti, about 64 km from Ancona. It is on the site of Camerinum (*Camertes Umbri*), whose inhabitants became allies of the Romans in 310/309 BC. The town is enclosed with ancient walls of medieval origin giving it its picturesque looks, which Folkes appreciated. Much of the old town was damaged in a 2016 earthquake.

4 There is a cross in the margin here.

Figure 32. Gianni del Bufalo, City of Ancona Coat of Arms. (CC BY-NC-SA), https://www.flickr.com/photos/bygdb/27597806065. This shows the armed knight on horseback that Folkes described.

Musone[1] which is sometimes considerable, the hill is high and the steepest we have passd, the town which seems not large has a high and strong wall so as tho it cannot be said to be strong, yet considering its scituation especially is not so defenceless as I have often heard it represented, neither is it so close upon the shore as to be so much exposd on that account. the Church which seems very large is very strongly built so as to have it self much of the look of a Castle, but it has a good front and pretty handsome cupola,[2] there is before it a [[~~xxx~~]] good square, with a well looking building on one side and the end opposite to the Church. in the middle is a very fine fountain of marble adornd with

[1] The river Musone drains eastward from its headwaters on the Marche ridge and then cuts across the Cingoli ridge to the Adriatic Sea south of Ancona. Wegmann and Pazzaglia, 'Late Quarternary', p. 139.

[2] Basilica della Santa Casa or the Basilica of the Holy House was built from 1468 onwards. As Marcia Pointon has noted: 'a number of distinguished architects worked on the Basilica and its adjacent buildings, most notably Giovanni da Maiano and Bramante (responsible for designing the elaborate marble casing constructed within the nave around the Santa Casa itself); Vanvitelli, who designed the campanile; and Guiliano da Sangallo, who was responsible for the cupola. The basilica opens onto a splendid piazza surrounded on two sides by the Palazzo Apostolico begun by Bramante and containing a fountain by Stefano Maderno.' Inside, was the 'walk-in relic of the Virgin's house, an architectural feature in its own right'. It was both relic and reliquary. The house contained a late 14th-century statue of darkened pine of the Madonna and Child, and worshippers would look through the window in which the angel of the annunciation was supposed to have travelled on its way to Mary. Pointon, 'Sacred Contagion', p. 131.

brass figures whence they have the water so high I have not yet learnt. on the steps of the Church is a noble brass statue of Sixtus the 5th, the pedestal finely adorned with figures of ffaith Justice Charity etc.[1] This is all I could see tonight. I am lodgd at the Liondoro seemingly a pretty good inn.

ffryday 23. I went the first thing I did to the Church which is very large and handsome, but what is principally remarkable is the Case of Marble over the holy house in the center of the Church, one of the finest pieces of carving that can be seen,[2] there are many prints of it,[3] so I can better refresh my memory by them than by any notes, only that it makes together a wonderfull fine appearance, we were in the holy house it self also and in the little treasury in the inner room all which are agreeable to the printed descriptions of them so [f. 73r] I set nothing down about them the jewels also gold lamps and images are stupendous, among the chief is the present of K James the 2^{ds} Queen,[4] an Angel of near 3 foot high of fine Gold^{+} holding a heart all coverd with ~~f~~ large diamonds. 19 Lamps all of Gold, an Angel of Silver holding a child all of gold the child 24 pounds weight the present of Lewis the 13 on the birth of his son,[5] but here also there are catalogues of in a book I bought of the history of the Holy house.[6] The Priest who showd it me told me the story of its being carried thro the air with great gravity, and made me take notice of several great Miracles about it, which consist in some pieces of wood work and her cloaks that came in her house which they say have never decayd. One of the Prebends of the Church sets over against its under a Canopy to watch the holy house all the time it is open, there is in it in a cupboard some of the earthenware said to to have been usd by the Virgin and our Lord which has been richly set in gold very finely chasd, there are tablets all over the church where the history of the Holy House is graved in all languages,[7] I read it in English

Sixtus V or Felice Peretti di Montalto, who was Pope from 1585 to 1590, fortified Loreto and gave it the privileges of a town. His statue, to which Folkes refers, was sculpted by Antonio Calcagni (1538–93).

The Marble Screen of Holy House was designed by Donato Bramante (1444–1514), and executed by Andrea Sansovino, Girolamo Lombardo, Bandinelli, Guglielmo della Porta, *inter alia*. 'The lavish sculptural decoration includes nine marble panels carved in a horizontal format in high relief depicting scenes from the *Life of the Virgin.*' See Bruschi, Grimaldi, and Olszewski, 'Santa Maria di Loreto', *Grove Art Online*, https://doi.org/10.1093/oao/9781884446054.013.90000369872 [Accessed 27 March 2023].

There were several prints of the marble walls of the Holy House to sell to tourists, including (inter alia) those by Hubert Vincent (1686) and those by Giovanni Battista de' Cavalieri (1567).

As Fissell has noted, James II's Queen, Mary of Modena, 'did pray to Our Lady [of Loreto] for a successful pregnancy; her mother made a pilgrimage to Loreto itself, and many sources [including Folkes's diary] also noted that a large offering was made to the statue'. Fissell, *Vernacular Bodies*, p. 232. In the right-hand margin next to the superscripted + was the marginalia: 'qu: Silver', indicating that perhaps the statue was solid silver, rather than gold as Folkes initially thought.

In 1683, in thanks for the birth of an heir, Louis XIII and Anne of Austria gave an ensemble of solid silver statues, including the Madonna and a likeness of the new Dauphin, to the Shrine of Loreto. This gift was accompanied by a present for the Pope consisting of books including an edition of the Pope's poems. Rietbergen, *Power and Religion*, p. 27.

Folkes owned the *Notizie della Santa Casa della Gran Madre di Dio Maria Vergine adorata in Loreto* (Venice: Federico Sartorj, 1728), indicated in the auction catalogue of his books. See *Catalogue of the entire and valuable library*, p. 36, lot 1306.

The first printed history of the origins of Loreto's shrine was published in Latin by Pietro di Giorgio Tolomei in 1472. 'Vicenzo Casale, the papal governor of Bologna, arranged for Tolomei's text to be translated into Italian, Slavonic, Greek, Spanish, French, German, and even Arabic so it could be made available to the full spectrum of Loreto's pilgrims. These translations, in abbreviated format, were eventually posted as explanatory displays for visitors to the Lauretan sanctuary.' Vélez, *Miraculous Flying House,* p. 51.

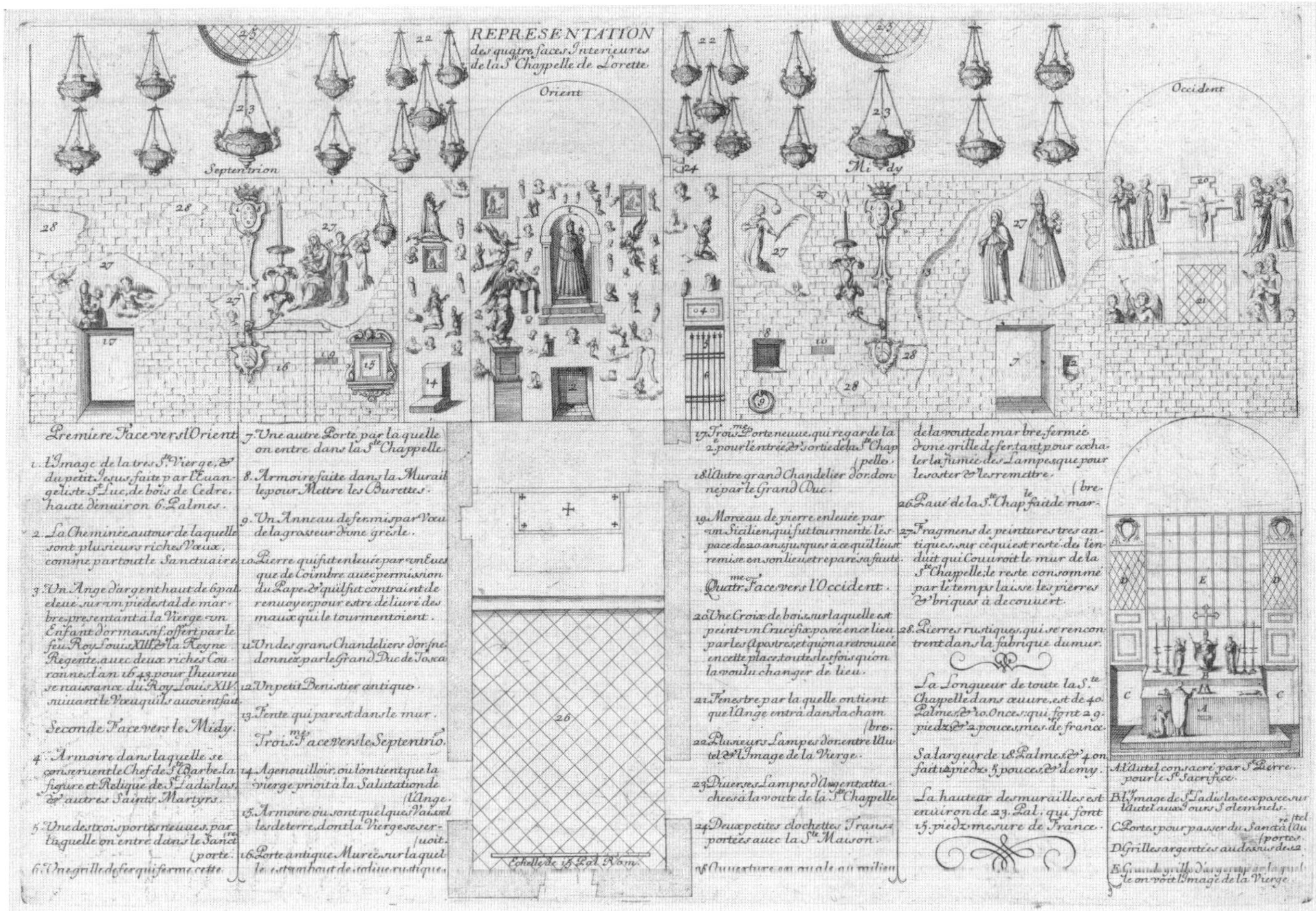

Figure 33. Stefano della Bella, *Représentation des quatre faces intérieures de la S.te Chappelle de Lorette* (1649–64), etching on paper, 23.4 × 34.5 cm. Rijksmuseum, Amsterdam. A view of the four internal walls of the Chapel of Loreto, with a floor-plan at centre. It is possible to see the statue of an angel that Louis XII donated to the shrine that Folkes mentioned in his diary. The Rijksmuseum entitled this etching 'Wanden van de Basilica della Santa Casa in Loreto'. Courtesy of the Rijksmuseum, Amsterdam.

and Scotch;[1] there are also confession chairs to the number of 24 in which the Jesuits attend, and there is wrote over them what nation they belong to, ffather Atkinson[2] the English Penitentiary came to me and offerd me all civilitys to shew me the treasure etc and accordingly he calld upon me in the afternoon with a companion not being to go out without. he went with me to the Treasure which is a large room beyond the Sacristy, filled with presses in which the riches are kept and the Quantity of Jewels exceeds all one can imagin it were endless to attempt any memorandums of them the cieling of the room is painted finely in compartments by Cav[alier] Pomarancio[3] of whom also here is a fine Crucifix at the upper end, there are also here 2 other fine pictures one the nativity of the BV.[4] by Annibal Carach[5] and the other a Madonna and Jesus by Raphael[6] finely preserved. A prebend of the Church also attends in the treasure and a petty canon shows it; I gave a cequin to the Petty canon, and another to him who showd me the holy house. There is in the church a picture said to be of Baroccio of the Annunciation,[7] but I suspect it not to be the original, the Gates of Corinthian brass are a very rich ornament of the Church they are finely carvd with stories of the bible.[8] the cupola is also said to be painted by Pomarancio it is a heaven but the house keeps one from any point of sight where to look on it. ffather Atkinson went with me afterwards to the *Epicerie* or Apothecarys office[9] [f 74r] belonging to the Holy House, where is a room filld with jars etc that were formerly painted for the Duke of Urbino by Raphael or his Scholars on his designs there are stories of the bible and also several of Virgils Aeneis etc. I was as well pleased with them as with

1 Possibly Gaelic for the benefit of Jacobite visitors.

2 Father James Atkinson (b. 1687) was from Worcestershire, the son of Mr George Atkinson and his wife, Lucy Withy. In 1703, he was admitted to the English College, Rome, and he left the College for the novitiate of Sant Andrea, Rome in February 1708. Atkinson spent approximately 30 years as English Penitentiary, partially in Loreto and partially in Rome, dying in the College of the Penitents at St Peter's on 24 March 1763. Foley, *Records of the English Province*, V, p. 847.

3 Cristoforo Roncalli (*c.* 1553–1626), an Italian Mannerist painter known as Pomarancio or Il Pomarancio after the place of his birth (Pomerance) near Volterra. 'In 1605 Roncalli won a commission for a major cycle of frescoes in the new sacristy of the basilica of S Maria at Loreto. Bands of gilt stucco divide the ceiling geometrically and frame elegant scenes from the *Life of the Virgin*, which glow with iridescent colour.' di Sorio, 'Roncalli, Cristoforo', *Grove Art Online*, https://doi.org/10.1093/gao/9781884446054.article.T073762 [Accessed 27 March 2023].

4 Blessed Virgin.

5 Annibale Carracci, *Birth of the Virgin* (1575–1600), commissioned by Cesare d'Este, Duke of Modena in 1605. The work was still unfinished by 1609, and by 1633, was hanging in the Santa Casa in Loreto. Taken by Napoleonic troops in 1797, it now is in the Louvre. For the catalogue entry, see Louvre, https://collections.louvre.fr/ark:/53355/cl010064435 [Accessed 14 July 2023].

6 A copy of Raphael's *Madonna of Loreto*, bequeathed by Girolamo Lottorio to the sanctuary of Loreto, where it hung from 1717 to 1797. The original is now in the Musée Condé in Chantilly. See Gould, 'Afterthoughts', p. 337.

7 Federico Fiori, known as Barocci, *Annunciation* (1582–84), Mvsei Vaticani. 'The work was painted by Barocci in the years 1582–1584 for the chapel of Francesco Maria II della Rovere, duke of Urbino, in the Basilica of Loreto. Transferred to Paris in 1797, it returned to Italy in 1816 and since 1820 has been part of the Vatican Art Pinacoteca.' The painting's provenance shows that Folkes may have been too suspicious in his appraisal of its authenticity. See Mvsei Vaticani, https://www.museivaticani.va/content/museivaticani/en/collezioni/musei/la-pinacoteca/sala-xi---secolo-xvi/barocci--annunciazione.html [Accessed 13 Feb 2023].

8 'The main entrance doors (1590–1610) to the church were cast in bronze by Guglielmo Lombardo and Antonio Lombardi; the doors depict scenes from the Old and New Testaments.' Bruschi et al., 'Santa Maria di Loreto', Grove Art Online, https://doi.org/10.1093/oao/9781884446054.013.90000369872 [Accessed 27 March 2023].

9 The *spezieria* of Santa Casa di Loreto, famous for its majolica jars made by the Fontana and Patanazzi workshops in Urbino. They were a 'dominant force' in the production of majolica in Urbino during the late 16th century and 17th century, primarily from 1580 to 1620. The Patanazzi decorated their pots with *istoriato* or narrative scenes, and grotesques on a white ground. See 'Patanazzi', *Grove Encyclopedia of Decorative Arts*, p. 194.

any thing I have seen here. ffat[h]er Atkinson recommended me to ffather Baker at Rome[1] who he says will show me all civilitys. at Loretto the beggars are an army and so insolent[2] it is really terrible walking ~~ab~~ a hundred yards in the street it goes beyond all human imagination, and is irksom beyond expression, the more one gives amongst em the worse and the very persons that have received are the most importunate in a minutes time for more. and so far from being thankfull that they have no sooner taken the money but they reproach you you have given no more. the whole little town is made up of little more than goldsmith shops or toy shops to sell things for presents to the Virgin.[3] The people creep round the holy house on their knees I saw several do it and one a good genteelish woman, the pilgrims also kiss the steps that go up to the church on the outside. tho the town is small ffather Atkinson tells me they reckon about 8000 souls.[4]

Saturday. 24. I left Loretto about 7 this morning, and coming down Loretto hill, passd a handsome Aquaeduct of brick connecting this and another hill, this work was made by Pope Paul the V and bring the water that serves the fountain in the Area before the Church.[5] and about 4 miles from Loretto on a very high hill is the town of Recanati a Bishoprick and walld with old walls and towers,[6] we passd through it it is a large town but now seems very poor tho it has formerly had good houses, it made a good appearance at a distance as all the towns here on the Hills do tho they surprise one a little at coming in to them.[7] About 3 miles further close by the bridg that goes over the river

[1] Father David Lewis, alias Charles Baker (*c.* 1671–*c.* 1741), a native of Monmouthshire, educated at St Omer's College and son of Richard and Mary Lewis. He was admitted to the English College in Rome on 25 October 1690, and left the college for the Jesuit noviate in February 1691 'without taking the college obligation'. He resided mostly in Rome, and in 1728 was Penitentiary at St Peter's. He was the nephew of the better-known Jesuit martyr, David Lewis, alias Charles Baker (1617–79), a victim of the Shaftesbury-Oates Plot. See Foley, *Records of the English Province*, V, p. 912, n. 35.

[2] As Vélez notes, 'the cadre of poor drifters dominated the scene at sanctuaries like Loreto to such a degree that one eighteenth-century French dictionary defined "pilgrimage" as "a mixed-up journey of devotion" undertaken by "beggars who, out of superstition idleness or debauchery, go to Our Lady of Loreto ... begging for alms all the way."' Vélez, *The Miraculous Flying House*, p. 90.

[3] Folkes's observations reflected the fact that Loreto had a particular reputation for vendors of souvenirs. As Pointon has noted, even in the early 19th century, Lady Morgan described Loreto as 'the holiest and poorest of cities', consisting almost entirely of 'little shops and vast ecclesiastical edifices: the former, the bijouteries of the Church are exclusively devoted to the sale of religious trinkets, rosaries of every quality, texture and value, from the string of wooden or glass beads, to rows of amber and other precious materials'. Pointon, 'Sacred Contagion', pp. 133–4. Lady Morgan (née Sydney Owenson) was a professional Irish traveller and travel writer, publishing a travelogue to Italy in 1821. See Badin, 'Lady Morgan in Italy', pp. 127–8.

[4] There is a cross written in the right-hand margin here.

[5] The aqueduct was built by Giovanni Fontana and Carlo Maderno (1556–1629) during the papacy of Paul V (r. 1605–21), and it stretched across the valley from Loreto to Recanati and supplied water from the Apennine Mountains to Loreto.

[6] Recanati in the province of Macerata in the Marche region, founded *c.* AD 1150. A free commune, it was allied to the Papacy during the Guelf-Ghibelline conflicts; in 1240, it was given the Cathedral of San Flaviano and made an Episcopal see in 1354. Recanati was known for its annual fair, and during the 15th and 16th centuries, it played an important role in commerce in the Adriatic. Men of letters arrived there, such as the humanist Antonio Bonfini, and jurists, such as Antonio da Cannara, and famous painters were commissioned for several works, such as Lorenzo Lotto, Guercino, Caravaggio, Sansovino, and Luigi Vanvitelli. By the time of Folkes's visit, Recanati had declined economically. The city remained independent until occupied by Napoleon's troops in 1798. See 'Cenni storici', *Recanati: Turismo e Cultura*, https://www.recanatiturismo.it/la-citta/cenni-storici [Accessed 14 July 2023].

[7] Many Grand Tourists to Italy depicted the juxtaposition between 'the grandiose remnants of antiquity and the presence and daily toils of the poor townspeople'. Syrjämaa, 'The Clash of the Picturesque', p. 204.

Potentia[1] are some roman Ruins, what is most compleat is an Amphitheater whose whole compass may be tracd and some of the Arches are very entire it is all of stone and well deserves notice, having passd the Potentia some 4 miles beyond is Macerata[2] without the gate of which we baited at a small inn and while dinner was getting I walkt into the town which is of some size but not handsome and the streets dirty, there are several churches some of which I lookt into and saw 2 almost round[3] that were pretty enough. there is a handsome gate[4] to the town and like recannati it makes a good figure at a distance, the ramparts which stand on a pretty high hill being

[f 74v] Clemens XI. Pont. Max.
Ruinorum montis Latus
Quinque Murii substratis munivit
Viamque ad Aetem Lauretanam
Negligentia temporum corruptam
Magnificentius ~~struxit~~ stravit
Curante etc Anno. 1708. Pontificatus 8°.[5]

[f. 75r] all facd with white stone of which also the town is built. without by the inn I were at are some little oratorys[6] in the manner of the Calvarys[7] in Germany these are the first of the sort I have met in Italy, but they are here also neglected. about a furlong[8] from the town is the view of one of the finest vales I have seen[9] and our Landscapes in General to-day have been good the grounds well cultivatd and planted with vines Olives, etc, we

[1] The river Potenza, whose course is 95 km. The river's source is in the territory of Fiuminata and the river flows into the Adriatic Sea at Porto Recanati. The Roman ruins to which Folkes referred, were formerly the town of Potentia, just north of the main Roman bed of the river. The territory in this area had belonged to the Picene peoples who were subsumed into the Roman Republic following Roman victories in the central Adriatic region, completed in 268 BC. The foundation of the Roman colony of Potenza occurred in 184 BC, marking a 'first phase of real urbanization'. 'According to literary sources (Livy 41.27.1, 10–13) ... the colony probably soon had, in 174 BC, a circuit wall with three arched gates, a street network with sewers, an aqueduct, a temple to Jupiter and a portico with shops enclosing the forum.' Vermeulen, et al., 'Potentia', p. 205.

[2] Like Potenza, Macerata was first a Picene city (named Ricina). The city, on a hill between the Potenza and Chienti rivers, was after Romanization called Recina and Helvia Recina. After the destruction of Heliva Recina by the Visigoths, the city was rebuilt as Macerata in the 10th and 11th centuries. A picturesque medieval-walled town, it was a commune (*libero commune*) in the 12th century and *c.*1445, became part of the Papal States under the protection of Pope Eugenius IV.

[3] One of these 'round churches', which so intrigued Folkes, was probably Santa Maria delle Vergini. The church was begun in 1554 and consecrated in 1573. Due to its plan of a large domed Greek cross, it was attributed to Donato Bramante who had produced a similar design for St Pietro Nuovo, but was actually designed by Galeazzo Alghisi da Carpi, who worked on the fortification of Loreto. See 'Chiesa di S. Maria Delle Vergini', Regione Marche Giunta Regionale: Cultura, https://www.regione.marche.it/Regione-Utile/Cultura/Catalogo-beni-culturali/RicercaCatalogoBeni/ids/65216/Chiesa-di-S-Maria-delle-Vergini [Accessed 17 March 2023].

[4] The gate, no longer extant, was apparently built 'like a triumphal arch'. Smith, *Sketch of a Tour*, II, p. 310.

[5] 'Pope Clement XI repaired the flank of a hill that had collapsed by constructing five supporting walls underneath it, and paved the road to Aetes in Laurentum, which had fallen into disrepair through long neglect, with greater splendour than ever. Supervised by etc. In the year 1708, the 8th year of his Pontificate.' This inscription is a reference to one of the building projects of Pope Clement XI (1649–1721). See below, p. 153, n. 7.

[6] Small chapel or shrine.

[7] The calvary is a series of representations, in a church or chapel, of the scenes of the Passion; it can also be a life-sized representation of the Crucifixion, on raised ground in the open.

[8] A furlong (one-eighth of an English mile) is just over 200 metres.

[9] Vale of Foligno.

see the high tops of the Appennine before us and have been upon pretty high hills our selves, yet all good ground, I saw at a little distance another good town on a hill St[a] Maria de Cassiano[1] which we left to the right coming to Macerata, we have passd to day the beds of several torrents all which have run very deep and hollow. the bed also of the Potentia is full of white stones and looks like a torrent bed, as does also the Chienti[2] upon which stand Tolentino[3] where we have taken our quarters for to night. the town is on a steep hill over looking the River and walld with towers in the old manner, it is a dull poor town now but like the rest carries some appearance of former greatness, there is a fountain in the market, the ascent to the town is steep and difficult, my inn has the appearance of a deserted palace, the rooms have parquet work cielings, but with that so miserable otherwise it is next to lying in a stable. I have seen no villages, but scatterd houses as yesterday and churches here and there and some convents. Tolentino is about 10 or 12 miles from Macerata. I came in just at dusk so could see but little I minded an old brick arch a little before I came to the gate, but whether antient or not, I had not time to examin.[4] we are here at the entrance of the Appennines.

Sunday 24 I set out just at 6 a clock and keeping on hills overlooking the small River Chienti came to the side of the town of Belfortè.[5] This town seems large \is walled/ and stands on the top of a pretty high and steep mountain, we passd between that and the River at about half way the height of the Mountain where there is a good Causewau [*sic*] made and the side of the mountain supported by 5 substantial walls, one above another for a good way, an expensive work of Pope Clement the XI[6] as appears by an inscription on marble[7] put up towards the middle of it. forwards is more way of the same sort which has been made at great expense by Paul the 5[8] and Urban the 8th [9] as Arches set up with inscriptions testifie; we crosst several hollow currents on bridges and one very large one with a noble Arch over against a little town called Pieve[10] was built in Clement XI[s] time at [f. 76r] the Expense of the town of Camerino not farr off. the country[11] here is well peopled, and I saw abundance of towns mostly on the tops or near the tops of hills, every thing seems well cultivated that is capable of it even to the top of high hills, these town[s] are all built of stone and make very pretty appearances at a distance tho when one comes

[1] Now called the town of Montecassiano, about 8 km north of Macerata.

[2] The river Chienti: is one of the main rivers in the Marche Region and it enters the Adriatic Sea south of the town of Civitanova Marche.' Crisanti and Taffetani, 'Diacronic Analysis', pp. 41–2.

[3] Tolentino is on the river Chienti, 223 metres above sea-level, and was once a fortified town of great strength.

[4] This arch is probably the quadrangular flag tower on the Ponte del Diavolo, a bridge built in 1268 and and entry point for the town of Tolentino. The flag tower has Guelph crenellation and a pointed arch door with double brickwork arched lintel. See Vivi Tolentino, 'Ponte del diavolo', https://www.vivitolentino.com/it/ponte-del-diavolo/ [Accessed 17 March 2023].

[5] Belforte del Chienti, a commune about 50 km south-east of Macerata, part of the Papal States.

[6] Pope Clement XI (1649–1721), born Giovanni Francesco Albani, who served as Pope from 1700 until his death. He was known for his condemnation of Jansenism, his architectural projects, his art art patronage and as a great benefactor to the Vatican Library. See Clough, 'The Albani Library', pp. 15–16; Heilbron, *Incomparable Monsignor*, pp. 89–93.

[7] On f. 74v, Folkes wrote down this inscription, referring to Clement XI's building project.

[8] Pope Paul V (1550–1621), born Camillo Borghese, was Pope and ruler of the Papal States from 1605 until his death.

[9] Pope Urban VIII (1568–1644), born Maffeo Vincenzo Barberini, was Pope from 1623 to his death in 1644.

[10] Pievebovigliana.

[11] There is a cross inked in the right-hand margin here.

to pass any of them one is surprised with the bad condition most of the houses are in. I coasted the River all day and passd through the little town of Valicimara[1] from whence the Mountains encreast all the rest of the day and grow rocky and consequently but little capable of cultivation however they are mostly full of trees, and wherever there is a vale it is made the most of. I dind very sorrily at a[2] very sorry little house on the place called Ponte de Trave,[3] and had by the way several very romantic prospects caused by the hollow beds of the torrents that run into the river Chienti and that of the River its self with old bridges, castles, cells chappels, etc among the rocks. soon after dinner I went through the small town of Muccia[4] but that seemd a little better within than most of the little towns hear abouts here I crosd the Ch[i]enti and traveld on causeways overlooking it on the right for some way that I crosst it again coasting it on the left to Serravalle[5] a very long village or rather town, but excessively poor seeming to consist of little but beggars, we are at the Port house but miserably quarterd and shall I believe fare very indifferently. the hills this afternoon are the highest we have seen in this country but in no ways equal to those of the Tirol, tho the eye does not carry much the difference in things so large, I observe there are frequently here habitations to the very top which is never in the very high ones of the other place, some of these have however clouds hung about em, one particularly had its head hid in one all this day. as we now tracd the Chienti to almost a kennel[6] I expect to lose ~~her~~ it early tomorrow, and my map shows the next rivers all run into the Mediterranean, whence I judg this is the highest ground between the 2 seas in this part of Italy. The journey to day was about 28 miles

Monday 26. getting out a little before 6, we immediately passd a formerly defended pass of the mountains where I supposed Serravalle takes its name, in about half a mile I came to a sort of pond whence the Ch[i]enti takes its first rise, and going up a pretty steep hill left a little town on a hill called Pupola[7] to the left hand, and came to another called Colle Fiorito[8] which we went through. [f. 77r] here the people lookt neat and tydy, and a good woman offerd us a dish of just churnd butter that lookt very clean and good a thing I had not seen of sometime; just without this town is another pond or morass on the right from whence rises immediately a little river which runs to Fuligno[9] where it joins another and later the name of Topino[10] running away for the mediterranean. so that the little town and hill of Colle ffiorito is just between 2 springs one that sends its waters to the Adriatick and the other to the Mare inferum or mediterranean, we crosst this little river twice pretty soon, going in between the mountains in an intense a cold as most I have felt; I say we seemd in vales between the mountains, but all at once the ground

1 Valcimarra, a little village near Caldarola.

2 There is a cross inked in the right-hand margin here.

3 Pontelatrave, with its 14th-century Romanesque bridge built upon Roman foundations dating from the 1st century AD. The town was an ancient staging post.

4 Muccia, a small commune about 40 km south-west of Macerata.

5 Serravalle di Chienti, about 80 km from Macerata, and, as its name implies, crossed by the river Chienti.

6 A 'kennel' was the surface drain of a street; the gutter, a later form of the word 'cannel'. See *OED*.

7 Popola is a hamlet in Foligno, Umbria, near the village of Cesi.

8 Colfiorito, formerly a Roman village called Plestia, in Umbria on the road from Macerata to the municipal seat of Foligno. The town is situated at a major road junction, which in ancient times already connected Umbria, Sabina and the Adriatic coast.

9 Foligno is 40 km south-east of Perugia, part of the Papal States until 1860 and an episcopal see.

10 The river Topino, the main tributary of the Chiascio, in the Tiber basin. The river was known by the Umbri as Tupunna and later in Latin as Tinia.

broke from us to the right and gave us a prospect into a vale so much deeper that shewd we were our selves upon a great height, this valley afforded one of the oddest views I ever saw little towns and several villages in perspective of venu [*sic*] d'oiseau,[1] the little river I have been speaking of tumbling at once from the rocks and making natural cascades[2] with the vale planted like a garden with vines and olives, we had this prospect for a good while as we coasted 2 or 3 mountains on a causeway cut about half way up and saw it as it were on all sides nothing I have seen here struck me more, our way tho by the side of a sort of precipice was very good and everywhere wide enough for two coaches to go abreast, it was mostly pavd and a great deal with as I take it is the remains of the old Roman pavement[3] we passd today several villages and well inhabited the ground everywhere made the most of even the hills where not rocky as the last we came at were almost entirely are plantd very completely. we came down at last by a pavd way from the mountains into a very large plain, and immediately found the climate, quite alterd a warm temperate sun, and passing again the river came to the town of Fuligni,[4] a very large walld town well inhabited and built, here we were at the George a very[5] large and good inn, and dind with our windows open after the excessive cold whe have felt in the morning. In this town I saw the Duomo or great Church a large handsome Church[6] with some not contemptible paintings both over the Altar and in ffresco round the church,[7] it has a Cupola and under it an Altar with a canopy supported by four gilt columns seeming a model of the high Altar of St Peters at Rome[8] – but one of the most remarkable things in this town is the altar piece painted by Raphael, in the convent of the Nuns of St Anne, which was given about 80 years since [f. 78r] by one of the house of Conti, and grandson to the person for whom Raphael painted the picture.[9] it is on board and well preservd, very graceful and well colourd tho somewhat of the stiffness he had in coming from the school of Perugin.[10] it represents above the Madonna and Bambino in the clouds and under them

1 Bird's-eye view of the picturesque Umbrian Valley.

2 There are several cascades in the area, such as the Menotre Cascade between Pale and Belfiore.

3 Via Flaminia, which lead from Rome over the Apennines to Rimini.

4 Foligno.

5 There is a cross inked here in the right-hand margin.

6 The Basilica Cattedrale di San Feliciano or the Duomo di Foligno, built in the first half of the 12th century on the site where its patron saint was said to be buried. Its facade is elegant, blending Romanesque and Gothic features. The dome is from the 16th century and was built by Giuliano di Baccio d'Agnolo.

7 The Chapel of the Sacrament features two large frescoes by Vespasiano Strada and Baldassare Croce featuring the life of San Feliciano. There is also, in the apse, a large fresco by Francesco Mancini of the medieval mystic Angela of Foligno, dating from 1719.

8 This is a baldachin over the main altar, commissioned by the family of Salvi Roscioli, and built by Antonio Calcioni; it is a reproduction of Bernini's baldacchino in St Peter's Basilica in Rome.

9 The *Madonna of Foligno*, painted on panel by Raphael, was executed *c.*1511–12. It was commissioned by the humanist Sigismondo de' Conti, the chamberlain to Pope Julius II. Originally on the high altar of Santa Maria in Aracoeli in Rome, where Sigismondo was buried, the painting was moved by his descendant, Anne Conti, to the monastery of St Anne in Foligno in 1565. Here the work remained for two centuries until the Napoleonic invasions of 1797, when it was moved to Paris following the terms of the Treaty of Tolentino. Whilst in Paris the painting was transferred from a panel to a canvas. The painting was returned to the Papal States after the Battle of Waterloo and now hangs in the Pinacoteca Vaticana of the Vatican Museum. Licht, 'Raphael', p. 18. See also 'Raffaello Sanzio, Madonna of Foligno', Mvsei Vaticani, https://www.museivaticani.va/content/museivaticani-mobile/en/collezioni/musei/la-pinacoteca/sala-viii-secolo-xvi/raffaello-sanzio--madonna-di-foligno.html [Accessed 21 March 2023].

10 Raphael was a pupil of Perugino.

a naked boy with wings holding a tablet as if designd for an inscription on one side is S[t] John the Baptist, and S[t] Antonio (as I take it) on the other S[t] Jerom and the person for whom the picture was painted kneeling. there is a Landscape in which is a town and what I at first took for a rainbow, but found upon more inspection to be more like balls of Fire carrying a war like bombs and somewhat like lig[ht]ning seems to break from the cloud on which the Virgin is.[1] I took here particular observations not recollecting to have seen the print of his design. it is a noble picture in all respects. as the roads were very bad this morning and my wife being frightend with the precipices we lost time in walking I did not get out again of Fuligno till past 3 a clock, and went through the same plain quite to Spoleto, throug[h] plantations and vines one of the finest spots of ground in the world having the mountains on my left, and I observd on the way the large town of Trevi on a mountain and all down one side of it making a noble appearance.[2] I saw also part of my way to the right the river Clitumnus[3] so famous in the old Poets,[4] for the white cattel that breed seems now to be got all over Italy and I have scarce for these 4 or 5 days seen an ox of any other colour.[5] Spoleto is a very large town and stands just at the entrance of the mountains again we passd over the bed of a prodigious torrent to the gates of the City and which washes the wall;[6] I got into the Port house, the Agnuolo doro,[7] a little after 8 o'clock a tolerable good inn as they go here. our way to day abt 35 miles

Tuesday. 27. The City of Spoleto which is very large and pretty well inhabited[8] stands on a small mountain and is surrounded with very good and high walls after the old manner; there is an ancient Castle Gothic I mean on the highest part of the hill being a craggy rock.[9] between this and a much higher mountain is a deep valley over which water is still brought to the city by an Aquaeduct built

[1] Sigismondo de'Conti commissioned the piece out of gratitude to the Virgin for having saved his house in Foligno, which had been struck by lightning. Raphael portrayed the fireball in the landscape in the background of the piece.

[2] Trevi is an ancient town and commune on the slopes of Monte Serrano, 10 km south-south-east of Foligno and 20 km north of Spoleto.

[3] River Clitunno, its supposedly deified waters celebrated in antiquity as the Clitumnus by authors such as Pliny the Younger. Clitumnus was a son of Oceanus and Tethys and the god of the river. Pliny the Younger wrote: 'Close by is a holy temple of great antiquity in which is a standing image of the god Clitumnus himself clad in a magistrate's bordered robe; the written oracles lying there prove the presence and prophetic powers of his divinity. All round are a number of small shrines, each containing its god and having its own name and cult, and some of them also their own springs, for as well as the parent stream there are smaller ones which have separate sources but afterwards join the river.' Pliny the Younger, *Letters*, 8.8. 5–6, pp. 24–5.

[4] It is evident Folkes was assiduously reading Addison for his travels, as Addison describes in detail the Clitumnus and the Chianina cattle of the regions, providing quotations from the relevant classical authors. See Hurd, ed., 'Remarks on several parts of Italy', *Works of Joseph Addison*, I, p. 410.

[5] Chianina cattle raised in Tuscany, Umbria, and Lazio since antiquity, first as draft animals, and now for lean beef production. It was thought that the cattle turned white by drinking the waters of the Clitumnus.

[6] There is a large and deep ravine outside Spoleto spanned by the 240 metres long Ponte delle Torri, an aqueduct built by the Romans, then rebuilt in the 12th century to provide the city with a water supply. The aqueduct serves as a bridge, uniting the town with Monte Luco.

[7] The inn of the golden lamb.

[8] There is a cross inked in the right-hand margin here.

[9] Each end of the aqueduct of Spoleto has a fortress, the Rocca Albornoziana on the city side, and the Fortilizio dei Mulini on the other.

f. 78v]

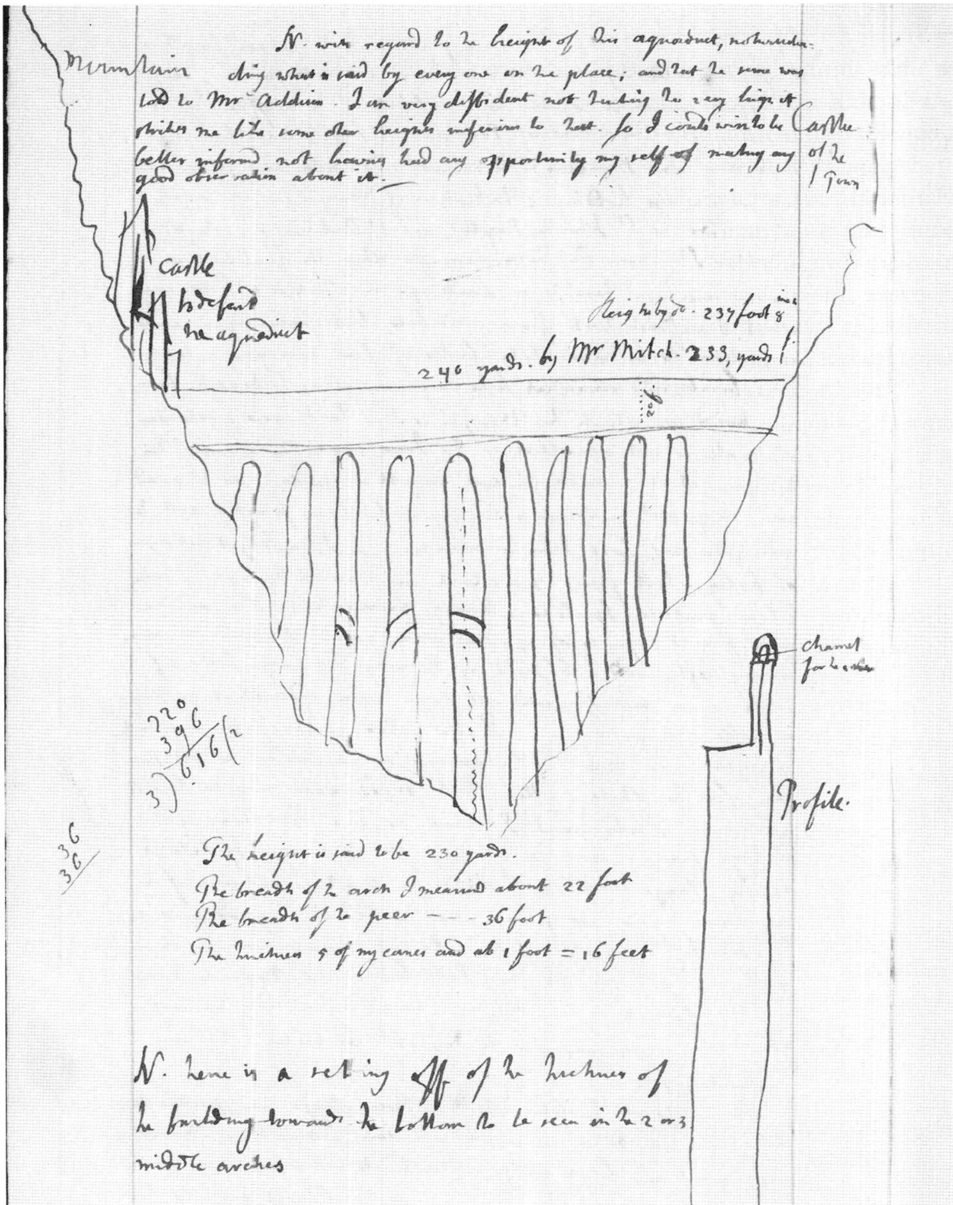

Figure 34. Folkes's drawing of Aqueduct of Spoleto, f. 78v, Bodleian Library, University of Oxford. Above it, Folkes wrote: 'N. with regard to the height of this aquaeduct, notwithstanding what is said by every one on the place; and that the same was told to Mr Addison. I am very diffident not thinking tho very high it strikes me like some other heights inferior to that. So I could wish to be better informd not having had any opportunity my self ~~of~~ making any good observation about it.' And below: 'Mountain | Castle | to defend the acquaduct heights by ad. 237 foot 8 \inch/| 240 yards by Mr Mitch 233 yards | 36 | 220 | 396] | 3)616/2 | The height is said to be 230 yards | The breadth of the arch I measurd about 22 foot | The breadth of the peer - - - 36 foot | The thickness 5 of my canes and ab 1 foot = 16 feet | Channel for the water| Profile | N. there is a setting off of the thickness of the building towards the bottom to be seen in the 2 or 3 middle arches'. Courtesy of the Bodleian Library, University of Oxford.

[f. 79r] built in the time of the latter empire, which is a work indeed very remarkable, it is of small stones and looks as of pale brick at a very small distance, by a gross pacing of it I take it to be 240 yards long on the top, the water runs in a channel on the top of all that has been formerly coverd but is now mostly open, about 20 or 25 foot as near as I could guess from the top is a bridg for people to pass over under which are ten Arches. I cannot be satisfied if they were all exactly equal but one I measured was about 22 foot wide and the peer[1] between that and the next was 36 foot, the thickness of the whole edifice was about 16 foot, and they say the height in the deepest place (for the arches are lower at each end and the mountain rising) is 230 yards, this I had no opportunity of examining but by my eye comparing at a distance the height with the breadth I believe not to be far from the truth, the deepest arch is the 5 from the left hand, that is the 6th from the city. and that and 2 more have had a half arch turnd about the mid way but which I take to be more modern than the rest of the work, it is surprizing to see how strait and upright the great edifice stands tho of no greater a thickness and of such antiquity, however, the plane of the whole is in no sort perfect but twisted whether it was always so I dare not determine but encline to think it was, tho I am not able to see how much it is out of the plain.[2] it still serves the use for which it was built and carries a channel of water which runs afterwards in a stone arch round part of the rock in which the castle is built and then delivers it self into a bason from the mouth of a vast antique head in the court before the castle, which has been within not very many ages repaird and adornd by the city, here the rest of it goes underground again, and is in channels carried the several cisterns and Condutts \to/ ~~p~~ different parts of the town, all which is well provided with this water; here are also in this town several good churches I was in 3 viz The Cathedral, a fine large Church with a Cupola but dark,[3] the Jesuits a pretty church,[4] but the best was that of St Philippo Neri,[5] a fine building and well adornd within, it has a dome also and the altars are [f. 80r] handsomely adorned; I went out of town just at noon, continued amongst the mountains almost all day, I think these are not the highest but they are the wildest part of the Apennines I have seen; I should have remarkt that the torrent I spoke of coming to Spoleto is the same as has its course through the arches of the Aquaeduct, we afterwards went a great way by the side of it. we passd also very near the top of one very high mountain our way difficult and uneasie prodigious precipices, (tho our way was sufficiently broad) and the road very rugged sometimes through loose stones and others over solid rock but grown uneven throug[h] time with deep ruts in almost marble ways, a great part of our Mountains

[1] Pier.

[2] Plane.

[3] Cathedral of Santa Maria Assunta, erected by Duke Theodelapius in 617, but restored, featuring a 13th-century facade. The choir features frescos by Fra Filippo Lippi (*c.*1406–59), completed after his death by Fra Diamante in 1470, and Lippi's tomb is at the left of the choir.

[4] Santa Maria della Piaggia (constructed 1594–1605).

[5] The construction of San Filippo Neri, based on a project by the Spoleto architect Loreto Scelli, began in 1640, but the building was consecrated only in 1724. The facade has stylistic characteristics typical of Roman churches of the early 17th century, and the interior has three naves with chapels housing 17th- and 18th-century altars; the transept is covered by a dome. The neoclassical aesthetic would have appealed to Folkes immensely. See 'Chiesa di San Filippo Neri', Comune di Spoleto, http://www.comune.spoleto.pg.it/turismoecultura/la-citta/chiese/chiesa-di-san-filippo-neri/ [Accessed 22 March 2023].

were absolute rock and showing little or no earth at all, in this high mountain, which I think is called Monte di Somma[1] was formally a fortifyd pass, this was upon the whole the most difficult way we have had, but at the same time the scenes very odd and entertaining many of the hills with hermitages upon them particularly that where the water comes to Spoleto; immense numbers of Olivetrees some seeming to stand on solid rock that one would wonder how they took any root, as in travelling mountains we generally follow the course of some river so we did today of a vast torrent that is the bed of it for it had no water tho it has raind all the last night and the greatest part of this day; we came out of the mountains into a fine vale a little before we got to Terni[2] where we arrived a little after Sun set, the way 20 miles. Terni is walld with an old wall and has been formally very large for there are great ffields and ruins within, yet they tell me it now contains about 8000 Souls.[3] [f. 80v] the way on the side of the mountain, is all almost in solid stone. some of it marble or what is very near it in hardness, it is not only narrow but very steep; and excessively rugged and uneven. I take it upon the whole to be the worst peice of way I have seen considering especially its length. The views are however extremely amusing and delight-full. one Sig^{r} Castelli[4] has a very pretty country house on the side of the river just over against Papino,[5] with Gardens and plantation [[~~xx~~]] very pretty as I lookt down upon them from the rocks.

Wednesday 28. We have been well lodgd at the Port house[6] the town of Terni is large but no ways handsome it stands in a plain and has been formally much more considerable than it now is, I was in the Cathedral[7] and several other of the churches but met with

[1] Monte Somma (1,132 metres).

[2] Terni – from the Roman name Interamna Nahars – was founded in 672 BC. It contains the remains of an amphitheatre from the age of Tiberius in the grounds of the episcopal palace, as well as Roman inscriptions in the Palazzo Publico. It is near the confluence of the Nera and the Serra rivers, the ancient city encircled by the rivers. See 'Roman Terni', Vivi Terni: Portale Turistico del Comune di Terni, https://turismo.comune.terni.it/en/content/roman-terni [Accessed 22 March 2023]. The Roman wall was replaced by a medieval wall with towers, buttresses and bastions, and in correspondence to the entrance gates, a bridge passed over the Nera and the Serra rivers that flowed into the town. See 'Medieval Terni', Vivi Terni: Portale Turistico del Comune di Terni, https://turismo.comune.terni.it/en/things-to-do/roman-terni [Accessed 26 March 2024]; See also 'Terni', *Princeton Encyclopedia of Classical Sites*, p. 411.

[3] Folkes is very roughly correct. In 1736, contemporary to when Folkes visited, the population of Terni was 6,486. 'Terni', *Enciclopedia Italiana*, https://www.treccani.it/enciclopedia/terni_%28Enciclopedia-Italiana%29 [Accessed 22 March 2023].

[4] The Castelli were a noble family of Terni, descended from the ancient Germanic counts of the Umbrian city. In addition to holding the lordship of Terni, which they governed with the title of count, they were counts of Melace and lords of Rocca Accarina (the contemporary Collestatte). Among the best-known members are Sant'Anastasio (co-patron of Terni together with Saints Procolo and Valentino) and Pope Celestine II, that is, Guido Guelfuccio di Castello. See Crollalanza, *Dizionario storico-blasonico delle famiglie nobili*, II, p. 256.

[5] Folkes was referring to the village of Papigno, a hamlet of medieval origin a short distance from the cascades and Terni.

[6] There is a cross in the margin here

[7] Cathedral of Santa Maria Assunta, which was, according to local legend, built on the foundations of a pagan temple in the 2nd century by the first Bishop of Terni, St Peregrine. Few elements survive from the 12th-century Romanesque cathedral, because modernization works were undertaken in the 16th and 17th centuries based on a project by Sebastiano Flori, a pupil of Giorgio Vasari. See 'La Cattedrale di Terni', Vivi Terni: Portale Turistico del Comune di Terni, https://turismo.comune.terni.it/da-vedere/cattedrale-di-santa-maria-assunta-duomo [Accessed 22 March 2022].

nothing remarkable in them, unless it be a small oval church with a dome on the Piazza,[1] which is on the foundation or at least in imitation of an old temple, it is in no ways well adorned yet has a genteel look, the frontispiece is begun but has never been finisht on the inside there are 3 altars on each side besides, the high altar and entrance which are on the extremitys, and between each are 2 Ionic Columns; before the church lies in the street an antient [f. 81r] Roman Altar with some figures on it in bass relief not fine but in a good taste; in the midst of the Piazza is a little obelisk of redish marble supported on an artificial rock on which have been 2 figures well enough done, it has been a fountain for is now decayd; the Mountains come very near the town on one side and at the foot and near it are vast numbers of little country houses of the Citadini that made a very delightfull appearance, I have indeed elsewhere in my journey seen several of these intermixed with convents and hermitages, that give a life to the Landscapes.[2] I got horses this morning to go to the Cascade in the neighbourhood,[3] most part of the way is up the side of a very high and steep mountain, seeing for the most the river Nar[4] by a precipice to the left, I should remark I passd it on a bridg as soon as I was out of the town, the vale before I came to the mountain was all vines, figtrees almonds etc and in the hill it self, infinite numbers of Olives, some way up the hill in a little walld town on a very steep rock calld Papino[5] a deep torrent or its bed markes a prodigious natural ditch, and the other side the rock is steep to the river, this river it self runs all the way in the tumbling manner of a torrent and in many places might be said to make sort of cascades if one were not in the way to see a much greater. from Papino we ascended continually by a narrow path to a vast height on the side of the mountain, I could make no estimate of the height, but I made \pretty large/ stones be thrown down, but which did witness reach the bottom yet I found several of these above nine seconds in falling, which should give a height of nearest 1300 feet. having ascended this height we turnd to the right and I immediately to my seeming found my self in a fine vale surrounded by mountains, coverd with fine green vines etc here I

[1] Santa Maria del Carmine, with its severely neoclassical aesthetic which appealed to Folkes. The church's construction was due to the presence of an image of the Madonna painted on the external wall of the Roman amphitheatre, which became an object of devotion. The works ended only in 1783 with the completion of the elegant brick facade, and the venerated image was detached from the wall and placed on the high altar. The church is now deconsecrated and serves as an auditorium. See 'Santa Maria Del Carmine', Vivi Terni: Portale Turistico del Comune di Terni, https://turismo.comune.terni.it/da-vedere/chiesa-del-carmine [Accessed 22 March 2022].

[2] The Valnerina valley with the river Nera running through it has long been a site for religious communities. In the 4th and 5th centuries the valley was populated by Hermetic movements, such as the first Syriac anchorites who fled from persecution, inhabiting isolated gorges with a water supply (Val Castoriana, Val di Narco, Val di Noce, Scentelle, Romita and Monteluco). In the 6th century, hermitages grew into monastic organizations, thanks to the followers of Benedetto da Norcia, the Benedictines. 'Cascata delle Marmore and Valnerina', UNESCO, https://whc.unesco.org/en/tentativelists/2031 [Accessed 23 March 2023].

[3] The Cascata delle Marmore is 165 metres high and an artificial construction, built in 271 BC by the Roman Consul [Manius] Curius Dentatus for flood control in the Sabine valley. These public hydraulic works directed the waters of the river Velino into the river Nera, away from the plains of Rieti. The Cascata was not only praised by Ancient Romans, but celebrated by Grand Tourists. 'Cascata delle Marmore and Valnerina', UNESCO, https://whc.unesco.org/en/tentativelists/2031 [Accessed 23 March 2023]. Addison remarked 'I think there is something more astonishing in this cascade than in all the water-works of Versailles'. Hurd, ed., 'Remarks on several parts of Italy', *Works of Joseph Addison*, I, p. 412. The waters still provide hydroelectric power.

[4] 'Nar' is Latin for the Nera.

[5] The hamlet of Papigno.

had the first view of the Cascade like a large smoke rising up a considerable height into the air.[1] I presently after saw somewhat of the water but lost it again and came upon it from behind for above a furlong, before I came at it I found of fine dew like very small rain in the air, and saw the Sun being behind me about nine in the morning a good rainbow but which imagination referred to the sky, when I got to the place I climbed down part of a rock that stood forward so to have a view of the fall is self, the water has 2 lesser ffalls just above the great one where it precipitates itself directly down a perpendicular rock; the height makes the body of water that falls

[f. 81v] The River that forms the Cascade is calld Velino.[2]

[f. 82r] look considerably narrower than it really is by the best information I could get and circumstances (for there is no coming, at least for one who only visits it by the by to take any exact measure of any thing) I judg it to be in breadth better than 40 foot I mean without 2 lesser cascades that they tell me are farr more considerable when the waters are high and sometimes almost joyn with the great one, this fall is with vast rapidity down among rocks below but the bottom is not to be seen looking like a cloud or thick mist that stops the sight and made by the broken parts of the water, this mist which I observd like a smoke at a distance, rises above as high again on the fall in some degree and in that and proper ~~cir~~ scituations [*sic*] are rainbows constantly seen, when I got in a due place I saw part both of the primary and secondary very distinctly,[3] and some of my servants who went there after I came away and saw them yet much finer the day proving very fine, and the sun being then placd to a better advantage for the Cascade. It is upon the whole \one of/ the finest things I have ever seen and the most grand; I attempted to discover the height of the fall in some degree by throwing down large stones, and upon several repeated tryals by my second watch, I found them to be better than 5 seconds and coming down whence the height should not be less than 400 feet, the water that falls runs immediately into the Nar, which seemd to have flowed pretty quickly till it receives this addition, but after that runs boyling with a White froth for above a mile; I got up above the cascade to see the river that forms it and found it to have a deep channel and to run pretty rapidly I should think about 60 foot over, or perhaps more it is really a very good river, and more considerable when the rains have swelld it, as it is now they tell me the lowest that has been known in memory; when the snows melt there is besides another \or 2/ that accompany it at some distance on either side and make the side cascades here unite with the great one. there was this morning a great [f. 83r] deal of Ice in the Mountain, and some high ones at a little distance were in the upper parts quite coverd with snow which they told me fell yesterday. there is all about the Cascade a beautiful verdure coming from the continual watering by that fine dew it throws up. these are the most material remarks I was able to make on a thing which would well deserve to be seen often and attentively to have a thorough idea of it; this however will leave me to refresh a little the impression one transitory view made ~~of~~ upon me. I got back between 12 and

[1] Addison made a similar observation about the mist of the falls. 'It is impossible to see the bottom on which it breaks, for the thickness of the mist that rises from it, which looks at a distance like clouds of smoke ascending from some vast furnace, and distils in perpetual rains on all the places that lie near it.' Hurd, ed., 'Remarks on several parts of Italy'. *Works of Joseph Addison*, I, p. 412.

[2] The river Velino joins the Nera and forms the Cascata delle Marmore.

[3] Folkes's interest in optics was ever present.

one, and lookt about Terni, I found nothing more there worth particular notice there ly about fragments of pillars etc but I could hear of no antiquity remaining of any consequence in the town.[1]

Thursday. 29. I was on the Coach before 5 and came by pretty good way through a pleasant Vale to Narni[2] one part, here we again enterd the mountains, the town itself \is walld and/ standing on one very uneven and rocky, especially on the outside next the Narr. the town is pretty large but for the most part very meanly built, I walkt on the outside the town next the river to see the remains of the old brigd [*sic*] built of Marble by Augustus[3] a very noble ruin,[4] it crosses the river and is without mortar. there is only one arch standing and that is nearer the town than the river, there is a rock fills up part of it so that I could not measure it, but I judg it at least 80 foot wide and between 2 and 3 times so high for it is formd more like the gate of a city, with uprights joins capt with a semicircle the next arch which is fallen down, must have been near as large again by the joins and beginnings of the Arch that remain, whether these have besides been one arch or 2 is doubtfull if 2 they were less than either of these, if one yet sensibly larger and this I take to have been the case, by the trimmings of the arch at each extremity, but there is a very substantial peer between, this I suspect to have been built after to hold up this immense arch upon some imperfection or giving way that was found in it. Wright has well represented this antiquity which is as noble a one as most I have seen,[5] from hence our road still lay by very high and steep mountains by the side of a very deep vale where

[1] Folkes clearly was unaware of Terni's archaeological riches, including the Fausto Amphitheatre.

[2] Narni, originally the Roman Narnia, overhangs a gorge of the river Nera in the province of Terni.

[3] The Ponte d'Augusto or Bridge of Augustus. As Bertolesi, et al., have noted: 'in the Augustan age the bridge started to be considered as a monument, a tangible expression of unity and power of the Roman Empire. From this period is the Bridge over the Nera river at Narni on the Via Flaminia ... The bridge was probably built in 27 BC during the seventh consulate of Augustus, during a restoration of the Via Flaminia.' Bertolesi, et al., 'Augustus Bridge', pp. 717–18, 721. The bridge once had four arches, 'the second arch, 32 m in span, ranking as one of the largest in Roman bridge building'. 'Narnia', *Princeton Encyclopedia of Classical Sites*, p. 608. But it is now a ruin having collapsed in the 11th century, with only one arch remaining; the final collapse of the second arch's pier was in 1885. The bridge has an outer coating made of masonry blocks of travertine with a filling of Roman concrete.

[4] Addison wrote of the bridge: 'From the Nar our next town on the road receives the name of Narni. I saw here abouts nothing remarkable except Augustus' bridge, that stands half a mile from the town, and is one of the stateliest ruins in Italy. It has no cement, and looks as firm as one entire stone. There is an arch of it unbroken, the broadest that I have ever seen, though, by reason of its great height, it does not appear so. The middle one was still much broader.' Hurd, ed., 'Remarks on several parts of Italy', *Works of Joseph Addison*, I, p. 414.

[5] Folkes is, again, referring to Wright. Wright wrote that as they came out of Narni: 'we went out of our way a little further on, to see the Remains of what is usually call'd *Augustus's Bridge*. Writers differ in their Opinion of it; some will have it to have been a Bridge, others an Aqueduct; and possibly it might have been both; as the *Pont du Garde* in *Languedoc*, I have been told, is. Certain it is, that, if we may judge by the present Condition of the river, the Arches are much higher than what had been necessary to a Fabrick that was intended as no other than a Bridge over it; for there is now a Bridge, on which we stood to view, and where I took a Sketch of those Ruins: The Arches of the modern Bridge are by many Degrees lower than those of the antique ones, and yet sufficient for any Height of Water ... the number of Arches may have been four. It is all built of Marble: the Pieces are very large, and join'd without any Cement, that we could discover; as several other antique Buildings are. I have been the more particular in my Account of this Piece of Antiquity, because it is called by a celebrated Writer [Addison] one of the stateliest Ruins in Italy'. Wright, *Some Observations*, I, pp. 130–32.

Figure 35. Illustration of the Ponte d'Augusto in Edward Wright, *Some Observations Made in Travelling through France, Italy, &c. in the years 1720, 1721 and 1722*, 2 vols, London, 1730, I, between pp. 130 and 131. Wellcome Images, London.

[f. 84r] we saw the nar running vehemently among rocks for some miles, upon this causeway as we were walking to avoid the uneasiness of the precipice in the coach, I had the bad luck to lose my dutch Dog,[1] of which I have unable to hear no more tho I sent two servants back and offerd a very great reward, as it has really given me far more uneasiness than I think such a sort of thing should but I have that weakness and am really fretted and vext about it beyond measure. when we left the nar we turnd to the left hand and going still among the hills at last discovered the Tyber sometime after he had receivd the Nar he appeared immediately a pretty good river and carrying boats, I saw at some distance on the other side, the city of Orta[2] on the summit of a very high hill and not long after came to Otricoli[3] a little walld town where we dined, here stood formerly the Castle of Ochiricolum and there yet stand considerable remains of a Gothic Castle[4] that has supplyd the place of the antient one and is now it self converted into tenements; there are in all these towns fragments of Antiquity remaining I took notice of an Altar lying in the street a square stone with sacrificing instruments on the sides and an inscription ~~only~~ on the front only considerable as containing the name of the old city.

[1] Although it is thought to have had its origins in China: 'known in its earliest days in Britain as a "Dutch mastiff" or "Dutch dog," the pug retained its Dutchness throughout the long eighteenth century.' Howard-Smith, 'China Pugs', p. 67.

[2] Folkes meant Orte, which is situated in the valley of the Tiber on a high cliff. It originally was Etruscan and called Hurta, then under the Romans called Horta. After being governed by a series of empires (Ostrogoth, Byzantine, and Lombard) it was a medieval free state or commune with a residential bishop governed by an elected magistrature, before subsequently becoming part of the Papal States, its status when Folkes observed the settlement at a distance.

[3] Anciently named Ocriculum, the settlement was the: 'southernmost town of Umbria, on the left bank of the Tiber just N[orth] of the point where the Via Flaminia crosses'. 'Ocriculum', *Princeton Encyclopedia of Classical Sites,* p. 1014. The Roman site of the town was excavated in 1775 under the patronage of Pope Pius VI and extensive discoveries were made (baths, amphitheatre, basilica). Several finds are now in the Vatican, including the Otricoli Zeus, a head of Claudius, and a mosaic pavement of the *Sala della Rotonda*. See 'Ocriculum', *1911 Encyclopaedia Britannica*, https://penelope.uchicago.edu/Thayer/E/Gazetteer/Places/Europe/Italy/Umbria/Terni/Otricoli/Otricoli/Britannica_1911*.html [Accessed 30 Dec 2023]; See also Hay et al., *Ocriculum* for a complete archaeological survey of the site.

[4] Probably the fortifications in Poggio di Otricoli on the western flank of Mount San Pancrazio.

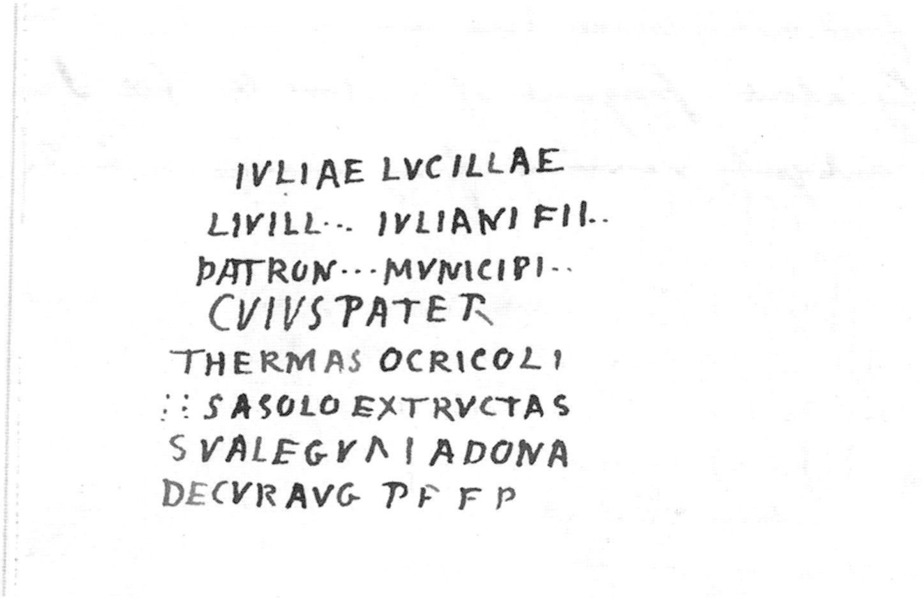

Figure 36. Folkes's notes about an inscription of an altar in Otricoli on f. 83v, Bodleian Library. Courtesy of the Bodleian Library, University of Oxford.

[f. 83v] IVLIAE LVCILLAE
LIVILL . . .IVLIANI FII . .
PATRON . . . MVNICIPI . . .
CVIVS PATER
THERMAS OCRICOLI
SASOLO EXTRVCTAS
SVALEGVΛIADONA
DECVRAVG PFFP[1]

[f. 84r *continued*] I walked from hence down to the side of the Tyber and at a miles distance are very considerable ruins of the ancient City, I had time only to take a very

[1] The inscription which Folkes notes on f. 83v refers to the 2nd-century Roman baths of Otricoli. The baths were built by the father of the lady to whom the statue was dedicated. The present location of the fragments of the altar is the crypt of the Collegiata di Santa Maria Assunta in Otricoli. For documentation about the altar, see EAGLE: The Electronic Archive of Greek and Latin Epigraphy, EDR177571, http://www.edr-edr.it/edr_programmi/res_complex_comune.php?visua=si&id_nr=EDR177571 [Accessed 24 March 2022]. My thanks to Dr Sophie Hay and Professor Andrew Wallace-Hadrill for this information and translation.

[Iuliae Luciliae,] | [L(uci) Iuli Iu]lian[i fil(iae),] | [patro]ni mu[nicipi(i),] | [cui]us pa[ter] | [th]ermas [Ocricola]= | [nis] a solo extructas | [sua pecunia dona]=[vit] | [Decur(iones), Aug(ustales), plebs]. | L(ocus) d(atus) d(ecreto).

[To Julia Lucilla, daughter of Lucius Iulius Iulianus, patron of the municipium, whose father donated with his own money the Baths of Ocriculum built from the ground up. Burial place granted by decree of the decurions (town council), Augustales and plebs].

Keysler also made note of the same inscription. Keysler, 'Letter LXI. Journey from Rome to Loretto', *Travels Through Germany*, III, p. 5.

cursory view but in that I observd a good part of an Amphitheater, an Aquaeduct and a round temple, beside several others that would well deserve more attention hath I have time or then disposition to give them, we continued our journey on the hills but grown more practicable from than before having the Tyber mostly in sight serpenting to our right, we passd the river Campano a very deep hollow torrent on a stone bridg and soon after the Tyber itself on a very noble one of 4 large Arches begun by Sixtus the V[1] as appears by an inscri[p]tion and finisht by his successor,[2] there is also on the same an Inscription to Urban VIII who repaired the banks of the river hereabouts and reducd it to its antient Channel,[3] we soon after went through a little walld town called Borghetto[4] and in about 5 miles more came to Citta Castellana[5] where we crossd the river Triglia[6] an excessive deep rapid hollow torrent and ascended a very steep hill to the town, where we lay in the worst quarters we have yet had. our journey to day 35 miles, pretty long.

[1] The Ponte Felice [Pons Felix], a bridge built by Pope Sixtus V (Felix Peretti), Pope 1585–90. The bridge is on the Via Flaminia, which crosses the Tiber between Borghetto near Magliano Sabina, and Otricoli. The bridge was built by Domenico Fontana, a noted architect and engineer of late 16th-century Rome. It was damaged in the Second World War and rebuilt on its original foundations.

[2] The successor of Pope Sixtus V was Pope Clement VIII. The inscription on the bridge says:
Sixtus V. Pont. Max. | Ut commeantes trajectionis molestiâ | Et vectigali sublevaret, | Pontem inchoavit | Ann. Sal. MDLXXXIX. Pontif. Sui V.
[For relieving travellers of the trouble and expense of a ferry crossing, his holiness Sixtus V began this bridge in the year 1589, and the fifth of his pontificate].

Opposite to this inscription was written this:
Clemens VIII. Pontif. Max | Pontem a Sixto V. Pont. Max. incaeptum | Opere magnifico absolvit, | Alveo excavato Iberim induxit, | Anno Sal. MDCIIII. Pontif. sui XIII
[Pope Clement VIII completed this bridge, which had been begun by Pope Sixtus V, with elaborate workmanship; he dug out a water-course and channelled the Tiber into it. In the year of our Salvation 1604, which was the the thirteenth year of his pontificate].
Keysler, 'Letter LXI. Journey from Rome to Loretto', *Travels Through Germany*, III, pp. 156–7.

[3] On the left-hand side of the bridge was a large stone with this inscription:
VRBANVS VIII. PONT. MAX. | Tiberim viâ publica eversâ | Veterem repetentem alveum, | Nova effossione | Ad dextram deductum, | Aggeris-objectu | Validâque compact. Lignorum | Sub ponte, quem declinabat, | Continuit, | Conservationi prospiciens | Peninsulam adjacentem | Attribuit, | Anno Salutis MDCXXVIII. | Pontif. sui sexto.
[In the year of our Salvation 1628, which was the 6th year of his pontificate, Pope Urban VIII confined the Tiber, which was now flowing towards its original course because of the collapse of a public road, beneath a bridge under which it had ceased to flow, by guiding it to the right by digging out a new channel, and constructing a bank strongly buttressed with stakes as a barrier. With a view to its future preservation he added an outcrop of land nearby].
Keysler, 'Letter LXI. Journey from Rome to Loretto', *Travels Through Germany*, III, p. 157.

[4] Borghetto, a village in Civita Castellana, Viterbo.

[5] Cività Castellana, known as Falerii Veteres in ancient Rome, situated 65 km north of Rome. It was part of the Papal States at the time Folkes visited and an episcopal see, its castle completed under Pope Julius II. The city was an essential site on the Grand Tour. As Insalaco has indicated, in 'the course of the eighteenth century, specific elements of the Civita Castellana landscape became stock images in the portfolio of artists who visited Italy. These were the cliffs and rivers which flowed through them: the Celle, Flaminia, Clementine and Terrano bridges; the convent of Santa Maria del Arco; the Sangallo Fortress and the entrance gates to the town. Many of these motifs became popularized by the so called "vedutisti" or view painters, who created prints for travel books that were distributed throughout Europe. Artists were often familiar with the Civita landscape before travelling to Italy.' Insalaco, 'Cozens', p. 34.

[6] The river Treja, also known as the Tregia or Triglia.

[f. 84v] #tho this is affirmd by all the modern inscriptions of the place where the town is called civitas veiorum, and S P Q Veientanus. I understand it is not allowd by the best Antiquarys who pretend to fix the site of this antient town where there are some ruins a pretty deal nearer to Rome

[f. 85r] Saturday 30. Citta Castellana which is \said to be/ the Old Veii#[1] is a pretty large handsome city, it is walld but stands besides on allsides upon almost inaccessible rocks with dreadfull precipices from it; the town has been adorned with several fountains, great reservoirs of water and other works of the like sort; the Duomo or Cathedral is a very good old Gothick Church[2] the front is handsome in its way and adorned with a colonade of granite Ionic pillars taken from some antient building, there is an old inscription on the ffront shows it to have been built about 1200,[3] is it adornd with a pretty deal of Mosaicks[4] in the same taste as the Confessor's tomb at Westminster and a very rich mosaic pavement;[5] there is a castle to the town of stone with bastions etc in the taste of the Castle St Angelo at Rome,[6] and one side of the ~~Castle~~ counterscarp is made of an Antient building of 2 kinds of Arches, but what its has been I have not skill to determine. without this are some remains of the Antient walls and such a precipice as I have scarcely seen, I observd here in walking about some very odd artificial grots[7] or caves in the rock very antient but to what purpose I cannot guess,[8] I was in another very pretty church on the market place one long Isle like Kings Coll Chappel at Cambridg.[9] I saw several fragments of Antiquity in the streets, I left the town about eleven a clock and came down a most prodigious and surprizing descent to another bridg over the river or torrent, and travelld through the hills being still very high the rest of the day. our way was mostly pavd, we have for some days had

[1] The # Folkes wrote at this point refers to the material on f. 84v. The ruins of the ancient Etruscan town of Veii (Veio in Italian) are not in Cività Castellana, but are situated near Borgo di Isola Farnese on a volcanic plateau between two streams. Veii was the most southern town of Etruria, the richest Etruscan city, and a traditional rival of Rome; it was defeated by Furio Camillo in 396 BC and was assimilated under Roman control as part of the lands of the *Ager Veientanus*. The site of Veii is now the Veio Archaeological Park. See 'Parco de Veio', www.parcodiveio.it [Accessed 27 March 2023].

[2] The cathedral, which was begun in the 7th century and rebuilt in the 12th and 13th centuries. In 1740, after Folkes visited, the cathedral was nearly completely renovated in the Baroque style, the interior designed by Gaetano Fabrizi.

[3] An inscription at the top indicates the portico's designer, 'IACOBUS CIVIS ROMANUS' (James, citizen of Rome) and the date of execution (1210).

[4] The mosaics on the facade are an example of the Roman Cosmati school of marble work and architecture who created mosaic work with classical motifs; members of the Cosmati signed the presbytery screens of Cività Castellana cathedral (*c.* 1230). Claussen, 'Drudus de Trivio', *Grove Art Online*, https://doi.org/10.1093/gao/9781884446054.article.T023736 [Accessed 27 March 2023].

[5] Folkes made a perceptive comment. Henry III (1207–72) held King Edward the Confessor (d. 1066) in great esteem, and he erected a new shrine for King Edward with Cosmati work in Westminster Abbey; the shrine was finished in 1269 and was a focus of the cult of St Edward. See 'Edward the Confessor and Edith', Westminster Abbey, https://www.westminster-abbey.org/abbey-commemorations/royals/edward-the-confessor-and-edith [Accessed 27 March 2023].

[6] This is another perceptive comment. Cardinal Rodrigo Borgia, who later became Pope Alexander VI, was responsible for the strengthening of the Castel Sant' Angelo in Rome. However, Alexander VI had also been Governor of Cività Castellana (1482–92), and administrated the construction of a fortress in much the same style (Forte Sangallo), to replace the medieval castle.

[7] Grottoes.

[8] The tufa cliffs around Cività Castellana were honeycombed with ancient Etruscan and Roman necropolises, tunnels, and caves.

[9] Probably the church of San Francesco, formerly the church of San Pietro.

a great deal of pavd way of the old Flaminia Via, but to day we have almost constantly travelld upon it, it is well describd in all books so I refer myself to em, we have almost surrounded and once at a very small distance Mount Soractè. ~~the Mount~~ now calld the hill of St Orestes;[1] I have seen it all day yesterday as one of the highest mountains hereabouts; yet is here some buildings on the very summit and there is a large town on a lesser mountain joyning to it.[2] our Country to day has been more barren than what I have seen for some days, and also less seemingly cultivated so that the Patrimony does not seem anyways equall to the Dutchy of Spoleto; I think we have past through but one town and that a very small one calld Rignano,[3] there is an antient statue that [f. 86r] of no consequence in the street, we came to some few scatterd houses belonging to a small town called castelnuovo,[4] where we took our lodging for this evening at a sorry inn but where we have had a good supper. here we have the first prospect of Rome, I have seen it with my glass[5] so as to discover St Peters Church[6] very distinctly; there are some fragments of Antiquity in the Church wall and the yard about it, one an Altar with some sacrificing utensils and a funeral description which I have copied having nothing better to do. there lie also some pillars in the street; I forgot to note a very fine twisted marble pillar I saw on the road yesterday not farr from Borghetto. there is a modern inscription on the inn I am at relating to some clearing the fflaminian way. I've gone today only 20 miles, they reckon as much more to Rome

[f. 85v] D.M
IVLIAE. ESQVILINAE
VIX. AN. XXVIIII. M. X
IVLIVS. POLEMONINUS
ET. IVLIVS
SYMPHEROS
CONIVGIS.SANCTISSIMAE
PECERVNT. ET
IVLIAE CLODIANAE
....RORIEIVS. V.A.X[7]

[1] The Mountain of Soracte (Monte Soratte) on the fringe of the Apennines, west of the Tiber, was mentioned by Virgil in the *Aeneid* (*Summe deum, sancti custos Soractis Apollo*, book 11, line 785). Over time, the name was erroneously written St Oracte and corrupted to St Oreste (Sant'Oreste).

[2] On the highest summit (691 metres) is the church of St Silvestro, and on the south-east slope is located the village of St Oreste.

[3] Rignano Flaminio, about 35 km north of Rome.

[4] Castelnuovo di Porto, now part of the Metropolitan City of Rome.

[5] Folkes had several telescopes, the auction catalogue of his collection listing: 'Four double concave and 2 object glasses, 2 conic glasses', 'a telescope in square tubes', 'a small reflecting telescope', 'a small silver reflecting telescope by Jackson', 'a 3-feet telescope mounted in silver', a '3-feet telescope with silver ends'. See *A catalogue of the . . . collection of instruments,* pp. 3–4, p. 7; lots 14, 16, 38, 45, 47, 63.

[6] St Peter's Basilica in Rome.

[7] The full epigraphy of the altar from Castelnuovo di Porto, which was outside the Church of San Giovanni, and the description of sacrificing instruments portrayed on it (a pitcher and a patera bowl) are as follows:

((:grypus) ((:candelabrum)) ((:grypus)) | D(is) M(anibus) | Iuliae Esquilìnae | vìx(it) an(nos) XXVIIII m(enses) X[---] | Iulius Polemonìnus | et Iulius | Sympheros | coniugi sanctissimae | fecerunt et Iuliae Clodianae | [sorori e]ius ⟨:quae⟩ v(ixit) a(nnos) X | ⟨:in latere intuentibus sinistro⟩ | ((:urceus)) | ⟨:in latere intuentibus dextro⟩ | ((:patera))

[Griffin | Candelabrum | Griffin | To the spirits of the departed | To Iulia Esquilina, who lived for 29 years 10 months | Iulius Polemonius and Iulius Sympheros constructed this for a most chaste wife, and for Iulia

Beatrix[1] Columna Anguillaria
Viam Flaminiam antea deviam
Spinis et terra alte obrutam
Labore et impensa oppidanorum
Castri novi purgatam apervit
Jussu Greg[orii]. XIII. Piissimi Principis
Pau[lus]. Bubalus Praef[ectus]. Viar[um]. Cur[andarum]. MDLXXX.

[f. 86r *continued*] Saturday. 31. we left Castro Nuovo just after 7 o'clock. and travelling along the fflaminian way through a hilly country had for the most part Rome in sight, all the country seems very barren and but indifferently cultivated nor could we see any number either of towns or villages among the hills; there were ever here and there bits of Antiquitys on the road as pieces of walls pillars, etc, I went into a pretty large square on the left hand with 2 gates which seem to have been the walls of a small City, at one corner is a building now turnd into a house, but that seems to have opend with a large arch every way like a temple of Janus[2] the facing has been taken away all but a part of a marble Cornish which has had an inscription but nothing of it now remains legible, at some distance on the right on a hillock in a vale is another antient large ~~stone~~ square building, built of black and white stones in rows; some way further on the left hand I took notice of a building buttrest as it seemd all round the facing gone and overgrown with thorn etc. I went to it, and found an Entrance it is a little rotunda the whole vault standing with a circular light in the middle the entrance and flooring is filled up a good way with Earth, a little further I observed a chappel built out of an old ruin, and soon after came to the bank of the Tyber, which here looks like a fine river but the water yellow and fowl;[3] we passd some torrents over stone bridges, and a little village calld prima porta,[4] one post

Clodiana her sister | who lived for 10 years | *On the left side as one looks at it* | *A pitcher* | *On the right side as one looks at it* | *A dish*]

EAGLE (The Electronic Archive of Greek and Latin Epigraphy), EDR 147295, http://www.edr-edr.it/edr_programmi/res_complex_comune.php?id_nr=EDR147295 [Accessed 30 Dec 2023]. The current location of the altar is unknown.

[1] This is translated as: 'Clarice Colonna Anguillaria cleared the Flaminian Way, which had previously been rendered impassable by brambles and buried deep under the soil, using the labour and expenses provided by the townsfolk of Castelnuovo, on the instructions of Gregory XIII, a most holy prince. Paulus Bubalus, Commissioner for Roads, 1580'. Folkes, however, mistranscribed here: 'Beatrix' should be 'Clarix'. Folkes was one of many travellers to record this inscription he saw in the wall of the Castelnuovo posthouse, still preserved *in situ* today in Castelnuovo. See Vistoli, 'La Via Flaminia', p. 18.

[2] The Temple of Janus, or: 'The Janus Geminus (to reflect his twin faces) was a small shrine that held an archaic bronze statue of the god, said to have been dedicated by Numa, Rome's second king (Plutarch, *Life of Numa*, XX.1–2) … The doors of the Janus Geminus were opened to indicate that Rome was at war and closed during times of peace.' Grout, 'Temple of Janus', *Encyclopaedia Romana*, https://penelope.uchicago.edu/~grout/encyclopaedia_romana/imperialfora/nerva/geminus.html [Accessed 28 March 2023].

[3] The Tiber had a long history of pollution; the sewers in ancient Rome, 'carried an estimated 100,000 lb of human waste every day out into the Tiber'. Bradley and Stow, *Rome, Pollution and Propriety*, p. 4. From the 14th century, official rubbish collectors in Rome 'disposed of everything along the system of the Tiber'. Rinne, 'Urban Ablutions', p. 186.

[4] Prima Porta or First Gate, is named after the section of Via Flaminia near Villa di Livia. The aqueduct that brought water to the villa formed a gate over Via Flaminia, which travellers saw as the first sign of having reached Rome.

from Rome, this village is oddly situated in a low rocky valley there are grots out in the rock which now are turnd into houses, a large [f. 87r] Arch that we passd under, and some ruins, particularly a part of the rock built up with a strong wall and buttresses to a great height, our way from here to Ponte Molle[1] was mostly by the side of a very steep rock and pretty high to our right, in it we saw several grots and on the top some buildings that look like antient sepulchres, one of those grots appearing somewhat more than the rest we stopt and went in and found it the Sepulchro de Nasonii[2] describd by bartoli,[3] but now entirely defaced, the plaster that was painted in a matter entirely gone, however one just sees it has been painted, I discoverd some of the compartments with 2 figures and a festoon between them, one of the hunting pieces, was distinguishable as also that of hercules and Antaeus, and the Judgment of Paris, these just remain enough to show Bartolis has been a faithfull representation of it. the Pediment over the door at entrance is also gone. I saw another little round tempietto to the left, but ~~who~~ was not in it. just afore I came at Ponte Molle I saw in the rock several Arches that made Grots together, an odd place I design to see again, I passd the Tyber at Ponte Molle the famous bridg where Constantine defeated the Tyrant Maxentius, of this there is a good representation in Corn. Meyer.[4] The Tyber is wide and Rapid and runs now on the ~~left~~ \right/ to rome, hence was the best view I had of the city before I came to it, from hence which is I judge about 2 mile there are nothing but vineyards of the Romans on either side so that we went to Rome between 2 walls in a sort of Street there are however handsome gates to most of the vignus and some pleasure houses with fountains and watering places under the walls, this is a grand avenue to the city but it is but narrow we enterd at the Porta del Popolo[5] and just where standst the great Egyptian Obelisk[6] and 2 churches front the

[1] Ponte Milvio, the bridge over the river Tiber in northern Rome, and the site of a battle in AD 312 when Emperor Constantine I defeated Maxentius.

[2] The ancient tomb of the Nasonii on the Via Flaminia, unearthed in 1674. The approximately 9 × 4.5 metres family tomb was decorated with colourful frescoes around AD 150. There was an erroneous assumption that the owner of the tomb was related to P. Ovidius Naso (Ovid), and that the poet was buried here. The onslaught of visitors resulted in subsequent deterioration of the images. See 'Le pitture antiche', Heidelberg Digitale Bibliothek, https://digi.ub.uni-heidelberg.de/diglit/bartoli1680 [Accessed 29 March 2023].

[3] Folkes was referring to the work of engraver and antiquarian, Pietro Santi Bartoli (1635–1700): *Le pitture antiche del sepolcro de Nasonii nella Via Flaminia, I. Disegnate, ed intagliate alla similitudine degli antichi originali da Pietro Santi Bartoli descritte, & illustrate da Gio: Pietro Bellori* (1680). See Figure 37. The engraver, draftsman and painter, Bartoli, created the copper engravings of the wall paintings in 1680 – probably on behalf of Cardinal Camillo Massimo. See 'Le pitture antiche', Heidelberg historic literature, https://digi.ub.uni-heidelberg.de/diglit/ bartoli1680 [Accessed 29 March 2023]. The original watercolour paintings by Bartoli are in the Special Collections of the University of Glasgow, MS Gen 1496, important in tomb reconstruction as the engravings are monochrome; the album may have been in the possession of Folkes's friend, Dr Richard Mead. See Pace, 'Pietro Santi Bartoli's Drawings', p. 1. The auction catalogue of Folkes's art collection showed that he owned two sets of prints by Bartoli, one of Raphael's Grotesques in the Vatican, and another of Raphael's 'Scripture prints'. See *A catalogue of the prints and drawings*, pp. 8, 12; lots 37, 48.

[4] Cornelius Meyer, *L'Arte di restituire à Roma la tralasciata navigatione del suo Tevere* (1685). See Figure 38. Folkes' library auction catalogue showed he owned this book in folio. See *A catalogue of the entire and valuable library*, p. 122, lot 4119.

[5] Porta del Popolo, a city gate of Rome's Aurelian Walls, built by Pope Sixtus IV (r. 1471–84) on the site of the Porta Flaminia. Lansford, *Latin Transcriptions*, p. 329.

[6] The Flaminian Obelisk in the Piazza del Popolo, where it remains today with its 18th-century embellishments. The obelisk is sometimes called the 'Flaminian obelisk' because it is first seen through the Porta Flaminia. Brier, *Cleopatra's Needles*, pp. 34–5.

Figure 37. Table I, in Pietro Santi Bartoli, *Le pitture antiche del sepolcro de Nasonii nella Via Flaminia. Disegnate, ed intagliate alla similitudine degli antichi originali da Pietro Santi Bartoli. Descritte, & illustrate da Gio: Pietro Bellori*, Rome, 1680. University of Heidelberg, https://doi.org/10.11588/diglit.3176#0001. This shows the destruction of the ancient tomb of the Nasonii on the Via Flaminia to which Folkes referred. Courtesy of the University of Heidelberg.

Figure 38. 'Veduta de Ponte Molle', Table 26, in Cornelius Meyer, *L'arte di restituire à Roma la tralasciata navigatione del suo Tevere: divisa in tre parti*. Bayerische Staatsbibliothek, Munich. Folkes referred to this representation of the Milvian Bridge by Meyer. Courtesy of the Bayerische Staatsbibliothek, Munich.

Figure 39. Giovanni Battista Piranesi, *Veduta della Piazza del Popolo*, or *The Piazza del Popolo, Vedute di Roma (1753)*, from *Views of Rome* (1750–78), etching on paper, 38 × 54 cm. Metropolitan Museum of Art, New York, Harris Brisbane Dick Fund, 1937. The view is roughly contemporaneous to Folkes's visit to the Piazza del Popolo in Rome. Courtesy of the Metropolitan Museum of Art, New York.

Figure 40. Giovanni Battista Piranesi, *Veduta della Dogana di Terra a Piazza di Pietra*, or *View of the Customs House in Piazza di Pietra* (*Vedute di Roma*) (1753), from *Views of Rome* (1750–78),etching on paper, 54.5 × 77.1 cm. Yale University Art Gallery, gift of Frank E. Brown. The view is roughly contemporaneous to Folkes's visit to the Dogana di Terra or customs house. Courtesy of Yale University Art Gallery.

entrance,[1] we were here stopt and directed to send our baggage which we had before put all upon one coach to the Doganna[2] to be searcht, but they were very civil and it only cost 12 peseti moyenant[3] which I had every thing in less than half an hour; I went to the Monte d Oro[4] on the Piazza d Espana[5] where I am for the present. It is soon to form any idea of the city yet I love to remember my first coup d'œil, it seems full of magnificent buildings yet has an uninhabited look there are few people and no bustle in the streets they are in no ways [f. 88r] regular and many narrow and but ill pavd, the houses are mostly plastered with white I mean the common ones which has rather a pretending than a real fine look. I was not so struck with the obelisk as I expected though it stands finely. I also saw as I went to the Custom house the Antonine pillar a noble antiquity and pretty entire;[6] the Custom house makes an odd appearance it is a modern building built close to a row of antient very fine but miserably decayd pillars which tho one admires one cannot help thinking no ways of a piece with new buildings joynd to em.[7] I receivd as I was just come in a compliment from Lord Harcourt[8] by Mr Bowman his Governour[9] and

[1] Santa Maria in Montesanto and Santa Maria dei Miracoli. The churches are called *le Gemelle*, or the Twins, from their resemblance (although not perfect symmetry) to each other. They were built in the 17th century on order of Pope Alexander VII who wished to create a focal point for those that passed through the square. As Sweet noted the: 'wide expanse of the Piazza del Popolo, dramatically reshaped for Alexander VII, opened up before the new arrivals and invariably made a positive first impression–as it was intended to do ... The visitor encountered a vast open space, flanked by two identical domed churches, from which three long, straight roads, the *tridente*, radiated out into the city of Rome, each one lined with grand and imposing palazzi. In the centre of the piazza was the obelisk, engraved with hieroglyphs, which Sixtus VI had erected in 1589.' Sweet, *Cities and the Grand Tour*, p. 99. See Figure 39.

[2] There was a Popolo Customs Office at Porta del Popolo which controlled traveller traffic. The Customs House or the Dogana di Terra to which Folkes referred was in the Piazza de Pietra.

[3] Moyennant, or 'for a fee', or for a consideration of'. In other words, a small bribe. The Act issued by Pope Clement XII which gave the city of Rome its own customs regime was not passed until 1738, after Folkes's visit. Wright described something similar: 'when we enter'd the City, the Postilion durst not set us down at the Inn but brought us strait to the Dogana, or Custom-house, to have our Bagge search'd for contraband Goods, or prohibited Books; but they gave us little trouble; a small gratuity made the Search very easy.' Wright, *Some Observations*, I, p. 133.

[4] Folkes's inn.

[5] The iconic Piazza de Spagna.

[6] Folkes commented on measuring the Antonine pillar in his article about the Trajan and Antonine pillars: 'The pillar of Antoninus Pius, now commonly called of Monte Citorio is now lying, along; the shaft of one piece, of red oriental granite, is about 48 English feet long, as I grossly measured it; the pedestal of one piece of marble was dug up not far from it.' Folkes, 'On the Trajan and Antonine Pillars', p. 121.

[7] In 1696, Pope Innocent XII commissioned Carlo Fontana to convert: 'the old temple of Neptune at the Ripa into a customs house for the commodities coming by sea, the Dogana di mare'. For overland goods, the Dogana di terra was constructed behind the colonnade of the temple of Hadrian in the Piazza di Pietra. Gross, *Rome in the Age of Enlightenment*, p. 23. See Figure 40.

[8] Simon Harcourt, 2nd Viscount (1714–77). He was a member of the Dilettanti (1736), an army officer, FRS (1753), Ambassador to France (1768–72) and Viceroy of Ireland (1772–77). He was in Rome on his Grand Tour with the antiquarian and traveller Walter Bowman from October 1733 until 22 April 1734. See Ingamells, *Dictionary*, p. 463.

[9] Walter Bowman (1699–1782), a well-known antiquary. Several of his volumes of manuscript notes survive: 'Memoirs for Travels in Italy' (1727), 'Travelling in Italy and France' (a companion volume), and 'Fugitive Mss' (1765) are in the Biblioteca Nazionale, Florence, and the 'Antiquitatum tabulae' (1764), an album of miscellaneous notes *c.*1720–50, is in the National Library of Scotland. Ingamells, *Dictionary*, pp. 113–14; See also Introduction, p. 27 above.

as soon as I had dind his Lordship and the same Gentleman did me the honour of a visit. as did also Mr Mace[1] and Mr Jeffreys[2] with whom I went out to see after some accommodation but yet have not done anything, whilst I was gone Mr Smith[3] an English Gentleman that is here calld to see me and spoke with my wife, he is I understand a Clergyman and has his lady with him. I also received a packet sent from Venice to Mssr Rizzi[4] Macirone on whom I have credit,[5] who also sent me a compliment. this evening Joseph,[6] whom I had sent back again in quest of my dutch Dog came in and brought him with him, the Dog I understand had walked back through all the places I've been and was taken up some miles on the other side of Terni towards Venice. what is pretty odd is he came all the way in the Coach and mostly slept at the bottom of it. the servant says he believes he has not been in any house and was almost starvd. the Dog moans still and seems to attempt at telling its misfortunes. I came today only 2 posts about 20 miles; we have [[*one word*]] these two days with a carriage of ten Malefactors that had been taken from a robbery and are carrying to the popes galleys at Civita Vecchia.[7]

Sunday, Nov. 1, I want to Mr Mace in the morning early, where I met Mr Feake,[8] Mr Hart and Mr Jeffreys, as also Mr Parker the Antiquary[9] who lives here; at my return I was visited by Mr Stevens[10] and Mr Michel,[11] as also by Mr Hart and Mr Parker. and

[1] Folkes encountered William Mace again in Rome, with whom he spent time in Venice (see above, p. 37, n. 2).

[2] Not much is known about Mr Jeffreys, other than he was travelling with William Hart of Bristol. Folkes encountered them several times in Rome in November. Ingamells, *Dictionary*, p. 471.

[3] Again, a shadowy figure. As Ingamells notes: 'On 11 July 1733 John Bagshaw reported from Genoa that "Mr Smith a Gentleman lately come from England has abjured his Religion he pretends to have had great Losses by the Charitable Corporation".' He was doubtless the clergyman Smith who had changed his religion and was in Rome with his wife in Oct–Nov 1733. Ingamells, *Dictionary*, p. 872.

[4] There is a cross in the margin here.

[5] A letter of exchange.

[6] Probably one of Folkes's servants.

[7] Civitavecchia fortress and port. In the pontifical navy, 'a majority of the oarsmen were convicts sentenced by papal courts to the galleys for various crimes or prisoners captured in the course of engagements with North African pirates. A small number of rowers (about 10 percent of the total) were unskilled free men who, for lack of any other employment opportunity, had volunteered for the galleys out of economic desperation. The volunteers received a meager wage and more and better food than did the prisoners.' Alvarez, *The Pope's Soldiers*, p. 8. There is a cross inked in the right-hand margin at this point.

[8] Samuel Feake, Chairman of the East India Company and member of the Dilettanti. See also above, p. 39, n. 6.

[9] Mark Parker (*c*.1698–1775), antiquary, who spent most of his life in Italy. Ingamells, *Dictionary*, pp. 739–40.

[10] Henry Stuart Stevens (d. 1760) FRS and mathematician. Stevens was nominated for membership in the Royal Society by Folkes in 1740, as well as Abraham de Moivre. RS/EC/1740/08. He may have been a contributor to the third edition of De Moivre's work on mathematical probability, *The Doctrine of Chances* (1756). Ingamells mistakenly identifies him as 'Celsius Stevens'. Ingamells, *Dictionary*, p. 895. Folkes left Stevens a gift in his will, a reflecting telescope made by renowned instrument maker James Short. PRO/11/809/301, Will of Martin Folkes, 3 July 1754; See also Bellhouse, 'Lord Stanhope's papers', pp. 184–5; and Roos, *Folkes*, p. 341.

[11] Most likely Andrew Mitchell. He and Stevens, 'left together for Bologna on 12 September 1734'. Ingamells, *Dictionary*, p. 895; See also Introduction, pp. 23–4 above.

afterwards by Mr Savage[1] M^r [*blank*][2] and Mr Dixon.[3] I then went out to St Peters Church, and in my way passd the Ponte St Angelo,[4] which is exceedingly magnificent, and the [f. 89r] size of Castle founded upon the old Mausoleum of Adrian is surprizing. the way from hence is but indifferent till one opens upon the area before St Peters,[5] which it first vastly surprizes as one has no idea it is near so large is it is; this has the effect to make one at first greatly deceivd in the size of the Church, whose front looks small till one approaches it when the plinths under the columns at the entrance are full 12 foot square, the beauty of this Colonade and of the Obelisk[6] and fountains is beyond all imagination whenever one may have fashiond to ones self from the prints, and so is indeed the magnificence and grandeur of the inside of the Church, with the many noble monuments of Popes within it all which with the statues of Saints apostles etc are masterpieces in their way, the Altarpieces are many of them made in Mosaick work like the paintings that were here and the rest are doing, these mosaicks are with an art beyond imagination, they absolutely represent paintings and the polish given to them lookes like a picture just varnisht out. I took no notice to day of particulars being surprized and struck with the tout ensemble of a place that may well furnish observation for several days. After dinner, I saw lodgings and agreed for the first floor at one Giuseppes in the Corso[7] where we are to go to morrow, and be provisioned much in the same manner we were at d'Henris at Venice.[8]

Monday. 2. I waited on Mr Hart and Mr Jeffreys by whose assistance I agreed for a Coach; I then went and just took ~~of~~ a slight view of the Pantheon and drove to the

[1] Samuel Savage, member of the Society of Dilettanti, was in Venice in May 1733, visiting at the same time as Samuel Feake. Savage saw Martin Folkes in Rome on 1 November, and on 26 March 1734 he left Venice with John Olmius and the Rev. John Swinton to visit Vienna. Ingamells, *Dictionary*, p. 842. In 1744, portraitist George Knapton (1698–1778) portrayed Savage in standard Venetian masquerade. Redford, *Dilettanti,* pp. 31–2. The painting is in the Society of Dilettanti in London.

[2] Folkes left a blank space to fill in later when he recalled the surname.

[3] Probably John Dixon, an antiquities dealer who spied on the Jacobites and was for a time in correspondence with Brinley Skinner at Leghorn. Ingamells, *Dictionary*, pp. 303–4.

[4] Ponte Sant'Angelo, formerly the Aelian Bridge or Pons Aelius. It was built over the Tiber by Hadrian and completed in AD 135. Its purpose was to connect the Campus Martius with Hadrian's mausoleum (renamed the Castel Sant'Angelo). The bridge has 7 stone arches, 5 main spans of 18 metres, supported on piers that are 7 metres high. Pope Clement IV installed an iron balustrade, and Pope Clement VII had statues installed of Sts Peter and Paul at the end of the bridge. See 'Ponte Sant'Angelo', *Encyclopaedia Britannica*, https://www.britannica.com/topic/SantAngelo-Bridge [Accessed 30 Dec 2023]. Between 1667 and 1672, ten statues of angels were mounted on the parapets, attributed to Gian Lorenzo Bernini (1598–1680), and the site is indeed magnificent. Weil, *History and Decoration*, pp. 41, 71–88.

[5] Folkes's awestruck appraisal of the approach to St Peter's Basilica is typical of Grand Tourists. Edward Wright noted in his *Observations*: 'S. Peter's alone has had Volumes written upon it in Folio. By the Prints of that noble Temple, frequent among us, it is very well seen after what manner it is built: and that for its general form, our S. Paul's agrees pretty much with it. It were to be wish'd that our's had such an Approach as that has, than which nothing can be more grand or magnificent.' Wright, *Some Observations*, I, p. 205.

[6] The Vatican Obelisk of red granite which is 25.5 metres high. The uninscribed obelisk was originally raised by Cornelius Gallus, Prefect of Egypt, during the reign of Augustus, in Forum Julium, a public square in Alexandria. It subsequently was brought from Egypt to Rome by Caligula, where it was, 'erected in the circus that Caligula was building to further his passion for horse racing'. After 1,500 years in this location (by the old St Peter's Basilica), Pope Sixtus V commissioned Domenico Fontana in 1585 to move it to its present location, a Renaissance engineering feat. See Brier, *Cleopatra's Needles*, p. 67.

[7] The Via Del Corso, the main street in the historical centre of Rome.

[8] There is a cross inked in the right-hand margin at this point.

Campo Vaccino[1] where I just saw the Antiquitys to know where they lie, and then removd to my new lodging in the Corso where I dind, after dinner I was visited by M[r] Smith and his Lady, he is a Clergyman who has taken it into his head to change his religion and come hither by what I can hear and observe he is himself a very imprudent man and I therefore intend to have no more acquaintance with him than I possibly cant help. I was visited also by Capt[n] Mawl[2] and Mr [*blank*],[3] and also by M[r] Mitchel, a very ingenious Scotch gentleman I have seen before, and D[r] Hamilton[4] formally Chaplain [f. 90r] the Duke of Ormond, a grave, discreet gentleman, who carefully I am told avoids giving any uneasiness to strangers and therefore whome every body converses without any difficulty. for M[r] Smith, I chiefly apprehend him for ~~ff~~ inconveniences that may arise from his acquaintance so fully resolved to avoid it all in my power tho he is so forward to offer to go about with my wife etc, and at the same time will talk in his own way, so that is shocking even to good manners.

Tuesday. 3. I made some more visits to the Gentlemen that had been so kind to come and see me particularly to D[r] Hamilton, who was the only one I found at home and where I met Mr Hart and Mr Jeffreys. ~~in the aft~~ there I saw some pretty small antiquitys the D[r] has collected; in the Afternoon Mr Mace calld on me and we went to the Latteran a very fine old church[5] but where I must go again to take notice of any thing this first being little more than a going to take a general view of things the front is rebuilding by Sig[r] Galileo[6] and the Portico will be very grand. the great Obelisk with egyptian Hieroglyphics[7] stands before the side door of this Church and before the palace built by Sixtus the V that joyns

[1] This is the Forum Romanum, at the heart of Ancient Rome. 'From the beginning the Forum Romanum was a market place with market booths on its long sides – later called Tabernae Veteres on the South and Tabernae Novae on the opposite side. The East and West sides were bordered by religious structures.' See 'Roma', *Princeton Encyclopedia of Classical Sites*, p. 763. In the 17th century, it was called Campo Vaccino, the 'field of cows', as it was the site of cattle trading. See Figure 41.

[2] Probably 'Captain Maule', who was the governor (tutor) of Simon Luttrell (1713–87) during his Grand Tour of 1733–4. Luttrell was born in Luttrellstown, Co. Dublin, a member of the Dilettanti (1735), and later Viscount and Earl of Carhampton (1781; 1785), and an MP variously for the Mitchell, Wigan, Weobly, and Stockbridge constituencies from 1755 until 1780. See Ingamells, *Dictionary*, p. 616.

[3] Folkes left a blank space to fill in later when he recalled the surname.

[4] The Rev. Ezekiel Hamilton, who served as the Anglican chaplain for the Pretender in Rome. Hamilton was previously with James III, at Avignon and Pesaro, and had since then been serving the Duke of Ormonde, as his chaplain and secretary. See Corp, *The Stuarts in Italy*, p. 196.

[5] On the slopes of the Caelian hill on the edge of an area enclosed by the Aurelian Wall, the Archbasilica of St John Lateran, or the Cathedral of Rome, was for nine centuries the principal residency of the Pope, long before the construction of the present Vatican complex. It was erected by Emperor Constantine in 315. The architect Francesco Borromini was then commissioned by Pope Innocent X in 1644–55 with the restoration and restructuring of the Basilica. See Bosman et al., 'The Lateran Basilica to 1600', pp. 1, 4.

[6] Tuscan neoclassical architect, mathematician, and engineer, Alessandro Galilei (1691–1737), a remote relation of the more famous astronomer, designed the facade of the Lateran Basilica for Pope Clement XII, as well as the Corsini family chapel inside. It was completed the year after Folkes's visit (1735) to ameliorate what neoclassical architects saw as the rococo excesses of Borromini, designed by Galilei using austere Vitruvian principles. As such, it would have had great appeal for Folkes. Kirk, *Architecture of Modern Italy*, p. 23.

[7] The Lateran Obelisk, the largest standing obelisk in the world. Brought to Rome by Constantius (AD 317–61), it was re-erected in the Circus Maximus. The obelisk was lost over the centuries then, after its rediscovery, on 3 Aug 1588, it was raised in the Piazza San Giovanni in Laterano, where it is now with a Christian cross on its apex. Brand, 'The "Lost" Obelisks', pp. 105, 111.

Figure 41. Giovanni Battista Piranesi, *The Forum Romanum,* or *Campo Vaccino, from the capitol, Vedute di Roma (1753)*, from *Views of Rome* (1750–78), etching on paper, 38 × 54 cm. Metropolitan Museum of Art, New York, Harris Brisbane Dick Fund, 1937. The view is roughly contemporaneous to Folkes's visit to see the antiquities in the Campo Vaccino. Courtesy of the Metropolitan Museum of Art, New York.

to it[1] as we came home we passd by St Mary Maggione [*sic*], where stands the pillar of the Corinthian order[2] brought from the Temple of Peace

Wednesd. 4. I made some more visits, and was at the church of St Carlo[3] where were about 14 or 15 Cardinals etc at High Mass. after dinner I was at the Villa Borghese[4] with Mr Mace and Mr Jeffries, this is one of the richest villas in antiques about Rome the whole outside is covered with bass reliefs some very good and considerable, the Curtius particularly as big as the life is famous;[5] ~~the po~~ the house stands in a very pleasant garden and park finally shaded and planted.[6] on the inside is also a vast collection of Antiques and some very celebrated modern statues. among the Antient are the faunes holding the Child,[7] the Gladiator,[8]

[1] The Lateran Palace, the residence of the popes until the 14th century when it was transferred to the Vatican.

[2] Santa Maria Maggiore, Rome, built at the top of Cispius, the highest peak of the Esquiline, which is the highest hill in Rome. Pope Liberius (352–66) founded the church, and the present structure was erected under Sixtus III (432–40). In the medieval tradition, the building site was indicated by the Virgin Mary, who left her footprints in a miraculous snow fall. *Oxford Dictionary of the Christian Church*, https://www.oxfordreference.com/display/10.1093/acref/9780199642465.001.0001/acref-9780199642465-e-6389 [Accessed 2 April 2023]. It was the premier Marian basilica in Rome and, when Folkes visited, in serious decay. The pillar outside Santa Maria Maggiore, the Column of Peace (Colonna della Pace), came from the Basilica of Maxentius, known as the Temple of Peace, and it was erected in 1613 at the behest of Pope Paul V (r. 1605–21). The inscription on the column indicated that the column was dedicated by Paul V to the Virgin Mary for, 'the Prince of true peace who was born of her womb, gave this gift' (EX CVIVS VISCERIBVS PRINCEPS VERAE PACIS GENITVS EST DONVM DEDIT).

[3] Francesco Borromini designed San Carlo alle Quattro Fontane, (Italian: 'Saint Charles at the Four Fountains', also called San Carlino), an influential Baroque church in Rome. The church was part of a small monastery for a community of Spanish monks. Commissioned in 1634, San Carlo was built during 1638–46, except for the tall facade, added about 1677. 'San Carlo alle Quattro Fontane', *Encyclopaedia Britannica*, https://www.britannica.com/topic/San-Carlo-alle-Quattro-Fontane [Accessed 30 Dec 2023].

[4] The Palazzo Borghese is near the Tiber in the Campo Marzio, the ancient monumental centre of Rome. The collection was formed in the early 17th century by Cardinal Scipione Borghese, nephew to Pope Paul V, and the collection was among the most significant in Rome. The paintings were largely early modern Italian works. The sculptures were mostly from antiquity, but included works of the young Gianlorenzo Bernini, and were exhibited mainly at the Casino Nobile. 'The Casino was the principal building on the grounds of the Villa Borghese, about a mile away on the Pincian Hill.' Paul, *The Borghese Collections*, p. 1.

[5] Bas relief showing Marcus Curtius throwing himself into the chasm, at the north facade of the Villa Borghese. 'A poem by Francucci (1613), written for Cardinal Borghese, indicates that the cardinal was compared with Marcus Curtius because of his own sacrifice in distributing alms to the Roman people during the 1606 flood of the Tiber.' Fiore, 'The Exhibition of Sculptures', p. 234.

[6] The garden would have appealed to an Englishman like Folkes. As Ehrlich noted: 'the garden-park design of the Villa Borghese in Rome both marks a profound change in a long tradition of villa-making in Italy and forms a bridge between Italian and English concepts of landscape design.' Ehrlich, 'Villa Borghese , p. 12.

[7] *Le Faune Borghese*, a Roman copy of the middle of the 2nd century AD after a Greek original by Lysippos (*c.*300 BC). The statue represents a scene from ancient Greek mythology, where the Satyr Silenus is carrying the infant Dionysus in his arms. It was among the Borghese marbles sold to Napoleon in 1807, and it is now in the Louvre. The statue was discovered by Carlo Muti before 1569 on his grounds near the Cassino Massimo, and subsequently became part of the Borghese Collection. The statue was displayed in the Stanza del Sileno (from which the room took its name) on the ground floor of the Casino Nobile. It was one of the most admired and copied statues in Rome. See Paul, *The Borghese Collections,* p. 300.

[8] The Borghese *Gladiator*, one of the most discussed antiquities at the Casino Nobile, which stood in the centre of the appropriately named Stanza del Gladiatore. It is a Hellenistic life-size marble sculpture created at Ephesus *c.*100 BC, portraying a swordsman, and like the *Faun* Borghese, it is now in the Louvre. Paul, *The Borghese Collections,* p. 194. At this point, Folkes inserted the names of two other statues in small print, the titles illegible.

\[[*two words*]]/ the Hermaphrodite[1] and the Narcissus all of the most celebrated things in Rome, here are also several others very much taken notice of I was pretty pleased with 2 boys one chained and crying[2] the other playing with a bird,[3] but the number of these must require several other [f. 91r] times going to take the notice they deserve. among the modern ones the Apollo and Daphne[4] and the David[5] of Bernini, and some family busts all masterpieces in their way. in the hall is a celebrated Antique bachus[6] lying along, now upon ~~the same~~ a pedestal on which is the famous bas relief of the death of Meleagre,[7] the antique pillars of the richest marbles herns etc in this house are almost infinite

Figure 42. Unknown sculptor, *Dionysos* (Roman), marble, 88 × 206 cm. Département des antiquités grecques, étrusques et romaines. Photo © Musée du Louvre, Dist. RMN-Grand Palais / Christian Larrieu. The 'celebrated Antique' Bacchus that Folkes saw in its original placement in the Galleria Borghese, now in the Louvre.

[1] 'The celebrated ancient marble statue of a sleeping hermaphrodite that stood before a shallow niche on the wall adjoining the Stanza del Gladiatore. This statue had been restored by Bernini, who also made its mattress, and it was exhibited in the Casino from at least 1638.' Discovered at Santa Maria della Vittoria near the Baths of Diocletian, it was absorbed into the Borghese Collection, and it is now in the Louvre. The ceiling of the room in the Casino Nobile in which the statue was displayed, was painted by neoclassical artist Nicola Buonvicini, illustrating the fable of Hermaphroditus and Salmacis (Ovid, *Metamorphoses* 4.313–92). See Paul, *The Borghese Collections,* p. 192.

[2] *Eros in Chains*, 2nd century AD, copy after Hellenistic original, Galleria Borghese.

[3] *Little Boy with a Bird*, 2nd century AD, copy after Hellenistic original, Galleria Borghese.

[4] Gian Lorenzo Bernini, *Apollo and Daphne* (1622–5), Galleria Borghese.

[5] Gian Lorenzo Bernini, *David* (1623–4), Galleria Borghese.

[6] One of the statutes was of Bacchus reclining on a sarcophagus. A drawing by Antonio Asprucci (1723–1808) shows the statue was in the Stanza del Sole of the Galleria Borghese. The work subsequently was sold by Prince Camillo Borghese to Napoleon in 1807 and it is now in the Louvre (inventory MS 1654). See Paul, *Making a Prince's Museum*, Fig. 25, p. 53, p. 117; Antonio Asprucci, *Elevation for the Stanza del Sole, Antonio Asprucci architectural drawings for the Villa Borghese* (*c.*1770–*c.*1793), pencil, pen and ink with gray and black wash on paper, between 1775 and 1780, The Getty Research Institute, Los Angeles; *Dionysos*, marble, Imperial Rome, 88 cm × 206 cm, Louvre, Département des Antiquités grecques, étrusques et romaines, https://collections.louvre.fr/en/ark:/53355/cl010250276 [Accessed 5 April 2022].

[7] The statue of Bacchus reclined on a separate sarcophagus, its bas-reliefs portraying Meleager, one of the Argonauts. He was killed by fire by his mother, Althaea, for murdering her brothers in an argument during a Calydonian boar hunt, a common motif for friezes on Roman coffers. The sarcophagus was in the Stanza del Sole of the Galleria Borghese. The busts that stood on the consoles in this room that Folkes mentions are lost. Paul, *Making a Prince's Museum*, Fig. 25, pp. 53, 117.

Figure 43. Attributed to Antonio Asprucci, *Elevation for the Stanza del Sole* (*c.*1775–80), pencil, pen and ink with gray and black wash on paper, 25.2 × 33.2 cm. The Getty Research Institute, Los Angeles. Sheet 13 of archive, Antonio Asprucci architectural drawings for the Villa Borghese, *c.*1770–*c.*1793. This is a drawing portraying the original placement of the statue of Bacchus and the antique sarcophagus below it that Folkes recorded seeing during his tour of the Galleria Borghese. Courtesy of the Getty Research Institute, Los Angeles.

Thursday, Fryday; Saturd. Sunday. besides returning several visits I went to take general views of the town and some of the principal things in it; the streets as I took notice before are narrow for the most part, pretty many are strait but none here have any regularity or beauty, tho they contain so many fine churches palaces fountains and another ornaments. I have been at the church of the Lateran to which a new front is now building by Sigr Galilei. The inside is fine but too low.[1] the Statues etc in it are well described by

[1] Folkes's comment about the ceiling height was perceptive. In 1646, Pope Innocent X had Francesco Borromini renovate the interior of the Lateran as the church had fallen into a decrepit state. However, the wooden coffered ceiling (made in 1562–67 by Flaminio Boulanger for Pope Pius V) was not to be altered. This was a constraint for Borromini, who had wished to create a monumental, coffered barrel vault for the Lateran. His contribution to the church was thus limited to the interior of the nave and four aisles. As Echols remarks: 'In the context of Borromini's crisply carved decoration, the forms of the ceiling seem heavy. The beams of the ceiling bear no relation to the pilasters and conflict with the regular rhythm of the elevation. That Borromini intended something other than the existing ceiling to cover the nave, something that would have unified and completed his design, seems implicit in the work that he did execute there.' Echols, 'A Classical Barrel Vault', p. 149.

Wright.[1] there are besides 2 magnificent pillars of Corinthian brass that support the Altar[2] that fronts the Entrance from the side of the palace. the High altar has an ancient arch of Mosaick curious for its Antiquity the Apostles Virgin etc are vastly larger than some saints of lesser note as if their size are answerable to their dignity.[3] the names are wrote in capitals standing one under another like Chinese writing.[4] there are 12 pretty old inscriptions belonging to the old church, one of which contains the account of some of their principal relicks, among which they affirm they have the Ark of the Covenant with the tables of the Law, the Table and Candlestick.[5] I enquird but they are not now to be seen, or I should have been glad to have known what sort of figure was given to these things some hundred of years since. I was also at St Mary Maggione before which has a fine column from the temple of Peace. The Columns of the Nave are Antique. the Church is not ill describd by Wright to whom I refer myself for it.[6] here is

[1] Wright wrote about the Lateran: 'The Nave of the Church is large, and finely adorn'd: The Twelve Apostles, twice as large as the Life, in white Marble, have a most magnificent Appearance: they are modern Performances, i. e. of the present Age, but by the best Masters in it, as Mons. le Grot [Pierre le Gros who sculpted St Thomas and St Bartholomew], Camillo Rosconi [Camillo Rusconi, who sculpted Sts Andrew, John, Matthew and James the Greater], &c. and some of them may justly be called very fine. They stand in spacious Niches, adorn'd on each side with Pillars of Verd antique; which were design'd by the Cavalier Boromini, who in some of his Works was a little particular in his Fancy, but in the main a great Master. Above these Statues are Basso Relievo's, fix out of the Old Testament, and as many on the opposite side out of the New, by way of Type and Antitype.' Wright, *Some Observations*, I, p. 216.

[2] Framing the altar of the Holy Sacrament at the opposite end of the transept are four gilded Corinthian columns dating from the early 2nd century AD which are 7.1 metres high and cast in bronze. These columns, likely made for a Roman temple or public monument, were repurposed for use in the old Constantine basilica and then installed around the altar in the Lateran in 1597–8. Claridge, 'Lateran Baptistry', *Rome,* p. 377.

[3] The Apse of the Triclinium of the Lateran Basilica. The Constantinian apse was rebuilt by the Franciscan Pope Nicholas IV (r. 1288–92), and the mosaic restored by Jacopo Torriti. Claussen, 'The Remodelling of San Giovanni in Laterano', p. 320.

[4] Folkes was correct about the relative sizes of the saints, a common iconographic technique in medieval mosaics. There are several inscriptions in the mosaic, including the saints' names. The insertion of the mosaic of the Saviour was especially noted in Grand Tour Guides: *sacrum vultum Salvatoris integrum reponi fecit in loco ubi primo miraculose populo romano apparuit quando fuit ista ecclesia consecrata* [He had the holy face of the Saviour replaced undamaged in the place where it first appeared to the Roman people when the church was consecrated]. See Claussen, 'The Remodelling of San Giovanni in Laterano', p. 322.

[5] 'The first mention of the Ark in the Lateran Basilica occurred in a tract from around 1100, later known as the *Descriptio lateranensis ecclesiae* (*Description of the Lateran Church* (hereafter *Descriptio*)). This text stated that the high altar of the basilica covered the Ark of the Covenant and included the rods of Aaron and Moses, the seven-branch candelabra, the tablet of the testament and a number of relics from the lives of Christ and the apostles.' The so-called Ark of the Covenant, the rod of Moses and the rod of Aaron were on display in an ambulatory chapel which also featured the table of the Last Supper, and the Ark, a decorated wooden chest covered with a cloth of silk. The Ark was exhibited above a glass box through which the rods were visible. The objects were displayed for veneration in the church for centuries. Whilst the table of the Last Supper was kept as a sacred object, the other objects were removed by Pope Benedict XIV in May 1745. No physical traces remain of these objects. See Oftestad, *Ark of the Covenant*, pp. xiii–xv.

[6] Folkes indeed takes his short note from Wright, who wrote: 'This Church is very noble and magnificent, as well as ancient. The back Front, which makes much the greatest Appearance, is modern: The Portico at the principal Entrance is ancient, supported by antique Pillars, and is adorn'd with old Mosaic Work. The Pillars which are on each side the great Nave are antique; there are several smaller, which support Tabernacles, towards the upper End of the Church; these are antique likewise, of beautiful Colours, and rare kinds of Marble ... The two great Ornaments of this Church, are the magnificent Chapels of Sixtus V. and Paulus V. on each side the Church, opposite to each other. These Chapels perfectly match one another, and are both surprisingly fine. The Cielings are of Stucco gilt; and the Walls perfectly cover'd over with Marble, Sculpture and Painting. In each

a Catafale[1] that has stood ever since the death of Charles II of the Spain, and they still say masses for him here. the 2 fine chappels here of Sixtus V. and Paul V. are of the finest and richest things in Rome. I was too at the Certosa a large Church built from the ruins of the Therms of Diocletian, it is plain and Spacious and there is Bianchinis Meridian Line seemingly in good order. I waited till noon and set my watch by it one may I think pretty easily determine the passage of the Suns Center to less than 2″ in time. I am told it is described in a book particularly. If I do not find it I will make some further memoranda about it.[2] The Piazza Navona is the fairest street I think in Rome the obelisk and fountain under it set up by Bernini is very fine and elegant more so I think than any other I have seen.[3] as there is a print it is useless to make any description of it.[4] The tomb of Cestius[5] is the completist thing that [f. 92r] I think now remains it has slightly been repaired, and is considerable as one is from it well enabled to form a just idea of the Egyptian Pyramids, tho this is sharper than they are. In the way to it I saw the palace Savelli whose wall is that of the old Theater of Marcellus.[6] I also saw the little round church of S[t] [*blank*][7] ~~here~~ in the wall of which are the old corinthian pillars supporting I

---

of them is the Monument and Statue of the Founder of the Chapel on one side of it, and on the opposite side is that of the Patron or Benefactor of the Founder.' Wright, *Some Observations*, I, pp. 219–20.

[1] This is the catafalque (bier to support a coffin) designed by the Venetian architect Domenico Rossi (1657–1737) for the funeral rites of Charles II in 1701 in Santa Maria Maggiore. Rossi's catafalque was in the middle of the nave, consisting of a stepped platform surmounted by two portraits of the deceased Hapsburg monarch, one facing the high altar and the other the basilica's entrance. It was draped in black and surrounded by silver candelabra and torchères. See Deupi, *Architectural Temperance*, p. 11.

[2] With the patronage of Pope Clement XI, the astronomer and antiquary Francesco Bianchini (1662–1729), FRS (1713), was responsible for the Clementine Meridiana at the church of Santa Maria degli Angeli e dei Martiri, which itself was built on the site of Diocletian's Baths in Rome. Bianchini's heliometer predicted the appearance of heavenly bodies and was used for Biblical chronology, as well as improving the accuracy of the calendar. The sun's image is projected onto Bianchini's line every solar noon. Heilbron, *Incomparable Monsignor*, p. 90. Folkes may have been alluding to Bianchini's book concerning the layout of his heliometer, the *De nummo et gnomon clementino* (Rome, 1703). The auction catalogue of Folkes's library shows that whilst he owned three other works by Bianchini, including his *Observationes circa planetam Veneris* (1728), he did not possess the book on the heliometer.

[3] A public open space in Rome, originally the Stadium of Domitian. As Sweet has noted, the Piazza Navona 'was widely praised by visitors'. Sweet, *Cities and the Grand Tour*, p. 135. Folkes is referring to the *Fontana del Quattro Fiumi* (Fountain of Four Rivers), sculpted in 1651 by Gian Lorenzo Bernini, topped by the Obelisk of Domitian brought from the Circus Maximus. The auction catalogue of Folkes's art collection showed that he owned a book of prints of Giovanni Battista Falda's *Le fontane di Roma*, published by Giovanni Giacomo Rossi. See *A catalogue of the prints and drawings* p. 6, lot 45.

[4] See Figure 44.

[5] Pyramid of Cestius in the Piazzale Ostiense, a 'monumental tomb, built *c.*18–12 BC (during the reign of Augustus) in the fork of the Via Ostiensis and a road which ran across to the Tiber (approximately where the Via della Marmorata does now)'. See Claridge, 'Pyramid of Cestius', *Rome,* pp. 398–400. Folkes also had a pronounced interest in Egyptian antiquities, so much so that his personal medal, which he had struck at Rome, featured the pyramid of Cestius. On this see Roos, *Folkes,* pp. 234–5. See also *Martin Folkes Medal*, object M.8466, British Museum, London. https://www.britishmuseum.org/collection/object/C_M-8466 [Accessed 1 Nov 2019].

[6] The Theatre of Marcellus was between the Circus Flaminius and the Temple of Apollo and by far the most important of Rome's three theatres, capable of holding 20,500 people.' In the 16th century, the theatre was used by the Savelli family to create their palazzo out of its ruins, and in 1712, additions to the palazzo were constructed by the Orsini family. Claridge, 'Theatre of Marcellus', *Rome*, p. 275.

[7] Folkes left a blank space here to fill in the name later and failed to do so. This is the Round Temple in the Piazza Bocca della Verità, recorded in 1132 as being converted into the church of St Stephen 'of the carriages'. It was re-dedicated in the 17th century to St Mary 'of the Sun'. Claridge, 'Round Temple', *Rome*, p. 287.

Figure 44. Giovanni Battista Falda, *Altra Veduta della Fontana in Piazza Navona* [Fontana dei Quattro Fiumi], Plate 20 of portfolio 'La Fontane di Roma nelle Piazze e Luoghi Publici' (*c.*1691), etching on paper, 28.7 × 21.2 cm. Metropolitan Museum of Art, New York, gift of D. W. Langton, transferred from the Library. This may have been the print Folkes was referring to in his travel diary, as he owned *La Fontane di Roma*. Courtesy of the Metropolitan Museum of Art, New York.

suppose an open coridor when it was the temple of Vesta or the Sun as they commonly say. over against #[1]

[f. 91v] # this is the Church of St Mary Cosmedin,[2] built as it is thought out of the ruins of the templum Pudicitiàe,[3] here are some ancient pillars and other fragments of Antiquity. in the porch in what they call bocca de la Verita,[4] a large head in relief of marble on a stone like a mill stone they tell several things about it, but it was probably only the causeway of a Cloaca where the water drained in.[5]

[f. 92r *continued*]. Wednesd 11. I was at the library of the Vatican consisting of several fine and some magnificent Galerys where the books are all kept in handsome cases adornd with the arms of the Several Popes that have filld up the parts, it was began by Sixtus V. and almost everyone since has enlargd it, the present pope has thrown in a long Galery and filld it with cases upon which is the large and noble collection of Etruscan vases of Cardinal [*blank*][6] bought by the pope and just presented here the cases also are in great measure filld with the printed books of Cardinal Quirini that gives in his life time his collection to the Library of which he is chief Librarian.[7] A stranger can remark little about the MSS but their quantity I was shown here are commonly shown to strangers and mentiond in 20 places. The ffamous septuagint and new testament is not so fair a book as the Alexandrian, it consists of three columns whereof the other consists but of 2. the ink is blacker and does not seemingly carry the marks of so great age. I read part of the Epistle of St John in the new testament where the text *there are three* etc is not.[8] this I

[1] There is #, a hashtag-like mark, in the margin to indicate his entry continued on the facing page of f. 91v.

[2] Santa Maria in Cosmedin or the Schola Graeca which was associated with the 'Greek community in Rome since the eighth century'. Claridge, 'S. Maria in Cosmedin', *Rome*, pp. 289–90.

[3] Folkes probably took this identification from Wright: 'The Templum Pudicitia Patriciae is a patch'd up old Temple, now turn'd into a Christian Church, by the Name of S. Maria in Cosmedin, or in Scholâ Græcâ.' Wright, *Some Observations*, I, p. 253.

[4] Bocca della Verità, a mask of Phyrgian purple and white marble on the left wall of the portico of Santa Maria in Cosmedin. It was the mouth of Truth that supposedly would bite off the hands of liars. It was really a drain-hole cover of the 2nd century AD, the head representing Oceanus, the god of the underground river believed to encircle the world. Claridge, 'S. Maria in Cosmedin', *Rome*, pp. 288–9.

[5] Folkes was absolutely correct in his assessment. Although he likens the Bocca della Verità to a millstone like Wright, he differs from Wright who thought it was a representation of the Nile. Wright, *Some Observations*, I, p. 253.

[6] Folkes left a blank space here to fill in the name later and failed to do so. Cardinal Gualterio of Orvieto's collection of Etruscan vases was acquired by the Vatican Library via the efforts of Pope Clement XII (r. 1730–40). The nucleus of vases in the Clementine Gallery of the Vatican Library dates from 1733, and they are now in the Gregorian Etruscan Museum. See *The Vatican Collections,* p. 183.

[7] Scholar and Cardinal Angelo Maria Querini (1680–1755), who became head of the Vatican Library in 1730, and became subsequently prefect of the Congregation of the Index. Weber, 'Angelo Maria Quirini', *The Catholic Encyclopedia*, http://www.newadvent.org/cathen/12614d.htm [Accessed 5 April 2023].

[8] Folkes is describing his visit to the Vatican Library to see the Codex Vaticanus. The 4th-century AD Uncial Codex Vaticanus is one of the most famous authoritative biblical texts, one of two of the oldest Biblical manuscripts in parchment that are still extant, and a cornerstone of the Septuagint and the New Testament. Although a few scholars had gained access to the Codex: 'it was not consulted properly until the end of the nineteenth century ... The Vatican seemed reluctant to allow the readings of a manuscript whose text differs so much from the textus receptus to be freely available ... particularly its omission of 1 John 5:7–8, where the Holy Trinity is averred.' De Jonge, 'Erasmus', p. 389. Despite this, many early modern scholars 'travelling through Europe sought out ancient manuscripts to check whether the passage was present'. Iliffe, *Priest of Nature,* p. 272. Folkes was no different. Folkes was a deist, and his mentor Newton an anti-Trinitarian, so this also may

only take notice of because by some mistake a late traveller has positively asserted contrary to all collationers himself to have read it here;[1] 2 other Gentlemen that are scolars read it with me and know this is fact.[2] The illuminations of D Julio Clovio in the life of the Duke of Urbin[o][3] and some missals is a curiosity everybody takes notice of. there is also a Natural history of Animals ~~of~~ by Petrus Candidus extracted chiefly from Pliny, but has the best miniatures of Animals of all sorts I have seen.[4] The Virgil is imperfect a quarto work in Capitals,[5] but this I design to go again and set down some observations upon by

---

have been a motivating factor to examine if the manuscript referred to the Trinity or not. The passage in question, the *Comma Johanneum*, which provides the clearest support for the doctrine of the Trinity, is considered in modern scholarship as an interpolation. For an analysis of Folkes's religious beliefs, see Roos, *Folkes*, pp. 124–40.

[1] A 17th-century Oratorian, Denis Amelot, was charged by King Louis XIV of France, 'with creating an official French translation of the New Testament from the Vulgate'. Although Erasmus stated that the *Comma Johanneum* was not in the Codex Vaticanus, Amelot claimed, in a publication of 1687/8, to have seen the *Comma* in 'the oldest manuscript in that library'. Amelot's claim was subsequently cited by others, such as David Martin and Richard Smalbroke, Bishop of Lichfield and Coventry, 'as evidence for the genuineness of the comma' in the Codex Vaticanus. See Macdonald, *Biblical Criticism*, pp. 149, 226. In Folkes's case, before he left for Italy, Jonathan Richardson the Younger, the son of Folkes's neighbour, the artist Jonathan Richardson the Elder, asserted that he indeed saw the *Comma Johanneum* in the Codex Vaticanum. This was the 'late traveller' whom Folkes referred to in his diary. Folkes subsequently advised Richardson to drop his assertion as a mistake and Richardson took umbrage. As Folkes himself was walking a thin line of being accused of publicly avowing anti-Trinitarianism and losing Richardson's friendship, an exchange of mollifying letters subsequently took place between Folkes and Richardson. Richardson admitted that he may have mistakenly seen another document in the Vatican Library, perhaps the one that Amelot had related, that did have the Comma. See Richardson to Folkes, 17 March 1736/7, RS/MS/790. Richardson Senior and Junior also published their observations of Italy in 1722 as, *An account of some of the statues, bas-reliefs, drawings and pictures in Italy, &c., with remarks*, London, 1722. Folkes owned a copy in his library, *Catalogue of the entire and valuable library*, p. 120, lot 4061.

[2] We do not know who these two men were. Folkes related to Phillips Glover in a draft letter: 'I must have had the curiousity to see my self the Vatican Manuscript, which I indeed did twice and each time with two Gentlemen of my acquaintance with me, and we all read the verse before and after which was however somewhat difficult \at least to me/ by reason of the character and age, but we made use of a common Greek Testament to compare it with spelling the words one after another.' Undated draft letter for letter of 4 May 1737, RS/MS/790/47.

[3] Giorgio Giulio Clovio (1498–1578), Italian miniature painter of Croatian birth, and possible collaborator with Pieter Bruegel the Elder. Clovio is best known for the *Farnese Hours* of 1546 (now in the Morgan Library, New York), done for his patron Cardinal Alessandro Farnese for whom he was in service from 1540 until 1578. Hawkins, 'Clovio, Giulio', *Grove Art Online*, https://doi.org/10.1093/gao/9781884446054.article.T018262 [Accessed 6 April 2023]. Described by Giorgio Vasari as a 'Michelangelo of small works', Clovio was said by Vasari to have done a set of illustrations for biographies of Federico da Montefeltro, the Duke of Urbino, and Francesco Maria della Rovere in the Vatican Library. See Hollingsworth, 'Art Patronage', p. 361.

[4] Das Tierbuch des Petrus Candidus (1460), a Vatican Library manuscript (MS Urb. Lat. 276) of the history of animals by the Lombard humanist Pier Candido Decembri (1399–1477). As Pyle has indicated, Candido, 'based himself nearly exclusively on the thirteenth-century *Liber de natura rerum* of Thomas of Cantimpré', though most medieval works of natural history took inspiration from Pliny's *Naturalis historia,* (books 8–11 for zoology). The illustrations were 'almost certainly added by the Mantuan artist, Teodoro Ghisi (1536–1601) in the last decade of the 16th century'. See Pyle, 'Art and Science', p. 268.

[5] The Vergilius Vaticanus, also known as the Vatican Virgil (Cod. Vat. lat. 3225). The manuscript is Roman, dated to within one or two decades of AD 400, and it is now in the Vatican Library (Vat. Lat. 3225). When it was intact, it had some 440 folios, written in rustic capitals, of which 75 now survive. 'They contain 50 illustrations, by three separate illustrators, whose figure style and framed composition are said to be influenced by wall painting and mosaic decoration, particularly the mosaics of Santa Maria Maggiore (432–40), although

themselves, there is another very old imperfect Virgil, which has been confounded with the first in from whence some of Bartoli's figures are[1] [f. 91v] taken. the Terence is also considerable, but the figures of one very like it are gravd in a book I had in England de Personis.[2] this is printing they told me with the figures gravd at Urbino. both this and another Terence in the same case[3] were collated by Dr Bentley.[4] I visited to day Sr Hugh Smithson[5] Mr Crow[6] and Mr Hamilton[7] gentlemen just come from fflorence, Mr Dugood[8] whome I knew in England is come with them.

---

they are somewhat later in style'. Nicholas Sparks, 'Vatican Vergil', *Oxford Dictionary of Late Antiquity*, https://www.oxfordreference.com/display/10.1093/acref/9780198662778.001.0001/acref-9780198662778-e-4947 [Accessed 22 Sept 2023]; Vatican Library, DVL Digivatlib, https://digi.vatlib.it/mss/detail/Vat.lat.3225 [Accessed 5 April 2023].

[1] Folkes continued his entry on f. 91v. He is referring to the Codice Virgiliano (Bib. Apostolica, Cod. Vat. 3867), which painter, engraver, and antiquary Pietro Santi Bartoli (1635–1700) reproduced in 55 plates (1677), a commission from Cardinal Camillo Massimo. Vatican Library, DVL Digivatlib, https://digi.vatlib.it/mss/detail/Vat.lat.3867 [Accessed 5 April 2023]. The auction catalogue of Folkes's library showed that he owned the 1725 edition of Bartoli's work: *Virgilii Opera, quae supersunt in Antiquo Codice Vaticano*. See *Catalogue of the entire and valuable library*, p. 139, lot 4639.

[2] Folkes owned two copies of the three-volume set of *Comedies de Terence, Traduites en Francois par Madame* [Anne] *Dacier* (1699), as well as the 1717 Rotterdam edition. See *Catalogue of the entire and valuable library*, p.15, lot 528; p. 145, lot 4801. Anne Dacier (d. 1720) was a scholar of ancient Greek and Latin and also published a critique of La Motte's opinion of Homer and his translation of the *Iliad* (1714).

[3] Folkes is referring to two manuscript editions of Terence's comedies, shelfmarks Vatican lat. 3226 and Vatican lat. 3868.

[4] *Publii Terentii Afri comoediae, Phaedri Augusti Liberti Fabularum Aesopiarum, Publii Syri et Aliorum Veterum Sententiae, Recensuit et Notas Richardus Bentleius* (Cambridge, 1726). Richard Bentley's important edition of Terence corrected the text in almost 1,000 passages, often on the grounds of proper classical comedic metre.

[5] Sir Hugh Smithson, 4th Baronet (1715–86), FSA, FRS, Dilettanti (1736), created Duke of Northumberland (1766) and subsequent trustee of the British Museum (1753–86). Smithson was in Italy with his tutor Benjamin Crowe in 1733–4, travelling in Florence and Venice. In November 1733 they came to Rome with William Dugood, the jeweller, and on 12 November they dined at William Holbech's with Martin Folkes. In 1736, Smithson was back in England, and he was elected fellow of the Royal Society and the Society of Antiquaries. Smithson was also a founding member of the Society of Dilettanti. Ingamells, *Dictionary*, p. 873.

[6] Benjamin Crowe (*c.*1689–1749), cleric. In 1716, he was appointed chaplain to the British Factory at Leghorn, where he assembled a fine collection of drawings, curiosities, and objects of vertu, visited by antiquaries such as Richard Rawlinson. He continued as chaplain until 1729, when he returned to England. As Ingamells noted: 'Crowe returned to Italy in 1733–4 as travelling tutor to Sir Hugh Smithson (later 1st Duke of Northumberland)', and he was a correspondent for several years with [the architect] Alessandro Galilei. Ingamells, *Dictionary*, p. 257.

[7] Ezekiel Hamilton. See above, p. 177, n. 4.

[8] William Dugood (fl.1728–67), jeweller, FRS (elected 18 April 1728) and spy. Dugood was nominated to the Royal Society by John Theophilus Desaguliers, Edmond Halley and Martin Folkes, and made magnetic measurements off the coast of Portugal in 1733. See 'William Dugood', *Past Fellows Directory*, Royal Society, London, https://catalogues.royalsociety.org/CalmView/Record.aspx?src=CalmView.Persons&id=NA5045&pos=1 [Accessed 29 Jan 2025]. Ingamells notes of Dugood that he was in Rome 1718–22, and then in Florence, Leghorn, Parma, and Rome from 1731 to 1734, and that he acted as a 'double agent in Italy while practising as a jeweller'. Ingamells, *Dictionary*, p. 317. Ingamells further noted that Dugood was arrested in Florence on 14 September 1733. 'Two weeks later he was released and went to Rome with Sir Hugh Smithson and Benjamin Crowe, only to be rearrested, the Pretender having asked "that he should be banished from the Papal States"'. Ingamells, *Dictionary*, p. 317. It was at this point Folkes made his comments about Dugood in his diary. See also Introduction, p. 28 above, for a discussion of Dugood.

[f.93r.] Monday 9. I was in the morning at the Vatican but was only in the open galery where Rafaels bible is painted,[1] these galerys go round 3 sides of a court that is open on the 4th, ~~have~~ the 52 that make the Bible of Rafael are in one side of the second galery, the other 2 being painted in immitation of it by other hands; they have many of them been lately as it seems touchd over yet is here enough remaining to give infinite pleasure to one who loves these things, they are but small the figures as I judg tho pretty high not above 20 inches or 2 foot at most, they are in compartments 4 to each arch, between are great variety of Ornaments mostly stucho painted, some of which are very considerable in their way and the work of John da Udine.[2] the galerys above are painted with maps and views of citys[3] I could not now get into them, in the lowest galery and the front of St Peters Church are built the cells of the Cardinals in a conclave. ~~In the front of~~ In the Portico of St Peters fronting the church is the bark of Giotto of Mosaick finely preservd,[4] the air of the head of St Peter is extreamly good and like a later master I was pleasd in the whole to see a piece in such perfection of a Master that livd so near the ~~rev~~ revival of painting. In the afternoon I was at the Capitol and further took a turn on the Campo Vaccino. The Capitol I only walkd about. ~~Tuesday 10~~. the two horses \and figures/ called Castor and Pollux are gigantick but besides that they are much broke and mended,[5] one I thoug[ht] bad. the Trophys of Marius as they are commonly supposd are grand and fine.[6] the

[1] At the invitation of Pope Leo X, Raphael was commissioned to decorate the walls and ceilings of the Logge, the Papal Palace. His plans for what are now termed Raphael's Bible included designing 13 ceiling vaults, 52 ceiling frescoes (largely Old Testament scenes – 12 bays, with only the 13th with New Testament iconography), and 28 pilaster frescoes. It is unlikely Raphael himself did the paintings, his studio painters being left under the general supervision of Giulio Romano who was foreman of the works. Davidson, *Raphael's Bible*, p. 39.

[2] Giovanni da Udine, also known as Giovanni Nanni (1487–1564), responsible for the decorative elements of Raphael's projects in Rome, including the stucco work in Raphael's *Bible* (1517–19). See Davidson, *Raphael's Bible*, p. 66.

[3] Folkes is describing the *Loggia della Cosmografia* on the third storey of the Logge. The Western Wing (the Bella Loggia) and the Northern, or Central Wing, are decorated with geographical maps. As Banfi explains: 'The Western wing, or *Bella Loggia*, consists of 12 arcades, in which there are 13 maps with legends indicating the special characteristics of each country represented.' Banfi, 'Cosmographic Loggia', pp. 23–5.

[4] The *Navicella* or the Bark of St Peter, of Old St Peter's Basilica, by Giotto di Bondone, originally made for the inside of the entrance wall of the open atrium of the old basilica. The mosaic (1298) was mostly destroyed in the 17th century due to the construction of the new St Peter's Basilica. Seemingly unbeknownst to Folkes, fragments were incorporated into a new work which, in 1628, was placed at the centre of the narthex of the new building by Pope Urban VIII. This may have accounted for its more modern appearance to Folkes. It is a large piece (7.4 × 9.9. metres). 'The mosaic shows Jesus who is walking on the waters of the Sea of Tiberias and inviting Peter to come to him. But Peter, letting himself be overcome by fear is beginning to sink; Jesus saves him and says "How little faith you have! Why did you falter?" (Mt 14:24–31).' This act inaugurated the established church. See 'Navicella', St Peter's Basilica, https://stpetersbasilica.info/Interior/Navicella/Navicella.htm [Accessed 11 April 2023]. Folkes also repeats the common Enlightenment view that Giotto revived painting via his naturalism and use of perspective.

[5] Also known as the Dioskouro in ancient Greek, and by the Romans as Gemini, the statues of the twin brothers Castor and Pollux were excavated in 1561 on the site of their temple (484 BC), which was next to the Circus Maximus in a south-eastern corner of the Forum. The statues are now at the top of the cordonata, the stepped ramp to Piazza del Campidoglio, where they serve as guardians of Rome. Claridge, 'Capitoline Hill', *Rome,* p. 265.

[6] 'On the balustrade to right and left of Castor and Pollux are two famous marble compositions known since the Middle Ages ... as the "Trophies of Marius": "trophy" meaning the display of captured enemy arms and armour mounted on a wooden framework which was carried as part of the triumphal procession.' Ibid., *Rome,* p. 265. Folkes was right to be sceptical, as the association with the Roman general Marius is spurious.

columna Miliaria[1] and 2 statues of Augustus merit notice, as does the fine M Aurelius of brass on horseback,[2] the horse is full of spirit but of a heavy make, the forelock is erect and represents pretty well an owl, which they say the work men did to shew his being an Athenian.[3] on each side the door are the collossal rivers of the Nile and Tyber.[4] going down the other way, are the 3 pillars of Jupiter Tonans,[5] and the whole Ionic portico of the temple of Concord.[6] All which things are well described in many prints.[7]

Tuesday 10. The Pope went out,[8] so I saw his cavalcade, he went in a crimson velvet Sedan, the Prelates of his household rode mostly on horses and mules about him one carried a large gilt cross, he was environd[9] with his Swiss halberdeers,[10] and followed by his horse gards [*sic*], those they called the light horse have instead of carbines[11] Lances and bandoliers on each, the 2 captains of his gard came in black Cloaks and large bands; the march was closd by his litter carried by 2 mules and his Coach of State with six grey horses, there were also the led horses richly caparisond.[12] the whole appearance was fine but there is a sort of [troubld?][13] air in [f. 94r] his guard. he was in crimson velvet with a

[1] Proximal to the trophies at the far ends of the balustrade are the *Colonna Miliaria*, or two milestones, 'from the Via Appia, that on the right marking the first mile (headed I), with an inscription recording repairs under Nerva in 91 AD, that on the left the seventh (VII), inscribed with the names and titles of Vespasian in 76 AD'. Claridge, 'Capitoline Hill', *Rome,* p. 266.

[2] Folkes is referring to the Roman equestrian monument of Emperor Marcus Aurelius (AD 181), which is a gilded hollow cast bronze that is over life-size. The antique statue was brought to the Capitoline Hill in 1538 and placed in the centre of the oval piazza on the axis of the entire architectural composition, by Pope Paul III. Ackerman, 'Marcus Aurelius', p. 73. The statue is now in the Palazzo dei Conservatori.

[3] Folkes is correct that the symbol of an Athenian is an owl. However, the statue of Marcus Aurelius in the medieval era had various identities, and the idea of an owl on the horse's head in the statue is a conflation of a medieval legend, the *Mirabilia urbis Romae.* The owl which supposedly stood on the horse's head was actually the gathered tuft of mane above its forehead. Stewart, 'The Equestrian Statue of Marcus Aurelius', p. 270.

[4] 'The two colossal, bearded males reclining semi-nude and holding cornucopia (horns of plenty) personify rivers, the Nile on the left, leaning on a sphinx, and the Tiber on the right (with wolf and twins). They came in 1518 from the Quirinal Hill, where they had decorated the Baths of Constantine ca 315 AD'. Claridge, 'Capitoline Hill', *Rome,* p. 267.

[5] Temple of Jupiter Tonans, now attributed as the Temple of Vespasian and Titus, completed by Domitian in the 1st century. Claridge, 'Temple of Vespasian and Titus', *Rome,* p. 82.

[6] Temple of Concordia Augusta, dedicated by Tiberius in AD 10 to replace an earlier temple of Concord 'which had been struck by lightning or otherwise burned down in 9 BC.' Claridge, 'Temple of Concordia Augusta, *Rome,* p. 80.

[7] A cross is inked in the right-hand margin here.

[8] Pope Clement XII (1652–1740), born Lorenzo Corsini, who served as head of the Catholic Church from 12 July 1730 until his death in February 1740. He was known (inter alia) for having the Trevi Fountain built, for a programme to restore the papal finances using lotteries, and, of relevance to Folkes, for issuing the first papal decree against the Freemasons (1738). The latter event made Folkes ultimately decide not to return and settle in Italy. Clement XII enlarged the collection of antique statues in the Musei Capitolini which were made publicly accessible for the first time to travellers like Folkes. The uneasiness Folkes sensed in the papal procession may have been due to Clement's physical debility and political inefficacy, as the Pope was often bedridden with gout, blind from 1732, and suffered from loss of memory from 1737. Kelly and Walsh, 'Clement XII', *A Dictionary of Popes.*

[9] Environed, or: 'To be in attendance upon (a person); to be stationed round (a person or thing) as guards'. See *OED.*

[10] The Swiss Guard.

[11] A carbine was a short-barrelled lightweight firearm used by cavalry.

[12] Caparisoned. A *caparison* is a cloth covering laid over a horse or other animal for protection and decoration. See *OED.*

[13] This word is unclear.

Figure 45. Lorenzo Filippo de' Rossi and Gasparo Massi, after Antonio David, *Portret van Paus Clemens XII, Gasparo Massi, naar Antonio David* (*Portrait of Pope Clement XII*) (1730–31), engraving on paper, 25.6 × 19 cm. Rijksmuseum, Amsterdam. Folkes described the Pope as 'led through the [Lateran] church being quite or very near blind, but is an old gentleman of a very good and reverend Aspect'. Courtesy of the Rijksmuseum, Amsterdam.

rich embroidered stole. the people all kneeld as he passd. we went before him to St John Lateran where he came and his devotion before 3 altars, he was led through the church being quite or very near blind, but is an old gentlemen of a very good and reverend Aspect. his Niece and some other Ladys met him there and receivd his blessing. Sigr Galilei the Architect[1] presented him and his retinue with new prints of the intended face of the Lateran Church.

they are filling up and adorning here a little chappel on the left hand of the great door, for his burying place,[2] and they are new polishing a large and fine Sarcophagus of Porphyry to contain him, it is the same that stood antiently in the portico of the Pantheon. As we

[1] Alessandro Galilei. See above, p. 177, n. 6.

[2] Giovanni Battista Maini (1690–1752) sculpted the bronze figure of Pope Clement XII for his tomb in the Capella Clementina in the Lateran from 1732–6, as well as a marble statue of Cardinal Neri Corsini, the Pope's uncle, which is placed directly opposite the funerary monument. The chapel was designed by Alessandro Galilei and executed by Francesco and Pietro Ceci. The sarcophagus to which Folkes refers is Roman, made of porphyry. See Figure 47.

Figure 46. Guiseppe Vasi, *View of the new façade of the Lateran Basilica, as built to the design of Alessandro Galilei of Florence, 1741,* in *Veduta e prospetto della nuova facciata della Sacrosanta Basilica Lateranense fatta erigere dalla santa memoria di Clemente XII*, Rome (1739), etching, 20 × 29 cm. Europeana. Biblioteca Nazionale Marciana, Venice. Folkes had seen the plans for its construction being presented to the Pope during his visit to Rome. Courtesy of the Biblioteca Nazionale Marciana, Venice.

Figure 47. Giovanni Battista Piranesi, *Remains of ancient buldings among which stands the sepulchral urn of Marcus Agrippa made entirely of porphyry, which serves today as the Tomb of Clement XII*, Plate 5 from the portfolio 'Prima Parte di Architettura, e Prospettive' (*c.*1800–9), etching on paper, 38.7 × 23.6 cm. Metropolitan Museum of Art, New York, Rogers Fund, transferred from the Library. The view of the urn which Folkes describes is roughly contemporaneous to Folkes's visit. Courtesy of the Metropolitan Museum of Art, New York.

came home we saw the casino of the palace of Rospigliosi,[1] it stands in a pleasant little garden full of Orange trees. and consists of an open Portico and a room at each end the whole front is adornd with antient bass reliefs, some of which are in the book of the Admiranda. in the cieling of the portico is the famous Aurora of Guido[2] gravd by ffrey[3] an incomparable piece, and round about are the beginyngs of Love by Tempest, pretty things,[4] but in no comparison with the other. in the room[5] at one end or two pictures of Domenichin,[6] Samson throwing down the temple,[7] and the triumph of David for the slaughter of Goliath both noble pictures, in the same room is also a fine Andromeda of Guido very like that that the D of Devon has in England.[8] In the other room is the celebrated Adam and Eve of Domenichin accountd one of his chief works, and the Rinaldo and Armida of Albano.[9] the Adam and Eve is the grandest of his paintings I have yet seen, there is a pretty deal of Landscapes and Animals the piece being

[1] Palazzo Pallavicini Rospigliosi, built on the site of the Baths of Constantine on the Quirinal Hill. The portico or summer house that Folkes refers to is the Casino dell'Aurora, which also served as a baroque art gallery. 'History', Casino dell'Aurora Pallavicini, http://www.casinoaurorapallavicini.it/eng/History.htm [Accessed 12 April 2023].

[2] *L'Aurora*, an exemplar of Baroque classicism, painted by Guido Reni between 1613 and 1614 for the ceiling of the Casino dell'Aurora.

[3] Johann Jakob Frey the Elder (1681–1752) was a Swiss engraver and regarded as one of the ablest artists in Rome. See Figure 48.

[4] These are friezes by Antonio Tempesta to decorate the two side walls of the Casino's central hall The right side features a picture of the 'Triumph of a Roman general being given a crown by a winged Victory'. On the left is the 'Triumph of Love', portrayed as a Dionysian procession. 'The Collection: Frescoes', Casino dell'Aurora Pallavicini, http://www.casinoaurorapallavicini.it/eng/Frescoes.htm [Accessed 12 April 2023].

[5] The Galleria Pallavicini Palazzo Pallavicini Rospigliosi, the collection assembled by Cardinal Lazzaro Pallavicini (1602–80).

[6] Domenichino [Domenico Zampieri] (1581–1641).

[7] 19th-century works written before the Pallavicini Rospigliosi collection was dispersed attribute the *Death of Samson* to Ludovico Carracci (1555–1619). See, for instance, Anonymous, 'Rome and Her Works of Art', p. 39.

[8] Folkes is referring to the painter Guido Reni (1575–1642) and his work, *Perseus and Andromeda* (1635–6), The Pallavicini Gallery, Palazzo Pallavicini Rospigliosi, Rome. 'As Reni felt increasing financial pressures from his gambling losses, he worked more quickly and sold more studio replicas, often retouching them himself only slightly. Consequently, there are multiple versions of many works from the late 1630s' such as the *Perseus and Andromeda*. Spear, 'Reni, Guido', *Grove Art Online*, https://doi.org/10.1093/gao/9781884446054.article.T071466 [Accessed 12 April 2023]. There are in fact two *Perseus and Andromeda* attributed to Reni, one in the Galleria Pallavicini at Rome, and a 'London' version, the latter in the collection of the Dukes of Devonshire from *c.*1713 until after 1854. Frank, *Melville,* p. 9. See also Pepper, *Guido Reni*, catalogue numbers 158A and 158B. Wright did a critical comparison of the two versions, writing: 'Within the Apartments of the Garden-House are, An Andromeda by Guido, the same as the Duke of Devonshire's; the Colouring is warmer than that of his Grace's: but I know not whether 'tis better for that or no, or whether a somewhat colder Colouring do not full as well suit a Figure in such a Situation; expos'd naked, chain'd to a Rock in the Sea, expecting every Moment to be devoured by a horrible Monster, which advances towards her with dreadful wide open'd Jaws: The Colour of the Sea is turn'd blackish.' Wright, *Some Observations*, I, p. 311. William Cavendish, 4th Duke of Devonshire (1720–64) held the title at the time Folkes wrote his diary.

[9] Folkes was probably referring to the ceiling fresco of Giovanni Baglione (1573–1664) in the Casino dell'Aurora, Palazzo Pallavicini-Rospigliosi, Rome, of scenes from *Orlando Furioso:* 'for which the artist was paid in 1614 ... Baglione himself describes the subject of the fresco as "la favola d'Armida, quando trovò Rinaldo addormentato, e sopra il suo incantato Carro il riposo" [The tale of Armida, when she found Rinaldo asleep, and resting upon his enchanted Chariot].' Baglione, *Rinaldo and Armida*, Metropolitan Museum of Art, https://www.metmuseum.org/art/collection/search/338116 [Accessed 24 April 2023]. See also Baglione, *Le vite de' pittori*, p. 403.

Figure 48. Jakob Frey (I), after Guido Reni, *Aurora leading the Chariot of Apollo* (*Aurora ledit de wagen van Apollo, Bonarum Atrium Cultoribus*) (1722), etching and engraving on paper, 44.5 × 90.2 cm. Rijksmuseum, Amsterdam. A reproduction of what Folkes called the 'aurora by Guido' which he referred to during his tour of the Palazzo Pallavicini-Rospigliosi in Rome. Apollo in the chariot of the sun, drawn by four horses, is usually preceded by Aurora holding a torch. Courtesy of the Rijksmuseum, Amsterdam.

large, the scene is after the eating the fruit Adam is gathering leaves which he gives to Eve crouching on the ground.[1] In the afternoon I was at the Pantheon and Piazza Navona.[2]

Wednesday 11. I was in the Capella Sistina at the Vatican where is the judgment of Michael Angelo[3] tho vastly spoiled and never of an agreeable contouring, nor a fine composition there yet remains enough in it to ~~super~~ \sur/prise and raise the admiration from the exquisite muscleing of the figures, I lookt at it above an hour with instruction, and am well assurd those who examin it [f. 95r] carefully will approve of the applause and reputation it has based in notwithstanding the cavils at its parts. the prints and the descriptions in books will suffice me without any further observations. the Prophets and Sibils are in the Cieling[4] and some other designs of Michael Angelo of which I had seen prints[5] but did not know where they were; some of these are incomparable tho too gigantick especially the Jonas[6] who fronts ~~of~~ over the judgment, he is quite a colossus. round about are historys of Perugin[o][7] no great matters after seeing the other. the hall

[1] *The Original Sin* (*c.*1621–2), Galleria Pallavicini, Rome. See Zeri, *La Galleria Pallavicini*, pp. 285–7.

[2] Piazza Navona, or the Stadium of Domitian. The stadium was inaugurated by the Emperor Domitian, *c.* AD 86, and designed for athletic contests in the nude Greek fashion; games in honour of Jupiter Capitolinus were held there, as well as public executions. The Stadium could also substitute for the Colosseum for gladiatorial contests. The obelisk on Bernini's *Four Rivers Fountain* (1651), in the centre of the piazza, had nothing to do with the Stadium, but it was made in the reign of Domitian, cut by Roman stonecutters offering a hymn to Domitian and the deified Vespasian and Titus. See Claridge, 'Piazza Navona', *Rome*, pp. 236–7.

[3] *The Last Judgement* by Michelangelo, Sistine Chapel, Rome (1536–41). The fresco on the wall above the altar of the Sistine Chapel had been 'planned by Clement VII as a "penitential memorial" of the terrible Sack of Rome of 1527, which many considered a divine punishment for the paganish excesses of the Renaissance. ... The immense fresco was painted between 1535 and 1541; it represents the late style of Michelangelo, who was beginning to be more deeply interested in religious problems than in art.' See *The Vatican: Spirit and Art*, p. 126. Michelangelo's aesthetic norms of mystical passion and *terribilità* were alien to Folkes who favoured neoclassicism, deism and Freemasonry. Folkes largely seems to echo Wright's appraisal. Wright wrote: 'In the Sistina, as soon as you come in, you see in front at the further end that Great, and so much noted Performance of his, The Last Judgment. The Design of this famous Piece, and the capricious Fancies that are in some parts of it, are so universally known by the Prints, and the Accounts of it in several Authors, that I need say nothing of that matter. The Execution is very bold and strong, but is hard in the Outline, as are the Works of most of the Florentine Masters. The excessive strong Expression of the Muscles even in the Women, and the youngest Figures, shew rather an Ostentation of his Knowledge of the Situation and Movement of those Muscles, than a just Thought of what was altogether fit to be done in such Subjects: But, he seem'd industrious in all his Works, that the World should know he was an Anatomist: And 'tis perhaps as learned a Piece, in that respect, as there is in the World. The Colouring seems never to have been very beautiful (tho' somewhat must be allow'd for Age) and the want of large Masses of Light and Shadow, makes the Whole less agreeable, tho' the particular Figures are exceeding masterly.' Wright, *Some Observations,* I, pp. 259–60.

[4] 'Along the entire length of the vault are painted, in alternation, figures of prophets and sibyls, seated on massive thrones. Among the most beautiful are the weeping Jeremiah, next to the scene of the first day of Creation, and the Delphic Sibyl, next to that of the Deluge.' *The Vatican: Spirit and Art,* p. 111.

[5] The auction catalogue of Folkes's art collection indicates he owned several prints of Michelangelo's work in the Sistine Chapel, including, 'four, of M. Angelo's last judgment, and 5 others'. *A catalogue of the prints and drawings*, p. 8, lot 50.

[6] *Jonah*. The fresco (*c.*1508–12) of the reclining and nearly nude prophet, his head thrown back to draw the viewer's attention to the ceiling, is painted above the high altar due to Jonah's prophetic significance in the Old Testament. The figure is directly above the *Last Judgement*. The whale that swallowed him is painted behind him.

[7] Perugino [Pietro di Cristoforo Vannucci] (*c.*1450–1523), who worked on the decoration of the recently restored Sistine Chapel. Pagden, 'Perugino [Vannucci, Pietro de Cristoforo', *Grove Art Online*, https://doi.org/10.1093/gao/9781884446054.article.T066629 [Accessed 5 Nov 2022].

before the chappel[1] is painted by Vasari and the best things I have seen of him, they have somewhat the appearance of tapestrys but are very fine. there is among others the wounding Admiral Coligni and the Massacre of Paris.[2] from hence we went to what is called the Popes superior apartment the hall or guardroom[3] as painted by paul bril[4] as indeed is a great part of the Apartment and with Landscapes that are exceedingly agreeable; the Martyrdom of St [*blank*][5] who was thrown into the sea with an Anchor makes the lower end of the hall. the cieling and the rest of the hall contains a great many things in perspective[6] very curious particularly a mural crown that deceives the eye as well as most things I have seen. the Pope is at the upper end baptising an eastern Prince I do not know the story,[7] and there turks that seem just converted, and an angel leading three little blacks excessively pretty to be christend. tho this is not accounted one of the great works of the Vatican I dare promise any body of good taste he being extremely pleased with it. There are some pictures in the rooms but not of the most capital. one is celebrated however of Romanelli[8] in the popes secret chappel or closet.

#[9]

[1] The *Sala Regia*, an antechamber to the Sistine Chapel, in the south-west part of the papal residence.

[2] Folkes is describing Giorgio Vasari's three paintings of the *Massacre of the Huguenots* in the *Sala Regia* which, 'unfolds as a drama in three acts: The Wounding of Admiral Coligny, the Death of Coligny, and Charles IX before Parliament, events which took place in Paris in August 1573 only several months before Vasari commemorated them in the Vatican'. The events were commemorated by the Papacy as the massacre was interpreted as an act of divine retribution and Coligny a threat to Christendom. Vasari used Italianate motifs and architecture in the paintings as he had never travelled to France. Howe, 'Architecture in Vasari's Massacre', p. 258.

[3] The *Sala del Concistoro*, with its richly carved and gilded ceiling, was frescoed by Cherubino Alberti and Paul Bril. It is adjacent to the *Sala Clementina* which serves as the vestibule to the papal apartments. *The Vatican: Spirit and Art*, p. 138.

[4] Paul Bril (1554–1626), born in Antwerp, painter, printmaker and draughtsman, brother of Matthiis Bril. Paul Bril did a number of commissions for the Vatican in Rome, executing the 'monumental seascape, the *Martyrdom of St Clement* (*c.*1600–02/3), painted with the brothers Giovanni and Cherubino Alberti in the Vatican Palace's Sala Clementina for Pope Clement VIII (*reg* 1592–1605)'. Courtright, 'Bril family [Brill]', *Grove Art Online,* https://doi.org/10.1093/gao/9781884446054.article.T011341 [Accessed 25 April 2023]. Pope Clement I (r. 88–99) or St Clement of Rome was imprisoned under Emperor Trajan and executed by being tied to an anchor and thrown into the sea.

[5] Folkes left a blank space here, as he could not recall the iconography. In the margin another hand wrote, 'St. Cleme\nt/ Pope and martyr'.

[6] The ceiling fresco is *The Apotheosis of St Clement* by Giovanni Alberti, an exemplar of *trompe-l'œil.* The fresco portrays an open sky, and the kneeling figure of St Clement in the centre who gazes up toward the Trinity. The real cornice of the room supports an illusionistic architecture of balustrades and broken pediments, as well as a fictional loggia. Abromson, 'Clement VIII's Patronage', p. 537.

[7] Folkes is referring to the *Baptism of Constantine* frescoed by Italian painters Cherubino and Giovanni Alberti and Baldassare Croce. The fresco is on the opposite wall to Brill's fresco of the *Martyrdom of St Clement.* The Emperor kneels to receive the sacrament from the hands of Pope Sylvester (r. 314– 55). Abromson, 'Clement VIII's Patronage', p. 537.

[8] Giovanni Francesco Romanelli (1610–82), born in Viterbo, a major Italian Baroque painter whose patrons were Pope Urban VIII and Cardinal Francesco Barberini. He was a pupil of Domenichino and was commissioned to paint the Nativity over the altarpiece of the pope's private chapel (Capella Pontificia, also known as the Capella Secreta) in the Palace of Sixtus V on 8 June 1637. This is probably the painting to which Folkes is referring. Witcombe, 'The Vatican Apartment', p. 18.

[9] Folkes wrote this mark in the text.

Thursday 12. I spent the morning in walking about the town, I was at the Church of St Mary Eqyptiaca, near the Tyber formally the temple of ffortuna Virilis[1] this temple was small but formerly extremely elegant, the filleted Ionic columns and Cornish being very beautifull it had a portico before it[2] but now not remaining and houses are built close upon it. from hence I went to the little arch of Severus[3] which has been full of bass reliefs but now also miserably defaced,[4] just by are the ruins of the former temple of Janus with 4 doors exactly agreeable to the prints of it. and adjoyning is a church built upon what was the temple of Acca Laurentia Romulus's nurse. In the Church of St Mary Cosmedin[5] hard by is one of the oldest Christian Inscriptions remaining, the letters are exceed [f. 96r] dingly barbarous and odd, there is a representation of somewhat like an Aqueduct on the same stone. I went from the temple of Janus by the side of the Palentine hill, where I went into the little round ~~temple~~ Church of St Toto or Theodor,[6] where the statues of Romulus and Remus were found, it is a pretty round building standing as it were under the side of the hill, I then passd the Arch of Constantine[7] to the Campo Vaccino where I climbd the Capitol hill to the portico of the temple of Concord fine Granite pillars of the Ionic order, their bases coverd.[8] and next to the 3 corinthians ones supposd to have belongd to the temple of Jupiter Tonans,[9] there are some of the richest here but covered with earth a 4th of their height very near, on the cornish is a good bassrelief of sacrificing Instruments etc all these are excellently described by Desgodetz,[10]

[1] Temple of Fortunas, formerly known as the Temple of Fortuna Virilis. It was converted into a church of Santa Maria *c.* AD 872. By the 16th century, the building was known as St Mary of Egypt, and it was given to an order of Armenian monks. Most of the decoration moulded in stucco on the frieze of the travertine entablature is lost, but it can be reconstructed from early modern drawings; it featured erotes, candelabra and bull's skulls. Claridge, 'Temple of Fortunas', *Rome*, pp. 285–6.

[2] Folkes was correct, the Ionic portico was raised on a very high podium to keep it clear of Tiber floods. Claridge, 'Temple of Fortunas', *Rome*, p 286.

[3] The Arch of Septimius Severus. The inscriptions on the arch indicate that it was awarded to the 'emperor Septimius Severus and his sons Caracalla and Geta in 203 AD for "having restored the Republic and expanded the dominion of the Roman people"'. Claridge, 'Arch of Septimius Severus', *Rome*, p. 79.

[4] Folkes was correct about the extensive damage to the arch. In the medieval era, it was made into a fortress with towers erected on top, and the arch has scars of other later structures. Ibid.

[5] For the church Santa Maria in Cosmedin in the Forum Boarium, see above, p. 185, n. 2.

[6] San Teodoro is at the foot of an ancient staircase, which leads down from the north-west foot of the Palatine Hill. The church was almost entirely rebuilt in the early 18th century, its unusual round shape appealing to Folkes. The Capitoline Wolf was kept in the church until the 16th century. See Claridge, 'San Teodoro', *Rome*, p. 297.

[7] The Arch of Constantine was dedicated on 25 July 315, the 10th anniversary of his reign, spanning his triumphal route. It is similar in size and design to the Arch of Septimius Severus on the Roman Forum. Claridge, 'Arch of Constantine', *Rome,* pp. 308–9

[8] For the Temple of Concordia, see above, p. 189, n. 6.

[9] For the Temple of Jupiter Tonans, see above, p. 189, n. 5. Folkes echoes Wright's description here. 'Just by [the Temple of Concord] are three noble Pillars, which seem to have been the Angle of a Portico to a temple, with part of their Entablature. They are so far buried by the Ruines of the old Capitol, which stood a little higher, that scarce half of them is above ground, and what is so, is almost hid by Trees.' Wright, *Some Observations* I, p. 256.

[10] Antoine Babuty Desgodetz's (1653–1728) publication, *Les édifices antiques de Rome dessinés et mesurés très exactement* (1682). Desgodetz was employed by the French minister, Jacques-Louis Colbert, to visit Rome to make engravings of Rome's monuments and antiquities for the use of French artists and architects, as well as their measures – of great interest to Folkes. See Figure 49. The auction catalogue of Folkes's library shows he owned a copy. See *Catalogue of the entire and valuable library*, p. 126, lot 4255.

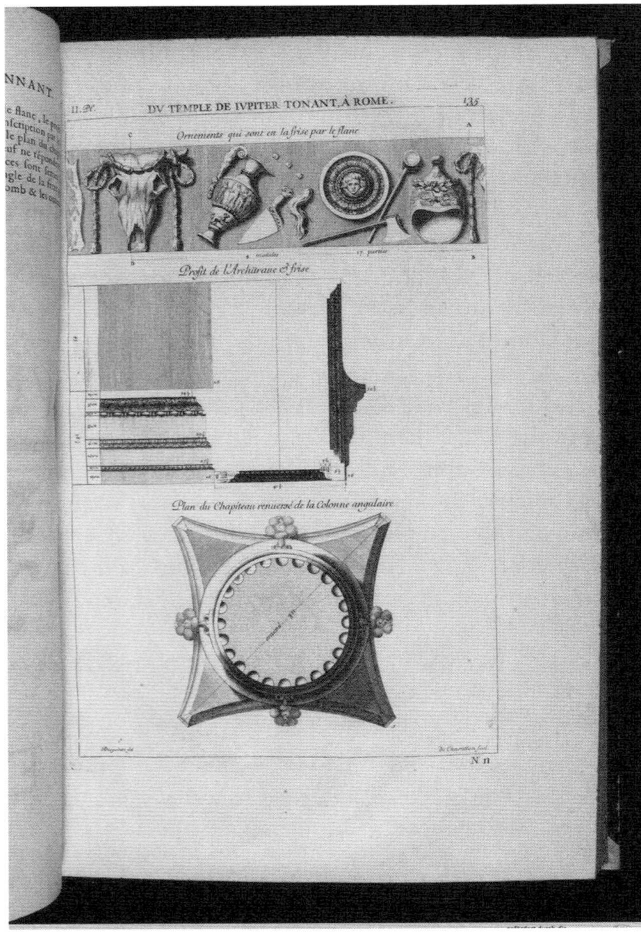

Figure 49. Plate showing the frieze decoration of the columns of Jupiter Tonans that Folkes admired, in Antoine Babuty Desgodetz, *Les edifices antiques de Rome dessinés et mesurés très exactement*, Paris (1682), p. 135. University of Heidelberg Digital Library, https://doi.org/10.11588/diglit.3251#0138. Courtesy of the University of Heidelberg.

as their present ruinous condition may be seen in Overbeeke,[1] which I look on as an excellent book to preserve the Idea of the general present appearance of things, as the other is to show what they have been by his accurate measures, and compleating the parts one from another.

[1] Bonaventura van Overbeek (1660–1705), a Dutch draughtsman and engraver. Folkes is describing his *Reliquiae antiquiae urbis Romae* (1708), a book of the major Grand Tour sites of Rome. It was translated into French as, *Les Restes de l'Ancienne Rome: recherchez avec soin, mesurez, dessinez sur les lieux, & gravez par feu Bonaventure d'Overbeke*, 3 vols (Amsterdam, 1709). The auction catalogue of Folkes's library shows he owned a copy of the 1708 Latin version. See *Catalogue of the entire and valuable library* p. 147, lot 4890.

I met at the Capitol,[1] Mr Mace and Mr Jeffryes, we went together to the Tarpeian Rock now probably much altered from its antient state however a precipice,[2] and to the Statue of Marsaris[3] in one of the courts of the Capitol, there is also just by a statue of the goddess Roma,[4] and 2 Egyptian Isises of Basaltin stone near 8 foot high each very entire and extreamly odd, there are hieroglyphicks on one and she holds the Lotus in her hand. I went with the same gentleman to Trajans Pillar,[5] and the remains of the old Sepulchre of Publicius now part of a house;[6] this I have only seen drawn in Overbeeke. I dind today at Mr Holbechs with Sr Hugh Smithson, Mr Crow and Mr Campbell. who gave us a very elegant entertainment. We staid there the best part of the afternoon. Mr Dugood[7] that came t'other day to town with with Sr Hugh etc. was yesterday put in prison by the Popes order and commanded to leave his dominions in 3 days he sent to us, but we have here no Protector and knowing nothing of his affairs can do nothing for his service.

[f. 97r. Letter from Folkes tipped into the diary, possibly to Madame Geoffrin (identified from the hand on the wrapper on f. 97v), including doodles of ratios based on William Oughtred's notation]

Cum venit[8] adhuc in Bibliotheca tabula antiqua Geografica quam ante 3 ad minimum saecula propria manu delineavit ipsius clarus RP Mauro[ensis] ejusdem cenobii, cosmographus inclytus; quae utilissima esse possit ad historiam Geografie recentioris docendam, ipsam publici juris facere praeposuimus, suasus praesertim universorum[9] aliquot qui tale negotium viris doctis gratum fore constanter asseruerunt.

[When he arrived, (there still remained) in the library an ancient geographical map which the famous cosmographer and Reverend Father Mauro of the same monastery[10] drew in

[1] Capitoline Hill.

[2] Folkes was again following Wright for the debates about the exact height of the Tarpeian Rock. Wright noted: 'I must not leave the Capitol without mentioning the Rupes Tarpeia [Tarpeian Rock] ... to convince us of the Mistake of Father Montfaucon, who says there is a little Precipice left; and of another very great Writer [Burnet], who represents it as what a Man might jump down without danger. What he shew'd is on your right hand, as you face the Capitol, and not far from the Palazzo Cassarelli; he affirms that he measur'd it, and found it to be eighty Palms [that is, sixty Foot] above ground, as it now is, besides what is hid of it with Rubbish at the bottom. Whether his measure is exact or not, I do not know; but it is manifestly so high, that no Man that was not quite mad, would take such a desperate Leap.' Wright, *Some Observations*, I, p. 326.

[3] Marsyas.

[4] In front of the staircase up to the Palazzo Senatorio in the Piazza Campidoglio were 3 statues assembled in the 16th century: the Nile, the Tiber, as well as Roma Triumphans, the seated statue in the central alcove. As Aiken noted: 'The seated Roma is also an ancient motif, and can be found on numerous Roman coins as well as in sculpture'. Aikin, 'Romae de Dacia', p. 588.

[5] For Trajan's Column and its importance to Folkes for his metrological studies, see Introduction, pp. 27–8 above.

[6] The tomb, at the base of the Capitoline Hill, is of C. Publicius Bibulus, who was a plebeian aedile c. 100 BC. Claridge, 'Tomb of Bibulus', *Rome,* p. 191.

[7] For Dugood, see Introduction, p. 28 above; see also p. 187, n. 8.

[8] There seems to be a verb missing after 'cum venit', which would then mean 'when he arrived there still remained ... '. Otherwise, 'cum venit' does not make sense. 'Tabula' must be the subject of this missing verb ('remained' or some such), as it is in the nominative.

[9] There seems to be an omission of 'a/ab' before 'aliquot', which is assumed for sense.

[10] RP Mauroensis; the medallion struck shortly after Fra Mauro's death calls him 'Maurus S. Michaelis Moranensis', that is, Maurus of (the monastery of) St Michael on Murano (an island in the lagoon)

his own hand at least three centuries previously. Since this could be most useful in teaching an account of modern geography, I have decided to place it in the public domain, being persuaded particularly by some of all those who declared unwaveringly that such an undertaking would be welcome to the academic world].

[Inside of back cover]
Diam. of the Farnesian Globe 2 ^{f}. 5 ^{Inch}. 1/10 English measure.[1]

[1] Folkes had a half-scale copy made of the Farnese Globe, 'constructed from linen stretched over a wooden armature', with the surface of the globe made of plaster applied to the linen support and polished. The globe is still extant in the British Museum. He also did a stereographic projection of the Roman Farnese Globe, relying on his knowledge of astronomy to date the globe by the positions of the constellations and stars it portrayed. Folkes is clearly noting the dimensions of the globe in his diary. As Lippincott has noted: 'The note is set above a cursory, unfinished sketch of an arch supported by columns or pilasters with simple, protruding capitals, which strongly recalls the arrangement of pilasters, capitals and arches found on the ground floor arcading of the interior *cortile* of the Palazzo Farnese.' Lippincott, 'A Chapter', p. 286; see also Introduction, p. 18 above.

APPENDIX A

Martin Folkes's Itinerary

The following pages are a transcription of Martin Folkes's itinerary across the Continent, written in his notebook 'Memoranda in case he should go abroad again, respecting his Estates, foreighn money & Places', presently held in the Norfolk Record Office in Norwich, NRS 20658. This title of his notebook was written by Folkes on rag paper, pasted onto the cover of the 48-page vellum-bound notebook and signed by him in 1734. It is thought the manuscript was begun early in 1733.

His record of the stops he made on his journey was utilized to create the map of his Grand Tour in this volume. The left column indicated time spent in the locality, each asterisk indicating one day. A cross indicated a prolonged period of stay. Folkes also indicated the 'posts' or milestones between each stopping place.

My Itinerary

	From London to Gravesend in the D. of Montagus[1] boat.
*	To Rotterdam on board the *Chandos* Sloop[2]
	To Delft
*	To the Hague
*	To Leiden
	To Harlem
*	To Amsterdam ..... to Naarden[3]
*	To Utrecht
	By Coach
	To Rhenen[4]
*	To Nimeguen[5] about 4 posts[6] from Utrecht on causeways
	To Cleeves[7] 2 posts short.
	To Somten 2 posts short

[1] John, 2nd Duke of Montagu (1690–1749), Folkes's friend and fellow Freemason, elected FRS in 1718.

[2] The boat to the Netherlands. A sloop was 'a small, one-masted, fore-and-aft rigged vessel, differing from a cutter in having a jib-stay and standing bowsprit'. *OED*.

[3] Naarden, a city and former municipality in the Gooi region in North Holland, Netherlands.

[4] Rhenen is in the central Netherlands, on the southernmost part of the chain of hills known as the Utrecht Hill Ridge where this meets the river Rhine.

[5] Nijmegen, located on the river Waal close to the German border.

[6] Milestones.

[7] Kleve, a town in the North Rhine-Westphalia near the Dutch border and river Rhine.

*	To Hoffstadt.[1] 2.
	To Nuyse. 2.
**	To Cologn.[2] 2.
	To Spieg.[3] 1.
	To Warth.[4] 1.
	To Weyerbus[5] 1 ½
	To Gulroth.[6] 1.
*	To Frailing.[7] 1.
	To Walmenoth.[8] 1.
	To Limburg.[9] 1.
	To Wirgas.[10] 1. ½
*	To Kuningstein.[11] 1. ½
*	To Francfort.[12] 1.
	To Hanaw.[13] 1.
	To Diettling.[14] 1.
*	To Besenbach.[15] 1 ½
	To Rottebrun.[16] 1.
	To Eyselbach.[17] 1.
	To Remlinghen.[18] 1.
*	To Wurtzburgh.[19] 1.
	To Kitzinghem.[20] 1 ½.
	To Posenheim.[21] 1.

[1] Hoffstadt, Lindlar, North Rhine-Westphalia.

[2] Cologne (Köln), spanning the river Rhine in North Rhine-Westphalia.

[3] Possibly Siegburg, a city in the district of Rhein-Sieg district in North Rhine-Westphalia, about 26 km from Cologne.

[4] Warth is a hamlet in North Rhine-Westphalia, situated nearby the hamlets of Schiefenthal and Halzemich.

[5] Weyerbusch, a town in the Rhineland-Palatinate.

[6] Gieleroth, a municipality in the Rhineland-Palatinate.

[7] Freilingen, a community belonging to a *Verbandsgemeinde* or collective municipality of Selters, in the Westerwald district in the Rhineland-Palatinate.

[8] Wallmenroth, a municipality in the district of Altenkirchen, in the Rhineland-Palatinate.

[9] Possibly Limberg, Asbach, a municipality in the Neuwied district, in the Rhineland-Palatinate.

[10] Wirges, a town in the Westerwald district in the Rhineland-Palatinate.

[11] Königstein im Taunus, now part of the Frankfurt Rhein-Main urban area.

[12] Frankfurt.

[13] Hanau, a town in Main-Kinzig district, Hesse, 25 km east of Frankfurt.

[14] Dettingen am Main, a municipality in Lower Franconia in Bavaria.

[15] Bessenbach, a community in the Aschaffenburg district, Lower Franconia in Bavaria.

[16] Rohrbrunn, in Aschaffenburg district, Lower Franconia in Bavaria.

[17] Esselbach, a community in the Main-Spessart district, Lower Franconia in Bavaria.

[18] Remlingen, a municipality in the district of Würzburg, Lower Franconia in Bavaria.

[19] Würzburg is a city in the region of Lower Franconia in Bavaria.

[20] Kitzingen is a town in Bavaria, capital of the Kitzingen district, Lower Franconia.

[21] Possenheim, a village in Lower Franconia, situated nearby to Markt Einersheim and Hellmitzheim.

	To Langfeld.[1] 1.
	To Farnbach.[2] 1.
***	To Nurenbergh.[3] 1.
	To Roth.[4] 2.
	To Pleinfeld.[5] 1.
	To Dietfurt.[6] 1.
*	To Monhaim[7] 1
*	To Donaweert.[8] 1.
	To Metinghan.[9] 1 ½
****	To Augsburg.[10] 1 ½
	Still in my Coach but voiture.[11]
	To Pruck.[12] 3 posts in value
**	To Munich.[13] 2.
*	To Wolfertshausen.[14] 2.
	To St. Benedicts Buren.[15] 2
*	To Wachen-Zee.[16] 1 ½
	To Mittewald.[17] 1 ½
	To Seefelde.[18] 1.
*	To [Kyrte?].[19] 1.
*	To Innspruch.[20] 1.

[1] Langenfeld, in a district of Neustadt (Aisch)-Bad Windsheim, Middle Franconia, Bavaria.

[2] Fahrnbach is a village in Bavaria.

[3] Nuremberg, the second largest city in Bavaria after Munich, located on the river Pegnitz, in Middle Franconia.

[4] Roth (formerly Roth bei Nürnberg) is a town in Bavaria, 25 km south of Nuremberg, in Middle Franconia.

[5] Pleinfeld, a municipality in the Weißenburg-Gunzenhausen district, Middle Franconia in Bavaria.

[6] Dietfurt is a town in the district of Neumarkt in Swabia, Bavaria, situated on the river Altmühl, and is 38 km west of Regensburg, and 30 km north of Ingolstadt.

[7] Monheim is a town in the Donau-Ries district, in Swabia, Bavaria. It is situated 15 km northeast of Donauwörth, and 27 km east of Nördlingen.

[8] Donauwörth is a town and the capital of the Donau-Ries district in Swabia, Bavaria.

[9] Meitingen, a market town in the district of Augsburg, in Swabia, Bavaria. It is situated on the left bank of the river Lech, 20 km south of Donauwörth, and 21 km north of Augsburg.

[10] Augsburg, Swabia, Bavaria.

[11] wagon.

[12] Possibly a hamlet called Prack.

[13] Munich, the capital of Bavaria.

[14] Wolfratshausen is a town of the district of Bad Tölz-Wolfratshausen, located in Upper Bavaria at the confluence of the Isar and Loisach rivers.

[15] Benediktbeuern, a municipality in the district of Bad Tölz-Wolfratshausen in Upper Bavaria.

[16] Walchensee is a small fishing village in the Kochel municipality, located on the western banks of a lake of the same name.

[17] Mittenwald is a town in the Alpine peaks of Upper Bavaria in the Isar river valley.

[18] Seefeld is a municipality in the Starnberg district of Upper Bavaria, lying on the western shore of Pilsensee.

[19] The transcription here is uncertain.

[20] Innsbruck, capital of Austrian Tyrol.

	To Schenberg.[1] 1.
*	To Steinach.[2] 1
	To Brenner.[3] 1.
*	To Stertzinghen.[4] 1.
*	To Brixen.[5] 2.
	To Colman.[6] 1 ½
*	To Bolzano.[7] 2.
	To Neimark.[8] 2.
	To Wa[[xxx]] Michael[9] 1
*	To Trent.[10] 1
*	To Roveredo.[11] 2.
	To Alla.[12] 2
	To Peri.[13] 1.
*	To La Chiesa[14] 1.
***	To Verona.[15] 1 ½
	To Torre de Confini.[16] 2.
*	To Vicenza[17] 1 ½
***	To Padua.[18] 2
+	To Venice by Burcello[19] about the value of 3 posts

[1] Schönberg im Stubaital is a municipality in the district of Innsbruck-Land, Tyrol, Austria. It is 9 km south of Innsbruck at the entrance of the Stubaital valley.

[2] Steinach am Brenner, a market town in the district of Innsbruck-Land in the Tyrol, Austria, which is located south of Innsbruck in the Wipptal at the river Sill.

[3] Brenner (Brennero), a municipality in the South Tyrol in northern Italy, 60 km north of Bolzano.

[4] Sterzing (Vipiteno), a municipality in the South Tyrol in northern Italy, the main settlement of the southern Wipptal; the river Eisack (Isarco) flows through the town.

[5] Brixen (Bressanone) is the oldest town in the South Tyrol, northern Italy, located about 40 km north of Bolzano.

[6] Colma (Kollman), on the lower Valle Isarco in the South Tyrol.

[7] Bolzano (Bozen), the largest city in the South Tyrol, known for its confluence of Italian and German-Austrian culture.

[8] Neumarkt (Egna), village in the South Tyrol, 25 km south of Bolzano.

[9] The transcription is uncertain here, but the town is probably San Michele all'Adige, about 15 km north of Trento.

[10] Trento (Trent), city in the province of Trento, South Tyrol region of northern Italy, governed by prince-bishops until the Napoleonic era.

[11] Rovereto, municipality in the province of Trento, South Tyrol.

[12] Ala, a village in the province of Trento, South Tyrol.

[13] Peri, a village in the municipality of Dolcè, in the province of Verona.

[14] La Chiesa in the province of Treviso.

[15] Verona, a city on the river Adige in the Veneto region, and a provincial capital.

[16] Torre di Confine, a locality in Verona province which belongs to the municipality of Gambellara, nearby the hamlets of Ciron and Fassacan-Locara.

[17] Vicenza, city in the Veneto region known for its Palladian architecture which Folkes must have appreciated.

[18] Padua, a major city and commune in the Veneto region.

[19] As noted by Wright in the mid-18th century: 'the Burcello is a large handsome boat; the middle part of which is a pretty room, generally adorn'd with carving, gilding and painting'. Wright, *Some Observations,* I, p. 43.

*	To Padua by burcello. Hence in my coach but voiture.
	To Monte-Felice,[1] about 14 miles.
	To Rovigo.[2]
	To Arqua.[3]
*	To Ferrara[4] abt 35 miles from Monte Felice.
*	To Argenta[5] abt. 20 miles
	To St Alberto.[6]
**	To Ravenna[7] abt. 35 miles from Argenta.
	To Cervia.[8]
	To Porto Cesannatico.[9]
*	To Rimini[10] abt. 36 miles from Ravenna.
	To Catholica.[11]
	To Pesaro.[12]
*	To Fano[13] abt 36 miles from Rimini
	To Sinigallia.[14]
*	To l Ulmo.[15] abt. 36 miles from Fano. To Ancona 3 miles.
**	To Loretto.[16] abt 20 miles.
	To Recanati[17]
	To Macerata[18]

[1] Monselice, which is on the road from Padua to Ferrara (*c.*20 km from Padua).

[2] Rovigo, capital of the Rovigo province, standing on the low ground known as the Polesine, about 40 km south-southwest of Padua, on the Adigetto canal.

[3] Arquà Polesine, about 7 km south-west of Rovigo.

[4] Ferrara at the time of Folkes's journey was a papal state and in the Emilia-Romagna region.

[51] Argenta is proximal to the Valli de Comacchio lagoon wetlands, about 30 km south-east of Ferrara.

[6] Sant'Alberto is in the hinterland of Ravenna, 14 km from the city.

[7] As the former centre of the Ostrogothic kingdom of Italy, Ravenna in the Emilia Romagna region was an essential stop for antiquaries on the Grand Tour for its architecture and antiquities.

[8] Cervia is on the Adriatic coast in the Emilia Romagna region.

[9] Cesenatico, a port town on the Adriatic coast.

[10] Rimini, on the Adriatic coast, in the Emilia Romagna region, at the southern tip of the Po Valley.

[11] Formerly a Roman settlement, Catholica is on the Adriatic coast, in the province of Rimini in the Emilia Romagna region. It is approximately 20 km from Rimini.

[11] Pesaro, part of the Papal States until the Napoleonic occupation. It is now the capital of the province of Pesaro and Urbino in the Marche region.

[13] Fano, a town 12 km south-east of Pesaro, a municipality of the province of Pesaro and Urbino in the Marche region.

[14] Senigallia, a port town on the Adriatic coast, in the province of Ancona in the Marche region. At the time of Folkes's visit, it was annexed to the Papal States legation of Urbino.

[15] The location of l'Ulmo is unknown, other than its proximity to Ancona.

[16] Loreto, Marche, a hill town in the province of Ancona, known as the Catholic pilgrimage site of the Basilica della Santa Casa.

[17] Recanati in the province of Macerata in the Marche region.

[18] Macerata, located on a hill between the Potenza and Chienti rivers. It is now the county seat of the province of *Macerata* in the Marche region.

*	To Tolentino.[1] abt. 30 miles or 35 from Loretto
	To Ponte di Trave.[2]
*	To Serravallè.[3] abt. 28 miles from Tolentino
	To Colle Fiorito.[4]
	To Fuligni.[5]
*	To Spoleto.[6] abt 35 miles from Serravalle.
**	To Terni abt. 20 miles. 3 or 4 to the cascade[7]
	To Narni[8]
	To Otricoli.[9]
	To Borghetto.[10]
*	To Civita Castellana.[11] abt 35 long miles from Terni
*	To Castelnuovo[12] abt 20 miles.
	To Prima Porta.[13]
	To Monte Ponte Mollè.[14]
+	To Rome abt 20 miles from Castelnuovo
	To Frascati by Grotta ferrata[15]
	To Albano,[16] and home by Frascati
	From Rome in Voiture chaises
	To Ponte Molle.

[1] Tolentino in the province of Macerata in the Marche region. It is on the river Chienti, 224 metres above sea-level, and was once a fortified town of great strength.

[2] Pontelatrave, a hamlet in Camerino, province of Macerata, Marche region. The town was an ancient staging post.

[3] Serravalle di Chienti, in the province of Macerata in the Marche region, about 80 km from Macerata. As its name implies, it is crossed by the river Chienti.

[4] Colfiorito, Umbria on the road from Macerata to the municipal seat of Foligno, of which it is a borough.

[5] Foligno, in the province of Perugia in east-central Umbria. It is 40 km south-east of Perugia.

[6] Spoleto is a city in the province of Perugia in east-central Umbria in the foothills of the Apennines. It is 29 km north of Terni.

[7] Terni is a city in the southern portion of the region of Umbria, in Central Italy. Terni is near the confluence of the Nera and the Serra rivers. The Cascata delle Marmore is 165 metres tall, the highest falls in Europe.

[8] Narni, a small medieval town in the province of Terni, Umbria region, is halfway between Rome and Assisi, and it overhangs a gorge of the river Nera.

[9] Otricoli is a town and municipality in the province of Terni, Umbria region.

[10] Borghetto, a village in Cività Castellana, province of Viterbo.

[11] Cività Castellana, in the province of Viterbo, situated 65 km north of Rome. It was part of the Papal States at the time Folkes visited and an episcopal see.

[12] Castelnuovo di Porto, now part of the Metropolitan City of Rome.

[13] Prima Porta or First Gate, is named after the section of Via Flaminia near Villa di Livia. The aqueduct that brought water to the villa formed a gate over Via Flaminia, which travellers saw as the first sign of having reached Rome.

[14] Ponte Milvio, the bridge over the river Tiber in northern Rome, and the site of a battle in AD 312 when Emperor Constantine I defeated Maxentius.

[15] Frascati is about 20 km south-east from Rome.

[16] Albano Laziale, 25 km from Rome, on the Alban Hills.

	To Monte Rossi[1] 3 posts from Rome.
*	To Ronciglione[2] 1 post
	To Viterbo[3] 1 post
	To Monte Fiasconè.[4] 1 post
	To Bolesenna[5] 1. post.
	To St Lorenzo[6] 1 post
*	To Aquapendente.[7] 1 post
	To Ponte Centino.[8] 1 post
	To Radi Cassani[9] 1 post
*	To St Zuirico.[10] 2 posts
	To Sienna[11] 3 posts.
*	To Castelloncello.[12] 1 post
	To Pongi-Bonzi.[13] 1 post
	To Tavernella.[14] 1 post
	To St Ascanio[15] 1 post
+	To Florence[16] 1 post
	To Pratolino[17] a post
	To La Trappe.[18]

[1] Monterosi is a municipality in the province of Viterbo in the Lazio region.

[2] Ronciglione, a municipality in the province of Viterbo, Lazio region, about 20 km from Viterbo

[3] Viterbo, a city in the Lazio region, the capital of the province.

[4] Montefiascone is a town in the province of Viterbo, in Lazio. It stands on a hill on the south-east side of Lake Bolsena, and it is 100 km north of Rome.

[5] Bolsena is a town in the province of Viterbo in northern Lazio region on the eastern shore of Lake Bolsena. It is 10 km north-north-west of Montefiascone.

[6] San Lorenzo Nuovo is located on the northern side of Lake Bolsena's crater rim.

[7] Acquapendente, a city in the province of Viterbo.

[8] Ponte Centino, a former Papal frontier station and custom house, on the banks of the river Elvella.

[9] Radicofani, a village in the province of Siena. It is located in the natural park of Val d'Orcia about 60 km south-east of Siena.

[10] San Quirico d'Orcia, a small municipality about 35 km south-east of Siena.

[11] Siena, capital of the province of Siena, Tuscany region.

[12] Possibly Castellina Scalo, a village within the municipality of Monteriggioni, in the province of Siena. It is on the road between Siena and Poggibonsi.

[13] Poggibonsi, a town in the province of Siena, Tuscany region, on the river Elsa in the Valdelsa valley.

[14] Probably Tavernelle Val di Pesa, about 12 km from Poggibonsi.

[15] San Casciano, a municipality in Val di Pese, now in the Metropolitan City of Florence.

[16] Florence, capital of Tuscany region and a major cultural and economic centre where Folkes stayed for some months.

[17] Villa di Pratolino, 12 km to the north of Florence, a Renaissance patrician villa built by Francesco de' Medici, and demolished in 1822.

[18] La Trappe, a convent near San Pietro a Sieve, a village within the municipality of Scarperia e San Piero, located in the Metropolitan City of Florence.

	From Florence in chaises Cambiatura[1]
	To Monte Lupo[2]
	To Epoli.[3]
*	To Pisa[4] about 40 miles from Florence.
+	To Leghorn[5] 38 miles
	from Leghorn to Pisa, to Lucca,[6] To Pistoia.[7]
	from Leghorn to Gibraltar and hence to London on board the Chitty fregate Captn. Brown Commr.[8]

The remainder of Folkes's notebook (not transcribed here) included not only his recording of the expenses of his journey, but also calculations of the costs of possibly returning to Florence to live, as he was finding it difficult to cope with his wife Lucretia's mental illness. He indicated he would travel with his son Martin only, leaving Lucretia in an asylum, and his daughters Lucretia and Dorothy to be cared for by a nanny or relatives. To accomplish these calculations, Folkes did a thorough enumeration of the value of his estates in Lincolnshire, London, Norfolk, Middlesex, costs of repairs, living expenses, and annuities he paid to family members such as his younger brother Harry. Like many of his contemporaries, including Newton, he owned South Sea Stock, highly lucrative after the South Sea Bubble and crash of 1720, its profits based partially on the transatlantic slave trade. On the back of page 16 and the recto of page 17 of his notebook, Folkes included 'Some remarks about foreign money. communicated by Mr Conduit, from acc[ts] transmitted to him. Ffrance'. This was probably John Conduitt (1688–1737), who was married to the half-niece of Sir Isaac Newton whom Conduitt succeeded as master of the Mint. The listing demonstrated Folkes's interest in coinage, also evinced in his continental travels. On page 22, there was a listing of the prices of different models of the watches made by a friend of Folkes's, horologist George Graham FRS. Folkes brokered purchases of these watches for friends, such as Girolamo Ascanio Giustiniani whom he met in Venice (see above, pp. 37–8, n. 8).

[1] Travelling *e cambiatura* or 'changing horses' was a system of changing chaises at every post. This involved leasing a carriage from a coachbuilder and hiring horses from each post house at every stage of the journey. Possibly at this point Folkes disposed of his coach, as he was on his way back to England.

[2] Montelupo Fiorentino, a municipality about 19 km south-west of Florence.

[3] Empoli, about 30 km south-west of Florence, to the south of the Arno in a plain formed by the river.

[4] Pisa, city in the Tuscany region, straddling the river Arno.

[5] Livorno or Leghorn, a port city on the Ligurian Sea on the western coast of Tuscany region.

[6] Lucca, a city on the river Serchio river in Tuscany region.

[7] Pistoia, the capital of a province of the same name, in Tuscany region.

[8] 'Chitty' referred to a member of the trading castes in southern India, particularly Madras, and a 'frigate' was a merchantman ship.

APPENDIX B

Venetian Coinage, 1700–1800

The Monetary System of Venice in the eighteenth century consisted of the following coinage minted by the Zecca [Venetian mint].[1] Nominal values in Venice were expressed in *denari* (d), *soldi* (s) of 12d, and *lire* (£) of 20s.

6 baggatini	=	1 bezzo
12 baggatini or denari	=	1 soldo
20 soldi	=	1 lira
30 soldi	=	1 lirazza
124 soldi	=	1 ducatone = 1 ducato
140 soldi	=	1 scudo = 1 tallero = 1 zecchino
2 scudi d'oro	=	1 doppia

Weights

zecchino	=	3.5 grams of gold
doppia	=	6.9 grams of gold

The design of the Venetian gold ducat or *ducati* remained unchanged for over 500 years, from its introduction in 1284 to the Napoleonic suppression of the Venetian Republic by the Treaty of Campo Formio in 1797.[2] As Michael remarked, many gold denominations were struck from the same dies, and therefore, it is more important to check the weight to determine the proper denomination of any specific piece.[3]

Mattia Viale has noted that gold coins in Venice, despite reference to them in governmental budgetary reports, were rare to find in circulation within the Republic and its dominions. Gold coinage was utilized for 'international trade and for domestic transactions of great value'. In fact, gold coinage progressively disappeared from circulation as it was hoarded, and 'the "commercial" role of gold coins was taken from high-value silver coins, and this, over the long run, had adverse effects on the ability of states to control the mass (and consequently the value) of fractional coins that was minted in alloys in which minimal quantities of silver was present'. For daily use, low-silver-content coins (billon) and copper coins dominated, forming a currency that was wide and heterogeneous. As the Republic was a centre of trade, there were multitudes of low-

[1] This table and the explanation of coin dies was taken from Michael, *Standard Catalogue of World Coins, 1701–1800*, p. 1104.

[2] Ives, 'Introduction', *Venetian Gold Ducat*, p. 1 in ANS Digital Library, http://numismatics org/digital library/ark:/53695/nnan28786 [Accessed 13 July 2021].

[3] Michael, *Standard Catalogue of World Coins, 1701–1800*, p. 1104.

value coins from foreign states as well. Although the 'circulation of these was illegal and forbidden', in practice it was impossible to stop, as in Venice there was a continual flow of merchants and travellers from Europe and the Mediterranean.[1]

By the reign of the last doge of Venice, Ludovico Manin (1725–1802), the Zecca or Venetian Mint produced coins of the following denominations, some tied officially to the lira of Venice and others floating against it according to supply and demand:[2]

- gold zecchino of 24 carats and multiples of 105, 50, 10, 9, 8, and 6 times the normal weight of 3.5 grams, as well as a half and quarter zecchino
- gold doppia of 22 carats and weight around 6.75 grams as well as its half and quarter
- silver scudo of .95 fineness and a weight of around 32 grams and its half, quarter, and eighth
- silver ducatone of .95 fineness and weight around 28 grams and its half, quarter, and eighth
- silver tallero of .835 fineness and weight around 28.5 grams and its half, quarter and eighth
- silver ducato of .825 fineness and weight around 23.4 grams and its half and quarter
- billon (alloy of gold and silver with copper) coin of 15 soldi of .39 silver fineness and weight around 3.7 grams and corresponding coins of 10 and 5 soldi
- billon soldo of .03 silver fineness and about 2 grams.

[1] Viale, 'Consumer Behaviour', p. 40, n. 21. All quotations in this paragraph were taken from this source.

[2] The coinage and their values and weights were taken from Stahl, 'Venetian Coinage and Money of Account', pp. 6–7; See also Stahl and Tonini, 'Zecca di Venezia', pp. 1217–24; Lane and Mueller, *Money and Banking*.

BIBLIOGRAPHY

MANUSCRIPTS

Baldassaroni–Corsini Archive, Florence

Letters of Antonio Cocchi. This is a private family archive without box or fonds numbers, but there is a published inventory: Anna Maria Magale Valente, *Le carte di Antonio Cocchi*, Milan, 1990

Biblioteca Nazionale Marciana, Venice

Letters of Giovanni Poleni, It. X, 135–136 (=6713–6714), It. X, 170 (=6572) and It. X, 173 (=6617)

Bodleian Library, Oxford

MS Engl. Misc. c. 444, Travel Diary of Martin Folkes

British Library, London

Add. MS 4222, ff. 22–56, Martin Folkes, President of the Royal Society, Memoirs of His Life: eighteenth century, manuscripts collected by Thomas Birch

Add. MS 22994, Letters of Richard Pococke

Add. MS 52362, Minute Book of the Egyptian Society

Add. MS 58318, Diary of Andrew Mitchell, Travels in Italy, Rome, 1733

MS Facs 589, Letters to and from Antonio Cocchi

Sloane MS 4058, f. 60, Martin Folkes to Hans Sloane, 1 October 1733

Keele University Library, Keele

MS 2806, Letters to Walter Sneyd including correspondence relating to the purchase of the Canonici manuscripts, Sneyd Family Papers. https://www.keele.ac.uk/library/specialcollections/sneydfamily/

MS 2807, Lists of Canonici MSS. purchased from Simoni Occhi by the Reverend Walter Sneyd, Sneyd Family Papers. https://www.keele.ac.uk/library/specialcollections/sneydfamily/

Museo Correr, Venice

Grandanigo Dolfin MS 49 (=191), Giovanni Grevembroch (Jan van Grevenbroeck)

Grevembroch, Giovanni (Jan), *Gli abiti de veneziani di quasi ogni età con diligenza raccolti e dipinti nel secolo XVIII*, or, *Customs of the Venetians*, 4 vols (18th century),

National Archives, The, Kew

PRO/11/809/301, Will of Martin Folkes, 3 July 1754

PRO/State Papers 99/63, f. 230 (Burges, 17 June 1733)

Norfolk Record Office, Norwich

NRS 8739, 21D4, William Folkes, Account of his brother's affairs while abroad

NRS 20658, Martin Folkes, 'Memoranda in case he should go abroad again, respecting his Estates, foreighn money & Places'

Northwestern University Library, Evanston, Illinois

Letters of Antonio Cocchi, Microfilm, 8 reels. These are microfilmed copies of the letters of physician Antonio Cocchi, the originals in the Baldassaroni-Corsini Archives in via S. Egidio n. 2, Florence, Italy.

Royal Society Library, London

RS/EC/1733/07, Election Certificate of Ludovico Riva

RS/EC/1733/08, Election Certificate of Father Giovanni Crivelli

RS/EC/1735/14, Election Certificate of Antonio Cocchi

RS/EC/1740/08, Election Certificate of Henry Stuart Stevens

RS/EC/1742/16, 'Certificate of Election'. Election Certificate of Walter Bowman

RS/EL/M2/61, Letter, Pietro Michelotti to James Jurin, 1723

RS/EL/M3/32, Letter, Pietro Michelotti to John Machin, 1733

RS/JBO/14, Responses to work on optics by Giovanni Rizzetti

RS/MM/20, Miscellaneous Manuscripts

RS/MS/213, Table of Contents of Journal Book, 1741–8

RS/MS/250, Correspondence of Martin Folkes

RS/MS/702, An Account of the Royal Society from its First Institution

RS/MS/704, Index of Papers Read Before the Royal Society, 1716–38

RS/MS/790, Correspondence of Martin Folkes

RS/MS/865, Letters of Charles Lennox, 2nd Duke of Richmond and Duke of Aubigny, to Martin Folkes, 1725–44

RS/RBO/19/5, Register Book Original, Paper of Ludovico Riva concerning meteors in Venice

RS/RBO/19/57, Register Book Original, Observations of the Aurora Borealis witnessed in London from September 1735 to April 1736 by Andreas Celsius

Society of Antiquaries of London

SAL/MS/264 B, vol. 2. Register transcribed by Joseph Ames of papers read or submitted to the Society

West Sussex Record Office

MS Goodwood House, MS 110, Martin Folkes to Charles Lennox, 8 July 1733

OBJECTS

Hogarth, William, 'Portrait of Martin Folkes', (*c.* 1740), oil on canvas, 73.5 cm × 62 cm. The Royal Society, London, image RS.9283, https://pictures.royalsociety.org/image-rs-9283

Manuscript globe belonging to Martin Folkes, British Library, London, inv. no. G 30

Martin Folkes Medal showing Egyptian motifs, including the Pyramid of Cestius, object M.8466, British Museum, London. https://www.britishmuseum.org/collection /object/C_M-8466

'To Martin Folkes esq.r, president of the Royal Society; Plan of the cities of London and Westminster and borough of Southwark; This plan of the cities of London and Westminster and borough of Southwark; this plan of the cities of London and Westminster and borough of Southwark, with the contiguous buildings is humbly inscribed', Harvard University, https://curiosity.lib.harvard.edu/scanned-maps/catalog/44-990102051310203941

PRINTED WORKS AND WEBSITES

Abromson, Morton C., 'Clement VIII's Patronage of the Brothers Alberti', *The Art Bulletin*, 60, 3, September 1978, pp. 531–47.

Ackerman, James S., 'Marcus Aurelius on the Capitoline Hill', *Renaissance News*, 10, 2, Summer 1957, pp. 69–75.

Agnew, David C. A., *Protestant Exiles from France, Chiefly in the Reign of Louis XIV. Or, The Huguenot Refugees and Their Descendants in Great Britain and Ireland*, 3 vols, 3rd edn, London, 1886.

Aikema, Bernard, *Pietro della Vecchia and the Heritage of the Renaissance in Venice*, Florence, 1990.

—, *Jacopo Bassano and his Public: Moralizing Pictures in an Age of Reform, ca. 1535–1600*, Princeton, NJ, 1996.

Aikin, Roger Cushing, '*Romae de Dacia Triumphantis*: Roma and Captives at the Capitoline Hill', *The Art Bulletin*, 62, 4, December 1980, pp. 583–97.

Ait-Touati, Frederique, Clouds and Meteors: Recreating Wonder on the Early Modern Stage', in Kirsten E. Shepherd-Barr, ed., *The Cambridge Companion to Theatre and Science*, Cambridge, 2020, pp. 188–202.

Alam, Ishrat, 'Diamond Mining and Trade in South India in the 17th Century', *Proceedings of the Indian History Congress*, 59, 1998, pp. 311–28.

Al-Bawwab, Rania Adel, 'The Zecca Mint: A Self-Enforcing Monetary Constitution in Historic Venice', *Economics of Governance*, 23, 2022, pp. 1–15.

Algarotti, Francesco, *Oeuvres de comte Algarotti traduit de l'italien*, 7 vols, Berlin, 1772.

Almagià, R., *L' 'Italia' di Giovanni Magini e la cartografia dell' Italia nei secoli XVI e XVII*, Naples, 1922.

Alvarez, David, *The Pope's Soldiers: A Military Victory of the Modern Vatican*, Lawrence, KS, 2011.

American Numismatic Society (ANS) Digital Library, http://numismatics.org

Amiani, P. M., *Dissertazione critico-lapidaria sopra l'antico Arco di Fano innalzato all'Imperadore Cesare Augusto*, Fano 1772.

Ammerman, Albert J., De Min, Maurizia, and Housley, Rupert, 'New Evidence on the Origins of Venice', *Antiquity* 66, 1992, pp 913–16.

Ammerman, Albert J., Pearson, Charlotte, L., Kuniholm, Peter I., Selleck, Bruce, and Vio, Ettore, 'Beneath the Basilica of San Marco: New Light on the Origins of Venice', *Antiquity*, 91, 360, 2011, pp. 1620–29.

Anderson, M.S., *Europe in the Eighteenth Century 1713–1789*, 4th edn, London and New York, 2000.

Andrews, Jonathan, and Scull, Andrew, *Undertaker of the Mind: John Monro and Mad-doctoring in Eighteenth-Century England*, Berkeley, CA, 2001.

—, *Customers and Patrons of the Mad-Trade: The Management of Lunacy in Eighteenth-Century England*, Berkeley, CA, 2003.

Anonymous, 'Rome and Her Works of Art', *The Art Journal*, 1, 1862, p. 39.

Anselm of Canterbury, *The Prayers and Meditations of Saint Anselm*, trans. and intro. by Sister Benedicta Ward, SLG, forward by R. W. Southern, New York, 1973.

Archaeology News Network, 'Roman Theatre of Fano Forgotten under Ex-factory', 19 March 2016, https://archaeonewsnet.com/2016/03/roman-theatre-of-fano-forgotten-under.html

Ariosto, Ludovico, *Orlando Furioso: A New Verse Translation*, edsed. David R. Slavitt and Charles S. Ross, Cambridge, MA, 2009.

Armogathe, Jean-Robert, 'Cartesianism and Eucharistic Physics', in Steven Nadler, Tad. M Schmaltz, and Delphine Antoine-Mahut, eds, *The Oxford Handbook of Descartes and Cartesianism*, Oxford, 2019, pp. 547–64.

Ashworth, William, 'Isaac Newton', Linda Hall Library, https://www.lindahall.org/about/news/scientist-of-the-day/isaac-newton-2/

Atlas of the Venetian Lagoon, http://cigno.atlantedellalaguna.it

'Attested Diplomatic Agents in Venice', in *Calendar of State Papers Relating To English Affairs in the Archives of Venice, Volume 1, 1202–1509*, ed. Rawdon Brown, London, 1864, pp. cli--clii. *British History Online*, http://www.british-history.ac.uk/cal-state-papers/venice/vol1/cli-clii

Avery, Charles, *The Triumph of Motion: Francesco Bertos (1678–1741) and the Art of Sculpture*, London, 2008.

Avery, Victoria, '"Belo et onorato per onor ancho suo" Alessandro Vittoria's Zane Altar Reconsidered', in Carlo Cosato and Debra Howard, eds, *Santa Maria Gloriosa dei Frari: Immagini di devozione, spazi della fede* [Devotional Spaces, Images of Piety], Padua, 2015, pp. 223–33.

Babitz, Sol, 'Treatise on Ornamentation: Guiseppe Tartini', *Journal of Research in Music Education*, 4, 2, Autumn 1956, pp. 75–102.

Bacchi, Andrea, 'Antonio Corradini e i Sagredo', in Paola Artoni, Enrico Maria Dal Pozzolo, Monica Molteni, and Alessandra Zamperini, eds, *Il tempo e la rosa. Scritti di storia dell'arte in onore di Loredana Olivato*, Treviso, 2013, pp. 130–35.

Badin, Donatella Abate, 'Lady Morgan in Italy: A Traveller with an Agenda', *Studi irlandesi: A Journal of Irish Studies*, 6, 2016, pp. 127–48.

Baglione, Giovanni, *Le vite de' pittori, scvltori et architetti*, Rome, 1642.

Bailey, Meryl, 'Punishment as Brotherly Love: Antonio Zanchi's "Expulsion of the Profaners from the Temple" and the Venetian "Conforteria"', *Artibus et historiae*, 37, 73, 2016, pp. 313–34.

Bamji, Alexandra, 'Health Passes, Print and Public Health in Early Modern Europe', *Social History of Medicine*, 32, 3, 2019, pp. 441–64.

—, 'The Materiality of Death in Early Modern Venice', in Suzanna Ivanič, Mary Laven, and Andrew Morrall, eds, *Religious Materiality in the Early Modern World*, Amsterdam, 2019, pp. 119–36.

Banfi, Florio, 'The Cosmographic Loggia of the Vatican Palace', *Imago Mundi*, 9, 1952, pp. 23–34.

Barbierato, Federico, *The Inquisitor in the Hat Shop: Inquisition, Forbidden Books and Unbelief in Early Modern Venice*, Farnham, 2012.

Barcham, William, 'Secular Commissions: Tiepolo as a Painter of History and Mythology and as a Decorator', in Keith Christiansen, ed., *Giambattista Tiepolo, 1696–1770*, New York, 1996, pp. 105–89.

Bardi, Alberto, 'Islamic Astronomy in Fifteenth-Century Christian Environments: Cardinal Bessarion and his Library', *Journal of Islamic Studies*, 30, 3, 2019, pp. 338–66.

Barocci, Federico, *The Entombment of Christ*, Art Institute of Chicago, https://www.artic.edu/artworks/100036/the-entombment-of-christ

Barone, Paola, Marino, Luigi, and Pignatelli, Olivia, *I trabocchi. Macchine da pesca della costa adriatica*, Caselle, 1999.

Barzazi, Antonella, *Gli affanni dell'erudizione: studi e organizzazione culturale degli ordini religiosi a Venezia tra Sei e Settecento*, Venice, 2004.

Basilica San Marco, https://www.basilicasanmarco.it

Bawden, Tina, et al., 'Early Visual Cultures and Panofsky's Perspektive als "symbolische Form"', *eTopoi: Journal for Ancient Studies*, 6, 2016, pp. 525–70.

Becherini, Marta, 'Staging the Foreign: Niccolò Manucci (1638–ca. 1720) and Early Modern Collections of Indian Paintings', PhD dissertation, Colombia University, 2017.

Bedini, S. A., *Clockwork Cosmos: Bernard Facini and the Farnese Planisferologio*, Vatican City, 1985.

—, 'A Marine Quadrant by Bernardo Facini', *Nuncius: Annali di storia della scienza*, 8, 2 1993, pp. 597–9.

Bellhouse, David R., 'Lord Stanhope's Papers on the Doctrine of Chances', *Historia Mathematica*, 34, 2, May 2007, pp. 173–86.

Benedict, Barbara, 'Collecting Trouble: Sir Hans Sloane's Literary Reputation in Eighteenth-century Britain', *Eighteenth-Century Life*, 36, 2, Spring 2012, pp. 111–42.

Benelli, Francesco, 'The Arch of Trajan in Ancona and Civic Identity in the Italian Quattrocento from Ciriaco d'Ancona to the Death of Matthias Corvinus', in Kathleen Christian and Bianca de Divitiis, eds, *Local Antiquities, Local Identities: Art, Literature and Antiquarianism in Europe, c. 1400–1700*, Manchester, 2019, pp. 37–56.

Benitez, Siobhan, 'Richness, Fatness and Salvation: The Movement of the Holy Spirit in the Blessings of the Holy Oils and Consecration of the Chrism during the Chrism Mass', *Horizons*, 46, 2019, pp. 1–23.

Berger, Tanya, ed., *Benedetto Marcello: Cassandra*, Middleton, 2016.

Berkeley, George, *Saggio d'una nuova teoria sopra la visione ...... ed un discorso preliminare al Trattato della cognizione*, trans. Giovan Bernardo Pisenti, Venice, 1732.

Bernabei, Mauro, Macchioni, Nicola, Pizzo, Benedetto, Sozzi, Lorena, Lazzeri, Simona, Fiorentino, Luigi, Pecoraro, Elisa, Quarta, Gianluca, and Calcagnile, Lucio, 'The Wooden Foundations of Rialto Bridge (Ponte di Rialto): Technological Characterisation and Dating', *Journal of Cultural Heritage*, 36, 2019, pp. 85–93.

Bernardino of Siena, *Opera omnia*, ed. P. M. Perantoni, 5 vols, Florence, 1950–.

'Bernardo Zendrini', Scientific Itineraries in Tuscany, https://brunelleschi.imss.fi.it/itineraries/biography/BernardinoZendrini.html

Bertarelli, Achille, *Le incisioni di Guiseppe Maria Mitelli, catalogo critico*, Milan, 1940.

Bertolesi, Elisa, Milani, Gabriele, Lopane, Fulvio Domenico, and Acito, Maurizio, 'Augustus Bridge in Narni (Italy): Seismic Vulnerability Assessment of the Still Standing Part, Possible Causes of Collapse, and Importance of the Roman Concrete Infill in the Seismic-Resistant Behaviour', *International Journal of Architectural Heritage*, 11, 5, 2017, pp. 717–46.

Betterton, Thomas, *The History of the English Stage from the Restauration to the Present Time*, London, 1741.

Biblioteca Nazionale Marciana, Venice Museum, https://www.venice-museum.com/biblioteca-marciana.php

Bibliotheca Phillippica New Series: First Part. Catalogue of the Celebrated Collection of Manuscripts Formed by Sir Thomas Phillipps, Bt. (1792–1872) ..., Sotheby & Co., auction catalogue, sale dates: 28–29 June 1965.

Binion, Alice, 'Algarotti's Sagredo Inventory', *Master Drawings*, 21, 4, Winter 1983, pp. 392–6.

Black, Christopher, *Church, Religion and Society in Early Modern Italy*, New York, 2004.

Black, Jeremy, 'British Foreign Policy in the Eighteenth Century: A Survey', *Journal of British Studies*, 26, 1, 1987, pp. 26–53.

—, *Italy and the Grand Tour*, New Haven, CT, 2003.

Blom, Philipp, *To Have and To Hold: An Intimate History of Collectors and Collecting*, New York, 2004.

Bolick, Laura, 'Culture, Humanism, and Intellect: Cardinal Bessarion as a Patron of the Arts', PhD dissertation, 2014, The Open University.

Boni, G., 'Trajan's Column', *Proceedings of the British Academy*, 3, 1–6, 1907, pp. 93–8.

Borean, Linda, 'Paolo Veronese Revisited: Art Collecting and Connoisseurship in Eighteenth Century Venice', in Andaleeb Badiee Banta, ed., *The Enduring Legacy of Venetian Renaissance Art*, London and New York, 2016, pp. 89–102.

Bosman, Lex., Haynes, Ian. P., and Liverani, Paolo, 'The Lateran Basilica to 1600', in Lex Bosman, Ian P. Haynes, and Paolo Liverani, eds, *The Basilica of Saint John Lateran to 1600*, British School at Rome Studies, Cambridge, 2020, pp. 1–5.

Bradley, Mark and Stow, Kenneth, 'Introduction', in Mark Bradley and Kenneth Stow, eds, *Rome, Pollution and Propriety: Dirt, Disease and Hygiene in the Eternal City from Antiquity to Modernity*, Cambridge, 2012, pp. 1–10.

Brand, Peter, 'The "Lost" Obelisks and Colossi of Seti I', *Journal of the American Research Center in Egypt*, 34, 1997, pp. 101–44.
Brier, Bob, *Cleopatra's Needles: The Lost Obelisks of Egypt*, London, 2016.
British History Online, http://www.british-history.ac.uk
Broughton, Thomas, *An Historical Dictionary of All Religions from the Creation of the World to this Present Time*, 2 vols, London, 1742.
Brown, Harcourt, 'Madam Geoffrin and Martin Folkes: Six New Letters', *Modern Language Quarterly*, 1–2, pp. 215–41.
Brown, Patricia Fortini, *The Venetian Bride: Bloodlines and Blood Feuds in Venice and its Empire*, Oxford, 2021.
Brundtland, Terje, 'Francis Hauksbee and His Air Pump', *Notes and Records: The Royal Society Journal of the History of Science*, 66, 2012, pp. 1–20.
Buchwald, Jed Z. and Fox, Robert, eds, *The Oxford Handbook of the History of Physics*, Oxford, 2013.
Bulman, Louisa M. Connor, 'Moral Education on the Grand Tour: Thomas Coke and his Contemporaries in Rome and Florence', *Apollo*, 157, March 2003, pp. 27–34.
Burke, Peter, 'Images as Evidence in Seventeenth-Century Europe', *Journal of the History of Ideas*, 64, 2, 2003, pp. 273–96.
—, *The Historical Anthropology of Early Modern Italy: Essays on Perception and Communication*, Cambridge, 2005.
—, *Popular Culture in Early Modern Europe*, Aldershot, 2009.
Burnett, Andrew, *The Hidden Treasures of this Happy Island: A History of Numismatics in Britain from the Renaissance to the Enlightenment*, 3 vols, London, 2020.
Burri, Renate, 'The Rediscovery of Ptolemy's Geography (End of the Thirteenth to End of the Fifteenth Century)', *Imago Mundi*, 61, 1, 2009, pp. 124–50.
Byrom, John, *The Private Journal and Literary Remains of John Byrom*, 2 vols, Manchester, 1854–7.
Ca' Sagredo, Venice, https://www.casagredohotel.com
Calendar of State Papers Domestic: William and Mary, 1691–2, ed. William John Hardy, London, 1900.
Calendar of State Papers Relating to English Affairs in the Archives of Venice, Volume 1, 1202–1509, ed. Rawdon Brown, London, 1864.
Campagnol, Isabella, *Forbidden Fashions: Invisible Luxuries in Early Venetian Convents*, Lubbock, TX, 2014.
Cantaluppi, Anna, 'Le carte del genovese Cladio Marini, ambasciatore del Re di Francia in Piemonte, nell'archivio della Compagnia di San Paolo', *Bollettino della Società Piemontese dei archeologia e belle arti*, 61–2, 2010–2011, pp. 309–30.
Cardarelli, François, *Encyclopaedia of Scientific Units, Weights and Measures: Their SI Equivalances and Origins*, trans. M. J. Shields, London, Berlin, and Heidelberg, 2003.
Carrington, Robert C., *Foreign Measures and their English Values Compiled from Official Sources*, London, 1864.
Casanova, Giacomo, *History of My Life*, ed. and trans. Willard R. Trask, 12 vols, Baltimore, MD, and London, 1970.
'Cascata delle Marmore and Valnerina', UNESCO, https://whc.unesco.org/en/tentativelists/2031
Casini, Paolo, 'The Reception of Newton's Opticks in Italy', in J. V. Field and Frank A. J. L. James, eds, *Renaissance and Revolution: Humanists, Scholars, Craftsmen and Natural Philosophers in Early Modern Europe*, Cambridge, 1997, pp. 215–29.
Casino dell'Aurora Pallavicini, http://www.casinoaurorapallavicini.it/
Cassibry, Kimberly, *Destinations in Mind: Portraying Places on the Roman Empire's Souvenirs*, New York and Oxford, 2021.
Castello de Monselice, https://www.castellodimonselice.it

Castello de Rocca Priora, *MuseoItaliana*, https://www.museionline.info/tipologie-museo/castello-di-rocca-priora

Catalogo generale de Beni Culturali, https://catalogo.beniculturali.it

Catalogue of the ... collection of coins, medals, and medallions ... of M. Folkes ... which will be sold by auction, by Mr. Langford ... the 27th of Jan. 1756, A, London, 1756.

Catalogue of the ... collection of mathematical instruments, gems, pictures ... etc. of M. Folkes ... which will be sold by auction, by Mr. Langford ... the 7th, and ... 8th, of May ... 1755, A, London, 1755.

Catalogue of the entire and valuable library of Martin Folkes, Esq. President of the Royal Society, and Member of the Royal Academy of Sciences in Paris, Lately Deceased ..., A, London, 1756.

Catalogue of an exhibition of drawings by Giovanni Battista Tiepolo (1696–1770) held at the Savile Gallery, London, during May 1928, London, 1928.

Catalogue of the genuine, entire, and curious collection of prints and drawings, bound and unbound, of the learned and ingenious Martin Folkes, Esq. late president of the Royal Society, deceas'd; which (by order of the executors) will be sold by auction, by Mr. Langford, at his house in the Great Piazza, Covent Garden, on Thursday the 15th of this instant January 1756, and the seven following evenings, A, London, 1756.

Catholic Encyclopedia, The, New York, 1912 http://www.newadvent.org/cathen

Cattaneo, Angelo, *Fra Mauro's Mappa Mundi and Fifteenth-Century Venice*, Turnhout, 2011.

Cattedrale di Ferrara, https://cattedralediferrara.it

Celsius, Anders, 'Observations of the Aurora Borealis Made in England by Andr. Celsius, F. R. S. and Secr. R. S. of Upsal in Sweden', *Philosophical Transactions*, 39, 441, 1735, pp. 241–4.

Cervia, https://www.turismo.comunecervia.it/en

Chamberlain, Edwin J., 'Long-Scale Slide Rules Revisited', *Journal of the Oughtred Society*, 13, 1, Spring 2004, pp. 23–42.

Chambers, Ephraim, *Cyclopædia: or, an universal dictionary of arts and sciences*, 2nd edn, London, 1738.

'Chiesa di San Filippo Neri', Comune di Spoleto, http://www.comune.spoleto.pg.it/turismoecultura/la-citta/chiese/chiesa-di-san-filippo-neri/

Chorus Venezia, https://chorusvenezia.org

Chrissochoidis, Ilias, '"true Merit always Envy rais'd": The Advice to Mr. Handel (1739) and Israel in Egypt's Early Reception', *The Musical Times*, Spring 2009, pp. 69–86.

Christensen, Keith, ed., *Giambattista Tiepolo 1696–1770*, New York, 1996.

Christie, Thony, The Renaissance Mathematicus, https://thonyc.wordpress.com

Churchill, Awnsham, *A collection of voyages and travels, some now first printed from original manuscripts, others now first published in English*, 6 vols, London, 1732.

Cibber, Colley, *An Apology for the Life of Colley Cibber, written by himself*, ed. Robert W. Lowe, ed., 2 vols, London, 1889.

Cipolla, Carl, *Money in Sixteenth-Century Florence*, Berkeley, Los Angeles, CA, and London, 1989.

Claridge, Amanda, with Toms, Judith and Cubberley, Tony, *Rome: An Oxford Archaeological Guide*, 2nd edn, Oxford, 2010.

Claussen, Peter C., 'The Remodelling of San Giovanni in Laterano by Pope Nicholas IV: Transept, Apse and Façade', in Lex Bosman, Ian P. Haynes, and Paolo Liverani, eds, *The Basilica of Saint John Lateran to 1600*, Cambridge, 2020, pp. 318–44.

Clough, Cecil H. 'The Albani Library and Pope Clement XI', *Librarivm*, 12, 1, May 1969, pp. 13–21.

Cocke, Richard, 'Exemplary Lives: Veronese's Representations of Martyrdom and the Council of Trent', *Renaissance Studies*, 10, 3, 1996, pp. 388–404.

—, *Paolo Veronese: Piety and Display in an Age of Religious Reform*, New York, 2017.

Coggins, Jay. S., and Perali, C. Federico, '64% Majority Rule in Ducal Venice: Voting for the Doge', *Public Choice*, 97, 4, 1998, pp. 709–23.

Cogo, Bruno, *Antonio Corradini: scultore veneziano: 1688–1752*, Este, 1996.

Coins of the Republic of Venice, 'The Root of all Evil', https://forgeofempiressite.wordpress.com/2017/07/18/ coins-of-the-republic-of-venice/

Colombo, Stefano, 'Baldassare Longhena's Funerary Monument to Doge Giovanni Pesaro', *Word & Image: A Journal of Verbal/Visual Enquiry*, 35, 1, 2019, pp. 48–65.

Cornaro, Flaminio, *Notizie storiche delle chiese e monasteri di Venezia, e di Torcello, tratte dalle chiese veneziane, e torcellane*, Venezia, 1758.

'Cornelis Cort', *British Museum*, https://www.britishmuseum.org/collection/term/BIOG23700

Corp, Edward, *The Stuarts in Italy, 1719–1766: A Royal Court in Permanent Exile*, Cambridge, 2011.

Cosmescu, Dragoş, *Venetian Renaissance Fortifications in the Mediterranean*, Jefferson, NC, 2016.

Costa, Thomas Dalla, Echols, Robert, and Ilchman, Frederick, *Tintoretto in Venice: A Guide,* Venice, 2018.

Cotton, Sir Evan, 'Editor's Note Book', *Bengal Past and Present: Journal of the Calcutta Historical Society*, 34, 1, July–September 1927, pp. 63–75.

Cotton, Jeff, Churches of Venice http://www.churchesofvenice.com

Cotty, Colonel, *Aide-mémoire a l'usage des officiers d'artillerie de France*, Paris, 1819.

Coulling, Ian, Images of Venice, https://imagesofvenice.com

Courtauld Gallery Collection Online, The, London, https://gallerycollections.courtauld.ac.uk/home

Coutu, Joan, *Then and Now: Collecting and Classicism in Eighteenth-Century England*, Montreal, 2015.

Craddock, Paul, *Scientific Investigation of Copies, Fakes and Forgeries*, New York, 2009.

Crisanti, M. A., and Taffetani, F., 'Diachronic Analysis of Variations Induced on the Flora and Vegetation of River Ecosystems by Actions Taken to Reduce the Risk of Flooding. Case Study of the River Chienti (Central Adriatic, Italy)', *Plant Sociology*, 52, 1 June 2015, pp. 41–64.

Crivelli, Giovanni, *Elementi di fisica*, Venice, 1731.

Croix, David de la and Vitale, Maria, 'Scholars and Literati at the University of Padua (1222–1800)', *Repetorium eruditorum totius Europae (RETE)*, 3, 2021, pp. 33–42.

Crollalanza, Giovan Battist di, *Dizionario storico-blasonico delle famiglie nobili e notabili italiane, estinte e fiorenti*, 3 vols in 1, Pisa, 1886–90.

Crouzet-Pavan, Elisabeth, *Venice Triumphant: The Horizons of a Myth*, Baltimore, MD, 2005.

Crowley, Roger, *Empires of the Sea: The Final Battle for the Mediterranean*, New York, 2008.

—, 'Bragadin's Defense', *MHQ: The Quarterly Journal of Military History*, 21, 3, Spring 2009, pp. 66–75.

Crucitti, Pierangelo, 'Distribution and Diversity of Italian Scorpions', *Redia*, 76, 2, 1993, pp. 281–300.

Da Canal, Vincenza, *Vita di Gregorio Lazzarini scritta da Vincenza Da Canal*, Venice, 1732.

Dal Prete, Ivano, 'Cultures and Politics of Preformationism in Eighteenth-Century Italy', in Raymond Stephanson and Darren N. Wagner, eds, *The Secrets of Generation: Reproduction in the Long Eighteenth Century,* Toronto, 2015, pp. 59–78.

Dandelet, Thomas James, *The Renaissance of Empire in Early Modern Europe*, Cambridge, 2014.

D'Andrea, David, 'Charity and Confraternities', in Eric Dursteler, ed., *A Companion to Venetian History, 1400–1797,* Leiden, 2013, pp. 421–47.

Davidson, Bernice F., *Raphael's Bible: A Study of the Vatican Logge,* University Park, PA, 1985.

Davidson, Linda Kay and Gitlitz, David Martin, eds, *Pilgrimage: From the Ganges to Graceland: An Encyclopedia*, 2 vols, Santa Barbara, CA, Denver, CO, and Oxford, 2002.

Davies, Jonathan, 'Violence and Italian Universities during the Renaissance', *Renaissance Studies*, 27, 4, 2013, pp. 504–16.

Davis, James C., *A Venetian Family and its Fortune 1500–1900: The Donà and the Conservation of Their Wealth*, Philadelphia, PA, 1975.

Davis, Paul K., *100 Decisive Battles from Ancient Times to the Present: The World's Major Battles and How They Shaped History*, Oxford, 1999.

Davis, Robert C., *The War of the Fists: Popular Culture and Public Violence in Late Renaissance Venice*, Oxford, 1994.

—, *Shipbuilders of the Venetian Arsenal: Workers and Workplace in the Preindustrial City*, Baltimore, MD, 2007.

De Blainville, Monsieur, *Travels through Holland, Germany, Switzerland, and other parts of Europe; but especially Italy. By the late Monsieur de Blainville, Sometime Secretary to the Embassy of the States-General, at the Court of Spain. Translated from the author's own manuscript, (never yet published) by William Guthrie, Esq.*, 2 vols, London, 1743.

Debus, Allen G., 'The Paracelsian Aerial Nitre', *Isis*, 55, 1, 1964, pp. 43–61.

De Grummond, Nancy Thomson, ed., *Encyclopedia of the History of Classical Archaeology*, New York, 2015.

De Jonge, Henk Jan, 'Erasmus and the Comma Johanneum', *Ephemerides Theologicae Lovanienses*, 66, 1980, pp. 381–9.

Deliyannis, Deborah Mauskopf, 'Bury Me in Ravenna? Appropriating Galla Placidia's Body in the Middle Ages', *Studi medievali*, 42, 2000, pp. 289–99.

Depaulis, Thierry, 'Bingo! A Material History of Modern Gaming', in Manfred Zollinger, ed., *Random Riches: Gambling Past & Present*, Abingdon and New York, 2016, pp. 36–56.

Desaguliers, J. T., 'An Account of some Experiments of Light and Colours, formerly made by Sir Isaac Newton, and mention'd in his *Opticks*, late repeated before the Royal Society by J. T. Desaguliers, F.R.S.', *Philosophical Transactions*, 29, 348, 31 December 1714, pp. 433–47.

—, 'Optical Experiments made in the beginning of August 1728, before the President and Several Members of the Royal Society, and Other Gentlemen of Several Nations, upon Occasion of Signior Rizzetti's Opticks ...', *Philosophical Transactions*, 35, 406, 1727–8, pp. 596–629.

Desgodetz, Antony Babuty, *Les édifices antiques de Rome dessinés et mesurés très exactement*, Paris, 1682.

Deupi, Victor, *Architectural Temperance: Spain and Rome, 1700–1759*, Abingdon and New York, 2015.

Devidutta, Soumen and Roy, Ambuj, 'St Vitus Dance', *Circulation*, 128, 19, 5 November 2013, p. 380.

Difederico, Frank R., *Francesco Trevisani: Eighteenth-Century Painter in Rome*, Washington, DC, 1977.

Di Giovanni, Serena, 'I dipinti murali giotteschi nella sala capitolare del complesso di Sant' Antonio a Padova: ricostruzione della vicenda conservativa', *Il Santo*, 61, 3, 2021, pp. 355–91.

Digital edition of the Fra Mauro Map1}, Biblioteca Nazionale Marciana, Venice.

Dizionario Biografico dei Friulani, https://www.dizionariobiograficodeifriulani.it

Dizionario Biografico degli Italiani, https://www.treccani.it/enciclopedia/

Donvito, Filippo, 'Storm of Steel', *Medieval Warfare*, 2, 5, 2012, pp. 17–21.

Doran, John, *'Their Majesties' Servants', or Annals of the English Stage*, London, 1865.

Dorfman, Rivka and Dorfman, Ben-Zion, 'The Jewish Community of Ancona, Italy', ANU Museum of the Jewish People, https://www.anumuseum.org.il/jewish-community-ancona-italy

Dursteler, Eric R., ed., *A Companion to Venetian History, 1400–1797*, Leiden, 2013.

EAGLE (The Electronic Archive of Greek and Latin Epigraphy), http://www.edr-edr.it/

Earle, Peter, 'The Commercial Development of Ancona, 1479–1551', *The Economic History Review*, 22, 1, 1969, pp. 28–44.

Echols, Robert, 'A Classical Barrel Vault for San Giovanni in Laterano in a Borromini Drawing', *Journal of the Society of Architectural Historians*, 51, 2, June 1992, pp. 146–60.

Edler, Florence, *Glossary of Medieval Terms of Business: Italian Series 1200–1600*, New York, 1934.

'Edward the Confessor and Edith', Westminster Abbey, https://www.westminster-abbey.org/abbey-commemorations/royals/edward-the-confessor-and-edith
Edwards, H. W., *Catalogue Number 114:Manuscripts IV*, Newbury, 1966.
'Egyptian Obelisks', NOVA, https://www.pbs.org/wgbh/nova/egypt/raising/rome.html
Ehrlich, Tracy L. The Villa Borghese and the rise of the Baroque Garden-Park', *Landscape: A Magazine of Human Geography*, 32, 2, 1994, pp. 12–19.
—, 'Pastoral Landscape and Social Politics in Baroque Rome', in Michel Conan, ed., *Baroque Garden Cultures: Emulation, Sublimation, Subversion*, Dumbarton Oaks, Washington DC, 2005, pp. 131–82.
Eisler, William, 'The Construction of the Image of Martin Folkes (1690–1754), Part I', *The Medal*, 58, 2011, pp. 4–29.
—, 'The Construction of the Image of Martin Folkes (1690–1754) Part II: Art, Science, and Masonic Sociability in the Age of the Grand Tour', *The Medal*, 59, 2011, pp. 4–16.
Enciclopedia on line, https://www.treccani.it
Encyclopaedia Britannica, https://www.britannica.com; 1911 *Encyclopaedia Britannica*, https://penelope.uchicago.edu/Thayer/E/Gazetteer/_Texts/EB1911/home.html
Engineering Historical Memory, Fra Mauro's Map of the World, https://engineeringhistorical memory.com/FraMauro.php
Epigraphic Database Heidelberg, https://edh.ub.uni-heidelberg.de
Ertman, Thomas, 'Opera, the State and Society', in Nicholas Till, ed., *The Cambridge Companion to Opera Studies*, Cambridge, 2012, pp. 25–52.
Europeana, https://www.europeana.eu
Evangelisti, Silvia, *Nuns: A History of Convent Life 1450–1700*, Oxford, 2007.
Evans, Joan, *A History of the Society of Antiquaries*, Oxford, 1956.
Evelyn, John, *The Diary of John Evelyn*, ed. William Bray, 2 vols, London, 1901.
Falchetta, Piero, *Fra Mauro's World Map: With a Commentary and Translations of the Inscriptions*, Turnhout, 2006.
—, *Biblioteca nazionale Marciana. Storia del Mappamondo di Fra' Mauro: con la trascrizione integrale del testo*, Rimini, 2016.
Fara, Patricia, '"Master of Practical Magnetics": The Construction of an Eighteenth-Century Natural Philosopher', *Enlightenment and Dissent*, 14, 1995, pp. 52–87.
Farquhar, Maria, *Biographical Catalogue of the Principal Italian Painters*, ed. Ralph N. Wornum, London, 1855.
Farrell, Sister Maureen FCJ, 'Rare Items Relating to William Whiston (1667–1752) in the Houghton Library', *Harvard Library Bulletin*, 24, 3, July 1976, pp. 349–59.
Favilla, Massimo, Rugolo, Ruggero and Meijers, Dulcia, 'Venetian Art 1600–1797', in Eric R. Dursteler, ed., *A Companion to Venetian History, 1400–1797*, Leiden, 2013, pp. 811–63.
Favilla, Massimo and Rugolo, Ruggero, *A Brief History of Palazzo Grimano dall'Albero d'Oro in Venice: From the Vendramin to the Marcello family 1449–1969*, Venice, 2021.
Ferrara terra e acqua, https://www.ferraraterraeacqua.it
Ferrari, Oreste, 'Luca Giordano', in Clovis Whitfield and Jane Martineau, eds, *Painting in Naples, 1606–1705, from Caravaggio to Giordano*, New York, 1982, pp. 168–80.
Ferrari, Oreste—, and Scavizzi, Giuseppe, *Luca Giordano*, 3 vols, Naples, 1966.
Ferraro, Joanne M., *Venice: History of a Floating City*, Cambridge, 2012.
Ferrone, Vincenzo, *The Intellectual Roots of the Italian Enlightenment: Newtonian Science, Religion, and Politics in the Early Eighteenth Century*, trans. Sue Brotherton, Atlantic Highlands, NJ, 1995.
Fiaccadori, Gianfranco, et al., eds, *Bessarione e l'Umanesimo: catalogo della mostra*, foreword by Marino Zorzi, preface by Giovanni Pugliese Carratelli, *Catalogue of an exhibition held in Venice at the Biblioteca Nazionale Marciana, 27 April–31 May 1994*, Naples, 1994.

Ficari, Maurizio, 'Una rilettura per la cattedrale di Civita Castellana e l'abbaziale di Santa Maria di Falleri al tempo di Innocenzo III', *Rivista di storia della Chiesa in Italia*, 2, 2017, pp. 427–57.

Filter, Susan, 'Historic Intent: Lodovico Ughi's Topographical Map of Venice: A Large Wall Map as a Historic Document, a Work of Art, and a Material Artifact', *The American Institute for Conservation*, 13, 1994. https://cool.culturalheritage.org/coolaic/sg/bpg/annual/v13/bp13-04html

Fiore, Kristina Herrmann, 'The Exhibition of Sculptures on the Villa Borghese Facades in the Time of Cardinal Scipione Borghese', *Studies in the History of Art*, 70, 2008, pp. 218–45.

Fissell, Mary E., *Vernacular Bodies: The Politics of Reproduction in Early Modern England*, Oxford, 2004.

Fletcher, Jennifer and Mueller, Reinhold C., 'Bellini and the Bankers: The Priuli Altarpiece for San Michele in Isola, Venice', *Burlington Magazine*, 147, 1222, January 2005, pp. 5–15.

'flor di battaglia/Fiore dei Liberi, Il', MS M.383, ca. 1410–20, The Morgan, https://www.themorgan.org/manuscript/77302

Folda, Jaroslav, 'Reliquary of Saint Marina', in Helen C. Evans and William D. Wixom, eds, *The Glory of Byzantium: Art and Culture of the Middle Byzantine Era, A.D.*, New York, 1997, pp. 496–7.

Foley, Henry, *Records of the English Province of the Society of Jesus: Historic Facts Illustrative of the Labours and Sufferings of its Members in the Sixteenth and Seventeenth Centuries*, 7 vols, London, 1879.

Folkes, Martin, 'An Account of the Standard Measures Preserved in the Capitol at Rome', *Philosophical Transactions*, 39, 442, 1735–6, pp. 262–6.

—-, *A table for English gold coins, from the eighteenth year of King Edward III. when gold was first coined in England. With their several weights, and present intrinsic values*, London, 1736.

—, *A table of English silver coins, From the Norman conquest to the present time with their weights, intrinsic Values, and Some Remarks upon the several Pieces*, London, 1745.

—, *Tables of English silver and gold coins: first published by Martin Folkes, Esq; and now re-printed, with plates and explanations, by the Society of Antiquaries*, London, 1761–1763.

—, 'On the Trajan and Antonine Pillars at Rome: Read 5 February 1735–6', *Archaeologia*, 1, 2, January 1770, pp. 117–21.

Fondazione Musei Civici Venezia, Museo Correr, https://correr.visitmuve.it

Fondazione Teatro della Fortuna, http://www.teatrodellafortuna.it

Ford, David Nash, 'Col. Elizeus Burges', Royal Berkshire History, http://www.berkshirehistory.com/bios/eburges.html

Fosi, Irene, *Papal Justice: Subjects and Courts in the Papal State, 1500–1750*, trans. Thomas V. Cohen, Washington, DC, 2011.

Foyster, Elizabeth, 'At the Limits of Liberty: Married Women and Confinement in Eighteenth-century England', *Continuity and Change*, 17, 1, 2002, pp. 39–62.

'Fra Mauro World Map', British Library, https://www.bl.uk/maps/videos/fra-mauro-world-map

Frank, Stuart M., *Herman Melville's Picture Gallery: Sources and Types of the 'Pictorial' Chapters of Moby-Dick*, Fairhaven, MA, 1986.

Franzoi, Umberto and Lugato, Franca, *Scuola Grande dei Carmini*, Ponzano, 2003.

Frascari, Marco, 'The Particolareggiamento in the Narration of Architecture', *Journal of Architectural Education*, 43, 1, Autumn 1989 pp. 3–12.

Frazer, Alfred, 'Architecture', in Kurt Weitzmann, ed., *Age of Spirituality: Late Antique and Early Christian Art,* New York, 1979, pp. 109–25.

Frischer, Bernard D., 'Ramsey's 'Enquiry': Text and Context', in Bernard D. Frischer and Iain Gordon Brown, eds, *Allan Ramsey and the Search for Horace's Villa,* Aldershot, 2001, pp. 73–104.

Fritsen, Angela, *Antiquarian Voices: The Roman Academy and the Commentary Tradition on Ovid's Fasti*, Columbus, OH, 2015.

Galilei, Galileo, *Sidereus Nuncius*, ed. Albert Van Helden, Chicago, IL, 1989.

Galleria Borghese, Rome, https://www.collezionegalleriaborghese.it/en

Gallerie Accademia, Venezia, https://www.gallerieaccademia.it

Galliazzo, V., *I ponti romani*, 2 vols, Canova, 1994.

Gambolati, Guiseppe and Teatini, Pietro, *Venice Shall Rise Again: Engineered UplifeUplift of Venice Through Seawater Injection*, London and Waltham, MA, 2013.

Gildon, Charles, *The Life of Mr Thomas Betterton: The Late Eminent Tragedian*, London, 1710.

Gilks, David, 'Attitudes to the Displacement of Cultural Property in the Wars of the French Revolution and Napoleon', *The Historical Journal*, 56, 1, 2013, pp. 113–43.

Gillgren, Peter, '*Una Dolcissima Estasi*: Performing *The Visitation* by Federico Barocci', in Mårten Snickare and Peter Gillgren, eds, *Performativity and Performance in Baroque Rome*, New York, 2017, pp. 157–79.

'Giocomo Soranzo', Archivio dei possessori, https://archiviopossessori.it/archivio/51-soranzo-giacomo

Gioielli Nascosti di Venezia rimarranno chiusi fino, I https://www.gioiellinascostidivenezia.it/en/the-jewels/chiesa-delle-zitelle/

Giustiniani, Bernardo, *Historia di m. Bernardo Giustiniano gentilhuomo vinitiano, dell'origine di Vinegia et delle cose fatte ca vinitiani ...*, trans. Lodovico Domenichi, Venice, 1545.

Glixon, Jonathan E., *Mirrors of Heaven or Worldly Theatres: Venetian Nunneries and their Music*, Oxford, 2017.

Gloria, Andrea, *I podestà e capitani de Padova dal 6 Giugno 1509 al 28 Aprile 1797*, Padua, 1861.

Gobel, Esther, 'The Sisters Sagredo: Passion and Patronage in Eighteenth-Century Venice', in Nebahat Avcioglu and Emma Jones, eds, *Architecture, Art and Identity in Venice and its Territories, 1450–1750: Essays in Honour of Deborah Howard*, New York, 2016, pp. 33–50.

Goff, Moira, *The Incomparable Hester Santlow*, Farnham and Burlington, VT, 2007.

Golahny, Amy, *Rembrandt – Studies in his Varied Approaches to Italian Art*, Leiden, 2020.

Google Arts and Culture, https://artsandculture.google.com/

Gordon-Lennox, Charles Henry, 8th Duke of Richmond and Earl of March, *A Duke and His Friends: The Life and Letters of the Second Duke of Richmond*, London, 1911.

Gori, Gigliola, 'Sporting Events Organized in Venice: Male Boating and the Amazing Case of Women's Rowing Contests', *The International Journal of the History of Sport*, 32, 4, 2015, pp. 584–96.

Gottardo, Ketty, 'Il gusto collezionistico di un eccentrico personaggio Veneziano: la raccolta di disegni di "Zotto" Sagredo', in Bernard Aikena, Rosella Lauber, and Max Seidel, eds, *Il collezionismo a Venezia e nel Veneto ai tempi della Serenissima*, Venice, 2005, pp. 239–58.

Gould, Cecil, 'Afterthoughts on Raphael's So-Called Loreto Madonna', *The Burlington Magazine*, 22, 926, May 1980, pp. 336–9, 341.

Goy, Richard J., 'To the Glory of God: Building the Church of S. Maria della Carità, Venice, 1441–1454', *Architectural History*, 37, 1994, pp. 1–23.

—, *Venetian Vernacular Architecture: Traditional Housing in the Venetian Lagoon*, Cambridge, 2010.

Gozzi, Count Carlo, *Memoirs of Count Carlo Gozzi*, trans. John Addington Symonds, 2 vols, London, 1890.

Grand Tour, The, Adam Matthew Digital, http://www.grandtour.amdigital.co.uk

Graziosi, Elisabetta, 'Women and Academies in Eighteenth-Century Italy', in Paula Findlen, Wendy Wassyng Roworth, and Catherine M. Siena, eds, *Italy's Eighteenth Century: Gender and Culture in the Age of the Grand Tour*, Stanford, CA, 2009, pp. 103–19.

Greaves, John, *A Discourse of the Romane Foot, and Denarius: from when, as from two principles, the Measures, and Weights, used by the Ancients, may be deduced*, London, 1647.

Greengrass, Mark, Hildyard, Daisy, Preston, Christopher D., and Smith, Paul J., 'Science on the Move: Francis Willughby's Expeditions', in Tim Birkhead, ed., *Virtuoso by Nature: The Scientific World of Francis Willughby FRS (1635–1672)*, Leiden, 2016, pp. 142–226.

Grendler, Paul F., *Jesuit Schools and Universities in Europe, 1548–1773*, Leiden, 2019.

—, *The Universities of the Italian Renaissance,* Baltimore, MD, 2022.

Grevembroch, Giovanni, *Gli abiti de' Veneziani di quasi ogni età con diligenza raccolti e dipinti nel secolo XVIII*, ed. Giovanni Mariacher, 4 vols, Venice, 1981.

Griffin, J. P., 'Venetian Treacle and the Foundation of Medicines Regulation', *British Journal of Clinical Pharmacology*, 58, 3, 2004, pp. 317–25.

Griffiths, Rhys, 'The Piraeus Lion', *History Today*, 66, 7, 19 July 2016, p. 70.

Gross, Hanns, *Rome in the Age of Enlightenment: The Post-Tridentine Syndrome and the Ancien Régime*, Cambridge, 2004.

Grossi, Gianluca and Polsinelli, Valentina, 'A Sustainable Management of Water and Soil: The Case of River Contracts', *GeoProgress Journal*, 7, i, 1, 2020, pp. 37–55.

Grout, James, ed., *Encyclopaedia Romana*, https://penelope.uchicago.edu/~grout/encyclopaedia_romana [

Grove Art Online, https://www-oxfordartonline-com

Grove Encyclopedia of Decorative Arts, The, ed. Campbell, Gordon, 2 vols, Oxford, 2006.

Grove Music Online, https://www.oxfordmusiconline.com/grovemusic

Gruter, Janus, *Inscriptiones antiqvae totivs orbis romani: in absolvtissimvm corpvs redactae olim avspciis Iosephi Scaligeri et Marci Velseri*, 2 vols with 2 parts each, Amsterdam, 1707.

Guerlac, Henry, 'The Poet's Nitre', *Isis* 45, 3, September 1954, pp. 243–55.

Guide to the Valle del Treja Regional Park, Rome, 2018.

Hahn, Cynthia, *Strange Beauty: Issues in the Making and Meaning of Reliquaries, 400--ca 1204*, University Park, PA, 2012.

Hall, A. Rupert, *All Was Light: An Introduction to Newton's Opticks*, Oxford, 1993.

—, Hall, Laura, Scott, James F., and Turnball, Herbert W., eds, *The Correspondence of Isaac Newton*, 7 vols, Cambridge, 1959–81.

Hansell, Sven Hostrup, 'Sacred Music at the Incurabili in Venice at the Time of J. A. Hasse – -I', *Journal of the American Musicological Society*, Summer 1970, pp. 282–301.

Hare, Augustus H. C., *Venice*, London, 1896.

Harris, Joseph, *An Essay Upon Money and Coins*, London, 1757.

Haskell, Yasmin, 'Group Therapy for Venetian Adolescents? Giannantonio Bernardi's "Prudence, a Didactic Prolusion" (Venice, 1709) and Jesuit Moral Counselling in Verse', *Journal of Jesuit Studies*, 4, 2017, pp. 186–208.

Hatton, Ronald G., and Hatton, Christopher H., 'Notes on the Family of Twysen and Twisden', *Archaeologia Cantiana, Kent Archaeological Society*, 58, 1945, pp. 43–67.

Haugen, Kristine L., 'Transformations of the Trinity Doctrine in English Scholarship: From the History of Beliefs to the History of Texts', *Archiv für Religionsgeschichte*, 3, 1, January 2011, pp. 149–68.

Hay, Sophie, Keay, Simon and Millett, Martin, *Ocriculum (Otricoli, Umbria): An Archaeological Survey of the Roman Town,* Rome, 2013.

Hayes, Richard, *The Negociator's Magazine or the Most Authentic Account yet purchased of the Monies, Weights, and Measures of the Principal Places of Trade in the known World*, London, 1777.

Healey, Robin, *Italian Literature Before 1900 in English Translation: An Annotated Bibliography, 1929–2008*, Toronto, 2011.

Hearne, Thomas, *Remarks and Collections of Thomas Hearne*, ed. Rev. H. E. Salter, et al., 11 vols, Oxford, 1885–1921.

'Hebrew Manuscripts from the Canonici Collection at the Bodleian', Digital Bodleian, http://bav.bodleian.ox.ac.uk/news/hebrew-manuscripts-from-the-canonici-collection-at-the-bodleian

Heffernan, Michael, 'The Scale of Two Cities: The Geography of Paris and London in the 1720s', *Notes and Records: the Royal Society Journal of the History of Science*, March 2024, https://doi.org/10.1098/rsnr.2023.0073forthcoming.

Heidelberg Digitale Bibliothek, https://digi.ub.uni-heidelberg.de

Heilbron, John, *The Incomparable Monsignor: Francesco Bianchini's World of Science, History and Court Intrigue*, Oxford, 2022.

Herrmann, Wolfgang, 'Antoine Desgodets and the Académie Royale d'Architecture', *The Art Bulletin,* 40, 1, March 1958, pp. 23–53.

Higgitt, Rebekah, 'Equipping Expeditionary Astronomers: Nevil Maskelyne and the Development of "Precision Exploration"', in F. MacDonald and C. W. J. Withers, eds, *Geography, Technology and Instruments of Exploration*, Basingstoke, 2015, pp. 15–36.

—, '""In the Society's Strong Box": A Visual and Material History of the Royal Society's Copley Medal, c.1736–1760', *Nuncius: Journal of the Material and Visual History of Science*, 34, 3, 2019, pp. 284–316.

Highfill, Jr, Philip H., Burnim, Kalman A., and Langhans, Edward A., eds, *A Biographical Dictionary of Actors, Actresses, Musicians, Dancers, Managers and Other Stage Personnel in London, 1660–1880,* 12 vols, Carbondale, IL, 1973–.

History of Cartography, The, Volume 4: Cartography in the European Enlightenment, edsed. Matthew H. Edney and Mary Sponberg Pedley, Chicago, IL, 2020.

Hoerschelmann, Emilie von, *Rosalba Carriera die Meisterin der Pastellmalerei: Studien und Bilder aus der Kunst-und Kulturgeschichte des 18. Jahrhunderts*, Leipzig, 1908.

Hollingsworth, Mary, 'Art Patronage in Renaissance Urbino, Pesaro, and Rimini, c. 1400–1550', in Charles Rosenberg, ed., *The Court Cities of Northern Italy: Milan, Parma, Picenza, Manuta, Ferrara, Bologna, Urbino, Pesaro, and Rimini*, Cambridge, 2010, pp. 325–68.

Homans, J. Smith, *A Cyclopedia of Commerce and Commercial Navigation*, New York, 1859.

Hood, William, 'The Narrative Mode in Titian's "Presentation of the Virgin"', *Memoirs of the American Academy in Rome*, 35, 1980, pp. 125–62.

Hopkins, Andrew, *Santa Maria della Salute: Architecture and Ceremony in Baroque Venice*, Cambridge, 2000.

Houghton, L. B. T., 'The Golden Age Returns: Virgil's Fourth Eclogue in the Political Panegyric of the Italian Courts', *Journal of the Warburg and Courtauld Institutes,* 78, 2015, pp. 71–95.

Houpt, Simon, *Museum of the Missing: A History of Art Theft*, New York, 2006.

Houston, R. A., 'Madness and Gender in the Long Eighteenth Century', *Social History*, 27, 3 2002, pp. 309–26.

Howard, Deborah, *Jacopo Sansovino: Architecture and Patronage in Renaissance, Venice and London*, 1975.

—, 'Venetian Architecture', in Eric R. Dursteler, ed., *A Companion to Venetian History, 1400–1797*, Leiden, 2013, pp. 743–78.

—, and Cosato, Carlo, 'Santa Maria Gloriosa dei Frari: Architecture and Community', in Carlo Cosato and Debra Howard, eds, *Santa Maria Gloriosa dei Frari: Immagini di devozione, spazi della fede* [Devotional Spaces, Images of Piety], Padua, 2015, pp. i–xviii.

—, and Moretti, Laura, *Sound and Space in Renaissance Venice: Architecture, Music, Acoustics*, New Haven, CT, 2009.

Howard-Smith, Stephanie, 'China-Pugs: The Global Circulation of Chinoiseries, Porcelain, and Lapdogs, 1660–1800', in Bärbel Czennia and Greg Clingham, eds, *Oriental Networks: Culture, Commerce, and Communication in the Long Eighteenth Century*, Lewisburg, PA, 2021, pp. 63–91

Howe, E., 'Architecture in Vasari's "Massacre of the Huguenots"', *Journal of the Warburg and Courtauld Institutes: The Warburg Institute*, 39, 1976, pp. 258–61.

Humfrey, Peter, 'Dosso Dossi: His Life and Works', in Peter Humfrey, Mauro Lucco and Andrea Bayer, eds, *Dosso Dossi: Court Painter in Renaissance Ferrara,* New York, 1998, pp. 3–16.

Humfrey, Peter and Sherman, Alison, 'The Lost Church of San Niccolò ai Frari (San Nicoletto) in Venice and its Painted Decoration', *Artibus et Historiae*, 36, 72, 2015, pp. 247–81.

Hunt, John Dixon, *Garden and Grove: The Italian Renaissance Garden in the English Imagination*, London, 1986.

Hurd, Richard, DD., *The Works of Joseph Addison*, 6 vols, London, 1893.

Hutton, Edward, *The Story of Ravenna*, London and New York, 1926.

Ilchman, Frederick, 'The Triumph of Tintoretto', *Apollo Magazine*, 188, 667, September 2018, pp. 27–8.

Iliffe, Rob, *Priest of Nature: The Religious Worlds of Isaac Newton*, Oxford, 2017.

—, and Mandelbrote, Scott, 'Newton and the Mint', University of Oxford, https://newtonandthemint.history.ox.ac.uk/

—, and Smith, George E., 'Introduction', in Rob Iliffe and George E. Smith, eds, *The Cambridge Companion to Newton*, Cambridge, 2016, pp. 1–33.

Iltis, Carolyn, 'Leibniz and the Vis Viva Controversy', *Isis*, 62,1, Spring 1971, pp. 21–35.

—, 'The Leibnizian-Newtonian Debates: Natural Philosophy and Social Psychology', *British Journal for the History of Science*, 6, 24, 1973, pp. 343–77.

Ingamells, John, ed., *A Dictionary of British and Irish Travellers in Italy, 1701–1800*, New Haven, CT, and London, 1997.

Insalaco, Maddine, 'John Robert Cozens (1752–1797), the Sublime, and Civita Castellana', *The British Art Journal*, 19 3, Winter 2018/2019, pp. 31–7.

Irvine, William, ed., *A Pepys of Mogul India, 1653–1708: being an Abridged Edition of the Storia do Mogor of Niccolai Manucci*, New Delhi, 1991.

Ives, Herbert E., *The Venetian Gold Ducat and Its Limitations*, New York, 1954.

Jacobi, Lauren, *The Architecture of Banking in Renaissance Italy: Constructing the Spaces of Money*, Cambridge, 2019.

Jeffares, Neil, ed., *Dictionary of Pastellists before 1800*, http://www.pastellists.com

Johns, Christopher M.S., 'Clement XI and Santa Maria Maggiore in the Early 18th Century', *Journal of the Society of Architectural Historians*, 45, 3, September 1686, pp. 286–93.

Johnson, Geraldine A., 'Approaching the Altar: Donatello's Sculpture in the Santo', *Renaissance Quarterly*, 52, 3, 1999, pp. 627–66.

Johnson, James H., 'Debt and Sincerity in Early Modern Venice', *Eighteenth-Century Studies*, 38, 3, Spring 2005, pp. 399–415.

—, *Venice Incognito: Masks in the Serene Republic.*, Los Angeles and Berkeley, CA, 2011

Jones, Emma Jane, 'The Business of Sculpture in Venice: 1525–1625', PhD dissertation, University of Cambridge, September 2016.

Joost-Gaugier, Christiane L., 'The Casa degli Specchi at Padua, its Architect Annibale da Bassano, Tito Livio, and a Peculiar Historical Connection', *Bollettino del Museo Civico di Padova* 72, 1983, pp. 113–24.

Kahr, Madlyn, 'The Meaning of Veronese's Paintings in the Church of San Sebastiano in Venice', *Journal of the Warburg and Courtauld Institutes*, 33, 1970, pp. 235–47.

Kaplan, Frederic, Second Digital Humanities Fall School in Venice: Case Studies, https://digitalvenice.wordpress.com/2014/10/01/second-digital-humanities-fall-school-in-venice-case-studies/

Kaplan, Paul H. D., 'Magi, Winds, Continents: Dark Skin and Global Allegory in Early Modern Images', in Maryanne Cline Horowitz and Louise Arizzoli, eds, *Bodies and Maps: Early Modern Personifications of the Continents*, Leiden, 2020, pp. 130–58.

Karet, Evelyn, *The Antonio II Badile Album of Drawings: The Origins of Collecting Drawings in Early Modern Northern Italy*, New York, 2016.

'Keeping the Books: The Rubinstein Collection of the Orsini Business Archive', University of Kansas Special Collections, https://blogs.lib.ku.edu/spencer/fall-exhibit-2022-keeping-the-books-the-rubinstein-collection-of-the-orsetti-family-business-archive/

Kelly, Jason, *The Society of Dilettanti: Archaeology and Identity in the British Enlightenment*, New Haven, CT, 2009.

Kelly, J. [John] N. D. and Walsh, Michael, eds, *A Dictionary of Popes*, 3rd edn, Oxford, 2015.

Keysler, John George, *Travels Through Germany, Bohemia, Hungary, Switzerland, Italy, and Lorrain. Giving a True and Just Description of the Present State of those Countries*, 4 vols, London, 1760.

Kirk, Terry, *The Architecture of Modern Italy: The Challenge of Tradition 1750–1900*, I, New York, 2005.

Kisby, Fiona, *Music and Musicians in Renaissance Cities and Towns*, Cambridge, 2005.

Klemenčič, Matej, 'Antonio Corradini, the Collegio dei Scultori, and *Neo-Cinquecentismo* in Venice around 1720', in Andaleeb Badiee Banta, ed., *The Enduring Legacy of Venetian Renaissance Art*, London and New York, 2016, pp. 103–20.

—, and Clemente, Maichol, 'Caritas Romana by Giusto Le Court and the Collection of Zaccaria Sagredo', paper presented at the Museum of Fine Arts, Budapest, Hungary, 9 October 2013.

Kolbe, George, 'Godfather to all Monkeys: Martin Folkes and his 1756 Library Sale', *The Asylum: Quarterly Journal of the Numismatic Bibliomania Society*, April–June 2014, pp. 38–92.

Korten, Christopher, 'Financial Policies in the Papal States, 1790s–1848: A Comparative Study of Napoleonic Europe', *Journal of Modern Italian Studies*, 23, 3, 2018, pp. 234–55.

Lane, Frederic C., *Venice: A Maritime Republic.*, Baltimore, MD, and London, 1973.

—, and Mueller, Reinhold C., *Money and Banking in Medieval and Renaissance Venice*, 2 vols, Baltimore, MD, 1985 and 1997.

Langé, Santino and ScolaScota, Angelo, *La dimora dei patriarchi: il Palazzo patriarcale di Venezia dopo i restauri del 2004–2007*, Venice, 2009.

Langins, Jânis, *Conserving the Enlightenment: French Military Engineering from Vauban to the Revolution*, Cambridge, 2004.

Lansford, Tyler, *The Latin Inscriptions of Rome: A Walking Guide*, Baltimore, MD, 2009.

Latorlonga, https://www.latorlonga.it

Laven, Mary, *Virgins of Venice: Enclosed Lives and Broken Vows in the Renaissance Convent*, London, 2002.

Lebowsky, Carlotta, *Bessarion's Library and the Biblioteca Nazionale Marciana, Six Early Inventories*, Rome, 1979, 1980 printing.

Lefèvre, Valentin, *Presentation of the Infant Christ at the Temple*, etching after Veronese, Wellcome Collection, https://wellcomecollection.org/works/w657mf48

Lehman, Robert and Szabó, George, *Eighteenth Century Italian Drawings, from the Robert Lehman Collection*, New York, 1981.

Licht, Fred, 'Raphael's Curtain Rod', *Notes in the History of Art*, 34, 1, 2014, pp. 18–23.

Lindberg, Sten G., 'Bibliotecket på Skokloster: Vårutflykt till vårt rikaste 1600-talsbibliotek', *Biblis*, 1976, pp. 133–72.

Lippincott, Kristen, 'The Neo-Latin Historical Epics of the North Italian Courts: An Examination of "Courtly Culture" in the Fifteenth Century', *Renaissance Studies*, 3, 4, 1989, pp. 415–28.

—, 'A Chapter in the "Nachleben" of the Farnese Atlas: Martin Folkes's Globe', *Journal of the Warburg and Courtauld Institutes*, 74, 2011, pp. 281–99.

Lippincott, Louise, 'Arthur Pond's Journal of Receipts and Expenses, 1734–50', *The Volume of the Walpole Society*, 54, 1988, pp. 220–333.

Litta, Pompeo, *Famiglie celebri di Italia. Giustiniani di Venezia*, Milan, 1840.

Lowe, Kate, 'Elections of Abbesses and Notions of Identity in Fifteenth- and Sixteenth-Century Italy, with Special Reference to Venice', *Renaissance Quarterly*, 54, 2, Summer 2001, pp. 389–429.

—, *Nun's Chronicles and Convent Culture in Renaissance and Counter Reformation Italy*, Cambridge, 2003.

Louvre, https://collections.louvre.fr

Lualdi, Alberto, 'Venetian Makers of Optical Instruments of the 18th--19th Centuries: Part 2 the Selva Family', *Bulletin of the Scientific Instrument Society*, 77, 2003, pp. 10–13.

Lubin, Matthew, 'Poisoning as a Means of State Assassination in Early Modern Venice', in Larissa Tracy, ed., *Medieval and Early Modern Murder: Legal, Literary and Historical Contexts*, Woodbridge, 2018, pp. 227–53.

Lugato, Franca, *The Frari Basilica Guide*, Venice, 2018.

Mac Tutor, St Andrew's University, https://mathshistory.st-andrews.ac.uk

McAndrew, John, 'Sant'Andrea della Certosa', *The Art Bulletin*, 51, 1, 1969, pp. 15–28.

—, *Venetian Architecture of the Early Renaissance*, Boston, MA, 1980.

McClintock and Strong Biblical Encyclopaedia, ed. John McClintock and James Strong, https://www.biblicalcyclopedia.com

McCluskey, Karen E., *New Saints in Late-Medieval Venice 1200–1500*, New York, 2019.

McConnell, Anita, 'The Flowers of Coral – Some Unpublished Conflicts from Montpellier and Paris During the Early 18th Century', *History and Philosophy of the Life Sciences*, 12, 1, 1990, pp. 51–66.

—, 'Instruments and Instrument Makers', in Jed Z. Buchwald and Robert Fox, eds, *The Oxford Handbook of the History of Physics*, Oxford, 2013, pp. 326–57.

McDonald, Grantley, *Biblical Criticism in Early Modern Europe: Erasmus, the Johannine Comma and Trinitarian Debate*, Oxford, 2016.

Machado, Carlos, 'Dedicated to Eternity? The Reuse of Statue Bases in Late Antique Italy', in Katharine Bolle, Carlos Machado and Christian Witschel, eds, *The Epigraphic Cultures of Late Antiquity*, Stuttgart, 2017, pp. 323–62.

McHam, Sarah Blake, 'Renaissance Monuments to Favourite Sons', *Renaissance Studies*, 19, 4, September 2005, pp. 458–86.

Mackie, Gillian Vallance, *Early Christian Chapels in the West: Decoration, Function and Patronage*, Toronto, 2003.

McKitterick, David, 'Anthony Robert Alwyn Hobson (1921–2014)', *Biographical Memoirs of Fellows of the British Academy*, 15, 2016, pp. 55–67.

McLeish, Simon, 'Monastic Provenances of Early Printed Books in Bodleian Collections: Case 1', The Conveyor, Bodleian Library, 5 March 2012, https://blogs.bodleian.ox.ac.uk/ theconveyor/ monastic-provenances-of-early-printed-books-in-bodleian-collections-case-1/

Macuglia, Daniele, 'Newtonianism and Information Control in Rome in the Wake of the Eighteenth Century', *Annals of Science*, 77, 2020, pp. 108–26.

Madden, Thomas F., *Venice: A New History*, New York, 2013.

Malamani, Vittorio, *Rosalba Carriera*, 2nd edn, Bergamo, 1910.

Manucci, Niccolao, *Histoire générale de l'empire du Mogol depuis sa fondation jusqu'à présent / sur les mémoires portugais de M. Manouchi* ..., ed. François Catrou, Paris, 1715.

'Mappamondo di Fra' Mauro, Il', Biblioteca Nazionale Marciana, Venice. https:// bibliotecanazionalemarciana.cultura.gov.it/la-biblioteca/il-patrimonio/patrimonio-librario/il-mappamondo-di-fra-mauro

Mapping our World: Terra Incognita to Australia, Canberra, 2013.

Margóscy, Daniel, 'Horses, Curiosities, and the Culture of Collection at Early Modern Germanic Courts', *Renaissance Quarterly*, 74, 4, 2021, 1210–59.

Marina, Areli, 'From the Myth to the Margins: The Patriarch's Piazza at San Pietro di Castello in Venice', *Renaissance Quarterly*, 64, 2011, pp. 353–429.

Marocchi, Massimo, *Castiglione delle Stiviere: un principato imperiale nell'Italia padana: (sec. XVI–XVIII)*, Rome, 1996.

Marra, Claudia, 'Venetian Affirmation and Urban Tradition in Sixteenth-Century Padua: The Palazzo de Podestà and its Façades on Piazza delle Erbe', in Brigit Blass-Simmen and Stefan Weppelmann, eds, *Padua and Venice: Transcultural Exchange in the Early Modern Age*, Berlin and Boston, MA, 2017, pp. 137–50.

Martin, Lillian Ray, *The Art and Archaeology of Venetian Ships and Boats*, College Station, TX, and London, 2001.

Märtl, Claudia, Kaiser, Christian and Ricklin, Thomas, eds, *Inter graecos latinissimus, inter latinos graecissimus: Bessarion zwischen den Kulturen*, Berlin and Boston, MA, 2013.

Martyn, Thomas, *An appendix to The gentleman's guide through Italy. Containing catalogues of the paintings, statues, busts, &c. By the author of the guide*, London, 1787.

Marzo, Mauro, 'A Theme, A Place: Defense of the Lagoon', in Mauro Marzo, ed., *Fortified Places in the Venetian Lagoon*, Parma, 2012, pp. 31–76.

Matthews, Albert, 'Colonel Elizeus Burges', *Transactions of the Colonial Society of Massachusetts*, 14, 1911–13, pp. 360–73.

Mauro, Maurizio, *Mura, porte e torri di Ravenna*, Ravenna, 2000.

Melton, James Van Horn, *Absolutism and the Eighteenth-Century Origins of Compulsory Schooling in Prussia and Austria*, Cambridge, 2003.

Mémoires concernant la vie et les écrits du Comte François Algarotti, Berlin, 1772.

Meneghin, Vittorino, *San Michele in Isola di Venezia*, 2 vols, Venice, 1962.

Merolle, Irma, *L'abate Matteo Luigi Canonici e la sua biblioteca: i manoscritti Canonici e Canonici-Soranzo delle biblioteche fiorentine*, Rome, 1958.

Metropolitan Museum of Art, New York, https://www.metmuseum.org/art/collection

Meyer, Cornelius, *L'arte di restituere à Roma la tralasciata navigatione del suo Tevere*, Rome, 1685.

Michael, Thomas, ed., *Standard Catalogue of World Coins, 1701–1800*, 7th edn, Iola, WI, 2016.

Michiel, Guistina Renier, *Origine delle feste veneziane*, 3 vols, Milan, 1830.

Middleton, Conyers, *Letter from Rome, showing an Exact Conformity between Popery and Paganism*, London, 1729.

Miller, David, '"Into the Valley of Darkness": Reflections on the Royal Society in the Eighteenth Century', *History of Science*, 27, 1989, pp. 155–66.

Miller, William, *Essays on the Latin Orient*, Cambridge, 1964.

Mioni, Elpidio, 'Bessarione bibliofilo e filologo', *Rivista di studi bizantini e neoellenici*, 5, 1968, pp. 61–83.

Miranda, Salvador, 'The Cardinals of the Holy Roman Church', https://cardinals.fiu.edu/cardinals.htm

Misericordia di Venezia, https://www.misericordiadivenezia.it/the-misericordia/heritage/

Misson, Maximilien, *A New Voyage to Italy: with curious observations on several other countries ....*, 4th edn, 2 vols, London, 1714.

Mitchell, John, 'Trevisan and Soranzo: Some Canonici Manuscripts from Two Eighteenth-Century Venetian Collections', *Bodleian Library Record*, 8, 3, 1969, pp. 125–35.

Mitelli, Guiseppe Maria, *The Martyrdom of St Sebastian, after Veronese. 1666, British Museum*, etching, https://www.britishmuseum.org/collection/object/P_W-9-84

—, St Sebastian Exhorting the Martyrs Mark and Marcellian not to be Swayed by the Prayers of their Families, after Veronese. 1666, etching, British Museum, https://www.britishmuseum.org/collection/object/P_W-9-85

Modena, Claudio, Caglioti, Benedetta, and Cescatti, Elvis, 'Monument of Ludovico Ariosto in Ferrara Italy: Conservation of Architectural Surfaces and Structural Consolidation', *WIT Transactions on the Built Environment*, 191, 2019, pp. 151–61.

Mohler, Ludwig, *Kardinal Bessarion als Theologe, Humanist und Staatsmann*, 3 vols, Paderborn, 1923–1942.

Molli, Giovanni Baldissin, 'Una "pella come un marmo": la riscoperta della tecnica dello stucco nella trattatistica rinascimentale, con particolare riguardo all'area veneta', *Bollettino del Museo Civico de Padova*, 78, 1989, pp. 93–101.

Monfasini, John, *Byzantine Scholars in Renaissance Italy: Cardinal Bessarion and other Emigrés*, London, 1995.

Moretti, Laura, 'How the Barbaro Brothers Created the Perfect Renaissance Villa', https://www.thebritishacademy.ac.uk/blog/how-barbaro-brothers-created-perfect-renaissance-villa

—,"'Quivi si essercitaranno le musiche": la sala della musica presso la "corte" padovana de Alvise Cornaro', *Music in Art*, 35, 1/2, 2020, pp. 135–44.

Morosini, Sergio Rossetti, 'New Findings in Titian's Fresco Technique at the Scuola del Santo in Padua', *The Art Bulletin*, 81, 1, 1999, pp. 163–4.

Muir, Edward, *Civic Ritual in Renaissance Venice,* Princeton, NJ, 1981.

Munby, Alan Noel Latimer, *Portrait of an Obsession: The Life of Sir Thomas Phillipps, the World's Greatest Book Collector*, New York, 1967.

Munman, Robert, 'The Last Work of Antonio Rizzo', *Arte Lombarda*, 47/48, 1977, pp. 89–98.

'Murano Glass Work', *Harper's Bazaar*, 18, 35, 29 August 1885, p. 563.

Muratori, Ludovico Antonio, *Della regolata divozione de' Cristiani*, Venice, 1747.

Murray, John, *Handbook for Travellers in Central Italy, including the Papal States, Rome and the Cities of Etruria,* London, 1843.

Mvsei Vaticani, https://www.museivaticani.va

Museo Castelvecchio in Verona, https://museodicastelvecchio.comune.verona.it

Museo Ferrara, http://www.museoferrara.it

Museo Galileo, Florence, https://catalogue.museogalileo.it

Museum of Fine Arts, Houston, https://emuseum.mfah.org

Muttoni, Pietro, 'Pietro della Vecchia', National Gallery of Art, Washington DC, https://www.nga.gov/collection/artist-info.6893.html

Nardone, Paola, 'Diplomacy and International Commerce in the 18th Century Adriatic: The Case of the French Consulate in the Port of Ancona', *The Journal of European Economic History*, 42, 1, 2013, pp. 165–207.

National Archives, *In Their Own Words 2: More Letters from History*, London, 2018.

National Gallery, The, https://www.nationalgallery.org.uk

Newton, Isaac, Sir, *Opticks: or, A Treatise of the Reflexions, Refractions, Inflexions and Colours of Light*, London, 1718.

—, *Chronology of Ancient Kingdoms Amended*, London, 1728.

—, *Observations Upon the Prophecies of Daniel and the Apocalypse of St John*, London, 1733.

Nordhagen, P. J., 'Mausoleum of Theodoric (Ravenna)', Grove Art Online, https://doi.org/10.1093/oao/9781884446054.013.90000369940

Norman, Larry F., *The Shock of the Ancient: Literature and History in Early Modern France*, Chicago, IL, 2011.

Norton, Marcy, 'Tasting Empire: Chocolate and the European Internalization of Mesoamerican Aesthetics', *The American Historical Review*, 111, 3, June 2006, pp. 660–91.

'Notable Buildings in Ravenna', https://edificistoricravenna.it/en/palazzo-della-prefettura-palazzo-del-legato-apostolico

Notizie della Santa Casa della Gran Madre di Dio Maria Vergine adorata in Loreto, Venice, 1728.

Nugent, Thomas, *The Grand Tour: Or, a Journey Through the Netherlands, Germany, Italy and France*, London, 1756.

Oberer, Angela, *The Life and Work of Rosalba Carriera (1673–1757): The Queen of Pastel*, Amsterdam, 2020.

O'Brian, Grant, *Historical Italian Metrology*, http://www.claviantica.com/ Historical_Italian_Metrology.htm

Ocean Space, https://www.ocean-space.org.

O'Connell, Monique, *Men of Empire: Power and Negotiation in Venice's Maritime State,* Baltimore, MD, 2009.

Oddy, Andrew, 'The Production of Gold Wire in Antiquity: Hand-Making Methods Before the Introduction of the Draw Plate', *Gold Bulletin*, 10, 1977, pp. 79–87.

Oftestad, Eivor Andersen, *The Lateran Church in Rome and the Ark of the Covenant: Housing the Holy Relics of Jerusalem*, Woodbridge, 2019.

Old Bailey Online, https://www.oldbaileyonline.org

Oliva, Marilyn, 'Review of Mary Laven, *Virgins of Venice: Broken Vows and Cloistered Lives in the Renaissance Convent*', H-Catholic, H-Net Reviews. July, 2006. https://networks.h-net.org/node/3595/reviews/4369/oliva-laven-virgins-venice-broken-vows-and-cloistered-lives-renaissance

Overbeek, Bonaventura van, *Les restes de l'ancienne Rome: recherchez avec soin, mesurez, dessinez sur les lieux, & gravez par feu Bonaventure d'Overbeke*, 3 vols, Amsterdam, 1709.

Owen, Lisa, *Carving Devotion in the Jain Caves at Ellora*, Leiden, 2012.

Owners of Incunabula, Consortium of European Research Libraries, https://data.cerl.org/owners

Oxford Classical Dictionary, The, https://www.oxfordreference.com/display/10.1093/acref/9780198606413.001.0001/acref-9780198606413

Oxford Dictionary of the Christian Church, The, https://www.oxfordreference.com/display/10.1093/acref/9780192802903.001.0001/acref-9780192802903

Oxford Dictionary of Late Antiquity, The, https://www.oxfordreference.com/display/ 10.1093/acref/9780198662778.001.0001/acref-9780198662778

Oxford English Dictionary, The [*OED*] Online, https://www.oed.com/

Pace, Claire, 'Pietro Santi Bartoli's Drawings after Ancient Roman Paintings in the Tomb of the Nasonii: Album MS Gen 1496 in Glasgow University Library', *Forma Urbis: itinerari nascosti di Roma antica*, 5, 4, 2000, pp. 1–40.

Padova medieval, http://www.padovamedievale.it

'Padua, Monastery of San Giovanni di Verdara', Bodleian Library, Oxford,

'Palazzo Ducale', Commune di Pesaro, http://www.comune.pesaro.pu.it/cultura/da-vedere/luoghi-da-vedere/edifici-storici-e-monumenti/edifici/palazzo-ducale

Panciera, Walter, 'The Industries of Venice in the Seventeenth and Eighteenth Centuries', in Paola Lanaro, ed., *At the Centre of the Old World: Trade and Manufacturing in Venice and on the Venetian Mainland*, Toronto, 2006, pp. 185–216.

Paoletti, John T. and Radke, Gary M., *Art in Renaissance Italy*, Upper Saddle River, NJ, 1997.

'Parco de Veio', https://www.parcodiveio.it

Park, Katharine and Pennuto, Concetta, 'Science and Medicine', in Michael Wyatt, ed., *The Cambridge Companion to the Italian Renaissance*, Cambridge, 2014, pp. 364–85.

Past Fellows Dictionary, The Royal Society of London, https://catalogues.royalsociety.org/CalmView/PersonSearch.aspx?src=Calm.View.Persons

Pastorino, Cesare, 'The Early Modern Study of Ancient Measures in Comparative Perspective: A Preliminary Investigation', in Dmitri Levitin and Ian Maclean, eds, *The Worlds of Knowledge and the Classical Tradition in the Early Modern Age*, Leiden, 2021, pp. 118–41.

Patrignani, Giovanna, *Pesaro. La radio storia della città*, Pesaro, 2008.

Paul, Benjamin, *Nuns and Reform Art in Early Modern Venice: The Architecture of Santi Cosma e Damiano and its Decoration from Tintoretto to Tiepolo*, Boca Raton, FL, 2016.

Paul, Carole, *Making a Prince's Museum: Drawings for the Late-Eighteenth-Century Redecoration of the Villa Borghese*, Los Angeles, CA, 2000.

—, *The Borghese Collections and the Display of Art in the Age of the Grand Tour*, Burlington, VT, 2009.

Peck, Amelia, et al., *Period Rooms in The Metropolitan Museum of Art*, New Haven, CT, and London, 1996.

Pedersen, Eva de la Fuente, 'Salvator Rosa's Democritus and Diogenes in Copenhagen', *Perspectives*, March 2017, pp. 1–22.

Pepper, D. Stephen, *Guido Reni: A Complete Catalogue of his Works with an Introductory Text*, Oxford, 1984.

Perlove, Shelley Karen, 'Power and Religious Authority in Papal Ferrara: Cardinal Serra and Guercino', *Konsthistorisk Tidskrift*, 68, 1, 1999, pp. 19–30.

Perrault, Charles, *Paralèle des anciens et les modernes en ce qui regarde l'éloquence*, 4 vols, Paris, 1688–97.

Perreau, Stéphan, *Hyacinthe Rigaud 1659–1743: Catalogue raisonné de l'oeuvre*, http://www.hyacinthe-rigaud.com/catalogue-raisonne-hyacinthe-rigaud/portraits/1128-bielke-carl-gustav

Perry, Marilyn, 'The Statuario Publico of the Venetian Republic', *Saggi e memorie di storia dell'arte*, 8, 1972, pp. 75–150, 221–53.

Petaros, Anja, Škrobonja, Ante, and Bosnar, Alan, 'Mummified Saints of the Northern Croatian Littoral', *Acta medico-historica Adriatica*, 10, 1, October 2012, pp. 131–40.

Péter, Róbert, 'Admissions and Lodge Meetings' in Róbert Péter and, Jan A. M. Smoek, eds, *British Freemasonry, 1717–1813*, 5 vols, New York and Oxford, 2016, V, pp. 1–42.

Petkov, Kiril, 'Relics and Society in Late Medieval and Renaissance Venice: The Miracles of the True Cross at the Bridges of San Lorenzo and San Lio', *Cahiers de recherches médiévales et humanistes*, 19, 2010, pp. 267–82.

Pezzolo, Luciano, 'The Venetian Economy', in Eric R. Dursteler, ed., *A Companion to Venetian History, 1400–1797*, Leiden, 2013, pp. 255–90.

Phillipps, Sir Thomas, *Catalogus librorum manuscriptorum in bibliotheca d. Thomae Phillipps, Bart.*, Middle Hill, Worcs, 1837.

—, *Bibliotheca Phillippica New Series: First Part. Catalogue of the Celebrated Collection of Manuscripts Formed by Sir Thomas Phillipps, Bt. (1792–1872) ......*, Sotheby & Co., auction catalogue, sale dates: 28–29 June 1965.

Pinacoteca di Brera, https://pinacotecabrera.org/collezione-online/

Pitteri, Matteo, *Per un confinazione 'equa e guista': Andrea Tron e la politica dei confini della Repubblica di Venezia nel '700*, Milan, 2007.

Pizzichi, Filippo, *Viaggio per l'alta Italia del ser. Principe di Toscana, poi Granduca Cosimo III. descritto da Filippo Pizzichi,* Florence, 1828.

Pliny the Younger, *Letters,* trans. Betty Radice, Cambridge, 1969.

Poggi, Francesco, *Origini e antichità de Fano*, Fano, 1895.

Pointon, Marcia, 'Sacred Contagion: Secular Jewellery and Votive Transvaluation at the Santa Casa, Loreto, 1720–1820', in Caroline Van Eck and Edward Winters, eds, *Dealing with the Visual: Art History, Aesthetics, and Visual Culture*, Basingstoke, 2005, pp. 127–48.

Poleni, Giovanni. *Exercitationes Vitruvianae ... Hoc est: Joannis Poleni commentarius criticus de M. Vitruvii Pollionis*. Padua, 1739–45.

—, *M. Virtuvii Pollionis architectura ... cum exercitationibus notisque novissimis Joannis Poleni ... additis nun primum studiis Simonis Stratico*, Udine, 1825–30.

Polichetti, M. Luisa., *San Ciriaco. La cattedrale di Ancona. Genesi e sviluppo,* Milan, 2003.

Pollens, Stewart, *Stradivari,* Cambridge, 2010.

Prado, Jeroìnimo [de], and Villalpando, Juan Bautista, *Hieronymi Pradi et Ioannis Baptistae Villalpandi e Societate Jesu in Ezechielem explanationes et apparatus urbis, ac Templi Hierosolymitani: commentariis et imaginibus illustratus. Opus tribus tomis distinctum,* Rome, 1596–1604, III, *Apparatus urbis ac templi Hierosolymitani,* Rome, 1604.

Present State of the Republick of Letters for July, The, 1730, Vol. VI, London, 1730.

Princeton Encyclopedia of Classical Sites, ed. Richard Stillwell, WilliamRichard, MacDonald, William L., McAlister, William L., and Maria Holland, Maria, eds, Princeton, NJ, 1976.

Pro Venezia, https://provenezia.ch

Prokop, Ellen, 'The Allegorical Frescoes of the Palazzo della Ragione, Padua', Frick Collection, https://www.frick.org/blogs/photoarchive/allegorical_frescoes_palazzo_della_ragione_padua

Publii Terentii Afri comoediae, Phaedri Augusti Liberti Fabularum Aesopiarum, Publii Syri et aliorum veterum Sententiae, Recensuit et Notas Richardus Bentleius, Cambridge, 1726.

Pugliano, Valentina, 'Pharmacy, Testing and the Language of Truth in Renaissance Italy', *Bulletin of the History of Medicine*, 91, 2, Summer 2017, pp. 233–73.

Puppi, Lionello and Ellero, Giuseppe, eds, *Le Zitelle: architettura, arte e storia di un'istituzione veneziana*, Venice, 1992.

Pyle, Cynthia M. 'The Art and Science of Renaissance Natural History: Thomas of Cantimpre, Pier Candido Decembrio, Conrad Gessner, and Teodoro Ghisi in Vatican Library MS Urb. Lat. 276', *Viator*, 27, 1996, pp. 265–321.

Queruel, Anne, *Andrea Tron (1712–1785): le maître de Venise*, Portet-sur-Garonne, 2008.

Quilici, Lorenzo, 'Land Transport, Part 1: Roads and Bridges', in John Peter Oleson, ed., *The Oxford Handbook of Engineering and Technology in the Classical World*, Oxford, 2012, pp. 551–79.

Raffa, Guy P., 'Bones of Contention: Ravenna's and Florence's Claims to Dante's Remains', *Italica*, 92, 3, 2015, pp. 565–81.

Raimondi, Sergio, 'The Bishop's Grand Palace,' *Ferrara: voci di una città*, 27, 2007, https://rivista.fondazioneestense.it/en/2007/item/44-il-gran-palagio-del-vescovo

Raines, Dorit, 'Book Museum or Scholarly Library? The "Libreria di San Marco" in a Republic Context', *Ateneo Veneto*, 9, 11, 2020, pp. 31–50.

—, 'Nobili Scaduti? The Return of Cretan Patriciates to Venice in the Seventeenth Century', in George Christ and Franz-Julius Morche, eds, *Cultures of Empire: Rethinking Venetian Rule, 1400–1700: Essays in Honour of Benjamin Arbel*, Leiden, 2020, pp. 153–75.

Rankin, Alicia, *The Poison Trials: Wonder Drugs, Experiment, and the Battle for Authority in Renaissance Science*, Chicago, IL, 2021.

Raper, Matthew, 'An Enquiry into the Measure of the Roman Foot', *Philosophical Transactions*, 51, 1759, pp. 774–823.

Ravenna Turismo, https://www.turismo.ra.it

Ravid, Benjamin, 'From Yellow to Red: On the Distinguishing Head-Covering of the Jews of Venice', *Jewish History*, 6, 1–2, 1992, pp. 180–210.

Rearick, W. R., 'More Veronese Drawings from the Sagredo Collection', *Master Drawings*, 33, 2, Summer 1995, pp. 132–43.

Recanati: Turismo e Cultura, https://www.recanatiturismo.it/

Redbourn, E. T., 'The History of Sweat and the Sweat Rash from Earliest Times to the End of the Eighteenth Century', *Journal of the History of Medicine and Allied Sciences*, 14, 2, April 1959, pp. 202–27.

Redford, Bruce, *Dilettanti: The Antic and the Antique in Eighteenth-Century England*, Los Angeles, CA, 2008.

Regione Marche Giunta Regionale, https://www.regione.marche.it

Religiana, https://religiana.com

Renaissance Duke, The. Catalogue to 2018 exhibition, Goodwood House, Sussex. https://www.goodwood.com/globalassets/venues/ goodwood-house/summer-exhibition-2017.pdf

Ribuffi, Gaspare, *Guida di Ravenna: con compendio storico della città*, Ravenna, 1835.

Ricci, Amico, *Memorie storiche delle arti e degli artisti della Marca di Ancona*, 2 vols, Macerata, 1834.

Ricci, Seymour de, *Catalogue raisonné des premières impressions de Mayence (1445--1467): avec une plance en phototypie*, Mainz, 1911.

Richardson, E. P., 'Walking Sticks of the Eighteenth Century', *Bulletin of the Detroit Institute of the Arts of the City of Detroit*, 23, 1, October 1943, pp. 6–8.

Richardson, Jonathan the Elder and Richardson, Jonathan the Younger, *An Account of some of the statues, bas-reliefs, drawings and pictures in Italy, &c., with remarks*, London, 1722.

Rietbergen, P. J. A. N., *Power and Religion in Baroque Rome: Barberini Cultural Policies*, Leiden, 2006.

Rigo, Antonio, 'Bessarione, Giovanni Regiomontano e i loro studi su Tolomeo a Venezia e Roma (1462–1464)', *Studi veneziani*, n.s. 21, 1991, pp. 49–110.

Rijksmuseum. Amsterdam, https://www.rijksmuseum.nl

Rinne, Katherine, 'Urban Ablutions: Cleansing Counter-Reformation Rome', in Mark Bradley with Kenneth Stow, eds, *Rome, Pollution and Propriety: Dirt, Disease and Hygiene in the Eternal City from Antiquity to Modernity*, Cambridge, 2012, pp. 182–201.

Ritzler, Remigius and Sefrin, Priminum, eds, *Hierarchia Catholica medii et recentis aevi,* Padua, 1913.

Rodda, Joshua, '"The Condition and State of a Scholar": Disputation in William Alabaster's Conversion Narrative', *British Catholic History*, 31, 3, May 2013, pp. 391–410.

Roeck, Bernd, 'Zu Kunstaufträgen des Dogen Agostino Barbarigo', *Zeitschrift für Kunstgeschichte*, 55, 1, 1992, pp. 1–34.

Roos, Anna Marie, *Salt of the Earth: Natural Philosophy, Medicine, and Chymistry in England, 1650–1750*, Leiden, 2007.

—, '"Only meer love to learning": A Rediscovered Travel Diary of Naturalist and Collector James Petiver (*c*.1665–1718)', *Journal of the History of Collections*, 29, 3, 2017, pp. 381–94.

—, 'Taking Newton on Tour: The Scientific Travels of Martin Folkes, 1733–1735', *The British Journal for the History of Science*, 50, 4, 2017, pp. 569–601.

—, *Martin Folkes (1690–1754): Newtonian, Antiquary, Connoisseur*, Oxford, 2021.

Rosand, David, 'Titian's *Presentation of the Virgin in the Temple* and the Scuola della Carità,', *Art Bulletin*, 58, 1976, pp. 55–84.

—, *Painting in Sixteenth-Century Venice: Titian, Veronese, Tintoretto*, 2nd edn, Cambridge, 1997.

Rossi, P. *Giovanni Marchiori alla Scuola Grande di San Rocco e le altre opera veneziane*, Venice, 2014.

Rossi, Vittorio, 'La biblioteca manoscritta del senatore veneziano Jacopo Soranzo. Appounti (continua) in – Il libro e la stampa', *Bulletino ufficiale della Società Bibliografica Italiana*, 1, 1907, pp. 1–8.

Rousseau, George and Haycock, David, 'Voices Calling for Reform: The Royal Society in the Mid-Eighteenth Century: Martin Folkes, John Hill, and William Stukeley', *History of Science*, 37, 188, pp. 377–406.

Royal Collection Trust, The, https://www.rct.uk/

Ruggeri, Ugo, *Lefèvre – —Valentin Lefèvre (1637--1677). Dipinti, disegni, incisioni*, Manerba del Garda, 2007.

Ruskin, John, *Stones of Venice*, 3 vols, New York, 1860.

—, *The Complete Works of John Ruskin*, ed.eds E. T. Cook and Alexander Wedderburn, 29 vols, London, 1903–1912, The Ruskin, University of Lancaster, https://www.lancaster.ac.uk/the-ruskin/the-complete-works-of-ruskin/

Rusnock, Andrea, ed., *The Correspondence of James Jurin (1684–1750)*, Amsterdam, 1996.

Saint Anthony of Padua, https://www.santantonio.org

'Saint Lawrence', The Courtauld Institute of Art, http://www.artandarchitecture.org.uk/images/gallery/a6b97aa5.html

St. Peter's Basilica, https://stpetersbasilica.info

Salerno, Luigi, *I dipinti del Guercino*, Rome, 1988.

Salmon, Thomas, *Modern History: or, the present state of all nations*, 33 vols, London 1729–39.

Salucci, Antonella, 'The Development, Recovery and Reuse of the Architectural and Urban Heritage: Former Barracks as University Location', in Enrique Castaño Perea, and Ernesto Echeverria Valiente, eds, *Architectural Draughtsmanship: From Analog to Digital Narratives*, Cham, 2018, pp. 1295–308.

Salvadore, Matteo, *The African Prester John and the Birth of Ethiopian-European Relations*, New York, 2016.

Salzberg, Rosa, 'Renaissance Lockdown: How Venice Tried to Control the Plague', *History Workshop*, https://www.historyworkshop.org.uk/venice-plague

Sami, Denis, Baldi, Elena, Booth, Anna, Chinni, Tania, and Toniolo, Luana, 'Ad-Novas-Cesenatico: From Roman Road to Late Antique Wooden Structures. An Interim Report on the Evaluation Test Pits and Excavation at Cà Bufalini', *The Journal of Fasti Online*, January 2014, pp. 1–20.

Samuel, Edgar, 'Diamonds and Pieces of Eight: How Stuart England won the Rough-Diamond Trade', *Jewish Historical Studies*, 38, 2002, pp. 23–40.

Sandys, John Edwin, *Latin Epigraphy: An Introduction to the Study of Latin Inscriptions*, Cambridge, 1919.

Sani, Bernardina, *Rosalba Carriera*, 2nd edn, Turin, 2007.

Sansovino, Francesco, *Venetia, città nobilissima, et singolare: descritta in XIIII. libri*, Venice, 1663.

Santarcangelo di Romagna, https://www.santarcangelodiromagna.info/en/chiesa-e-convento-dei-frati-cappuccini-25

Sapienza, Valentina, 'Notice sur le cycle de fresques de "Villa Bembo Valier alla Chitarra"', in Giuseppe Pavanello, ed., *Gli affreschi delle ville venete. Il Seicento*, Venice, 2009, pp. 21–4.

Save Venice, http://www.savevenice.org

Savoy, Daniel, 'Palladio and the Water-Oriented Scenography of Venice', *Journal of the Society of Architectural Historians*, 71, 2, June 2012, pp. 204–25.

Schaefer, B. E., 'The Epoch of the Constellations on the Farnese Atlas and their Origin in Hipparchus' Lost Catalogue', *Journal for the History of Astronomy*, 36, 2005, pp. 167–96.

Schaffer, Simon, 'Glass Works: Newton's Prisms and the Uses of Experiment', in David Gooding, Trevor Pinch, and Simon Schaffer, eds, *The Uses of Experiment: Studies in the Natural Sciences*, Cambridge, 1989, pp. 67–104.

Schaus, Margaret, 'Reliquary of the Hand of Santa Marina', *Feminae: Medieval women and gender index*, https://inpress.lib.uiowa.edu/feminae/DetailsPage.aspx?Feminae_ID=32224

Schoolman, Edward, *Rediscovering Sainthood in Italy: Hagiography and the Late Antique Past in Medieval Ravenna*, New York, 2016.

Schulz, Anne Markham, *Giammaria Mosca Called Padovano: A Renaissance Sculptor in Italy and Poland*, Philadelphia, PA, 2010.

Schulz, Juergen, *Venetian Painted Ceilings of the Renaissance*, Oakland, CA, 2023.

Scuola Grande di San Rocco, http://www.scuolagrandesanrocco.org/home-en/buildings/church/

Secondo, Joellen, 'The Mausoleum of Theodoric', *Sculpture Review*, 50, 4, 2001, pp. 32–6.

Selden-Goth, Gisella Collection, Library of Congress, https://memory.loc.gov/diglib/ihas/loc.natlib.scdb.200033734/default.html

Selfridge-Field, Eleanor, *The Music of Benedetto and Alessandro Marcello. A Thematic Catalogue, with Commentary on the Composers, Repertory and Sources*, Oxford, 1990.

—, *A New Chronology of Venetian Opera and Related Genres, 1660–1760*, Stanford, CA, 2007.

Selva, Lorenzo, *Esposizione delle comuni, e nuove spezie di cannocchiali, telescopj, microscopj, ed altri istrumenti diottrici, catottrici, e catodiottrici. Perfezionati ed inventati da Domenico Selva ottico* [...], Venice, 1761.

Sennert, Daniel, *Praticae medicinae liber quintus. De tumoribus*, Lyon, 1634.

Sensier, Alfred, ed., *Journal de Rosalba Carriera pendant son séjour à Paris en 1720 et 1721*, Paris, 1865.

Shank, Michael H., 'Regiomontanus and Astronomical Controversy in the Background of Copernicus', in Rivka Feldhay and F. Jamil Ragep, eds, *Before Copernicus: The Cultures and Contexts of Scientific Learning in the Fifteenth Century*, Montreal, Kingston, London, and Chicago, IL, 2017, pp. 79–109.

Shapiro, Alan E., *Fits, Passions, and Paroxysms: Physics, Method, and Chemistry, and Newton's Theories of Colored Bodies and Fits of Easy Reflection*, Cambridge, 1993.

—, 'Newton's Optics', in Jed Z. Buchwald and Robert Fox, eds, *The Oxford Handbook of the History of Physics*, Oxford, 2013, pp. 166–98.

Sheppard, Francis H. W., ed., *Survey of London: Volume 36, Covent Garden*, London, 1970, British History Online, http://www.british-history.ac.uk/survey-london/vol36/

Simane, Jan, *Grabmonumente der Dogen. Venezianische Sepulkralplastik im Cinquecento*, Sigmaringen, 1993.

Small, Alistair, and Small, Carola, 'South Italy, England and Elysium in the Eighteenth Century', *Antiquaries Journal*, 79, 1999, pp. 333–42.

Smith, James Edward, *A Sketch of a Tour on the Continent, in the Years 1786 and 1787*, 3 vols, London, 1793.

Smith, Jonathan Z., *Drudgery Divine: On the Comparison of Early Christianities and the Religions of Late Antiquity*, London, 1990.

Smith, Leonard, *Private Madhouses in England, 1640–1815. Commercialised Care for the Insane*, Cham, 2020.

Social Networks and Archival Context [SNAC], https://snaccooperative.org/view/3040193

Soren, David, and Soren, Noelle, *A Roman Villa and a Late Roman Infant Cemetery: Excavation at Poggio Gramignano Lugnano in Teverina*, Rome, 1999.

Sorensen, Lee, 'Zanetti, Antonio Maria Alessandro', *Dictionary of Art Historians*, https://arthistorians.info/zanetti

Sorrenson, Richard, 'Towards a History of the Royal Society in the Eighteenth Century', *Notes and Records of the Royal Society of London*, 50, 1, 1996, pp. 29–46.

Sperling, Jutta Gisela, *Convents and the Body Politic in Late Renaissance Venice*, Chicago, IL, 1999.

Stahl, Alan M., *Zecca: The Mint of Venice in the Middle Ages*, Baltimore, MD, 2001.

—, 'Venetian Coinage and Money of Account: An Introduction for Researchers', *News on the Rialto*, 35, 2016, pp. 1–8.

Stahl, Alan M.—, and Tonini, Camillo, 'Zecca di Venezia', in Lucia Travaini, ed., *Le zecche italiane fino all'unità*, 2 vols, Rome, 2011, pp. 1217–24.

Stearns, Raymond Phineas, ed., *Martin Lister, A Journey to Paris in the Year 1698*, Urbana, Chicago, IL, and London, 1967.

Stewart, Peter, 'The Equestrian Statue of Marcus Aurelius', in M. van Ackeren, ed., *A Companion to Marcus Aurelius*, Oxford, 2012, pp. 264–77.

Stone, Harold Samuel, *Vico's Cultural History: The Production and Transmission of Ideas in Naples*, Leiden, 1997.

Stone, Peter, The History of London, https://www.thehistoryoflondon.co.uk

Stukeley, William *The Family Memoirs of the Rev. William Stukeley: And the Antiquarian and Other Correspondence of William Stukeley, Roger & Samuel Gale, Etc*, 3 vols, London, 1882–7.

Subrahmanyam, Sanjay, *Three Ways to be Alien: Travails and Encounters in the Early Modern World*, Waltham, MA, 2011.

Sutton, Denys, 'Augustan Virtuosi', *Apollo*, 144, 1981, pp. 313–26.

Sweet, Rosemary, *Cities and the Grand Tour: The British in Italy, c.1690–1820,* Cambridge, 2012.

Syrjämaa, Taina, 'The Clash of Picturesque Decay and Modern Cleanliness in Late Nineteenth-Century Rome', in Mark Bradley and Kenneth Stow, eds, *Rome, Pollution and Propriety: Dirt, Disease and Hygiene in the Eternal City from Antiquity to Modernity*, Cambridge, 2012, pp. 202–22.

Tabacco, Giovanni, *Andrea Tron (1712–1785): e la crisi dell'aristocrazia senatoria a Venezia*, 2nd edn, Udine, 1980.

Talas, Sofia, 'New Light on the Cabinet of Physics of Padua', in Jim Bennett and Sofia Talas, eds, *Cabinets of Experimental Philosophy in Eighteenth-Century Europe*, Leiden, 2013, pp. 49–68.

Talbot, Michael, *The Vivaldi Companion*, Woodbridge, 2011.

Tazzara, Corey, *The Free Port of Livorno and the Transformation of the Mediterranean World*, Oxford, 2017.

Terpstra, Nicholas, *Cultures of Charity: Women, Politics, and the Reform of Poor Relief in Renaissance Italy*, Cambridge, 2013.

—, *Religious Refugees in the Early Modern World: An Alternative History of the Reformation*, Cambridge, 2015.

Terrall, Mary, *The Man Who Flattened the Earth: Maupertuis and the Sciences in the Enlightenment*, Chicago, IL, 2006.

Tesini, Federica, *Pesaro: guida storico-artistica*, Pesaro, 2000.

Thorne, E. H., 'Vincenzo Martinelli in England: 1748–1744', *Italian Studies*, 11, 1956, pp. 92–107.

Tieto, Paolo, *La meravigliosa Riviera del Brenta prolungamento del Canal Grande di Venezia*, Castelfranco Veneto, 2014.

Toaff, Ariel, 'Jews, Franciscans, and the First *Monti de Pietà* in Italy (1462–1500)', in Susan E. Myers and Steven J. McMichael, eds, *The Friars and Jews in the Middle Ages and Renaissance*, Leiden, 2004, pp. 239–54.

Tobie prélevant le coeur et le foie du poisson, in the collections du Musée des Beaux-Arts de Pau, https://mba-pau.opacweb.fr/r/bc8fca97-b20b-4cd2-bff1-29b1f0e56e10

Trapp, Joseph Burney, *Studies of Petrarch and His Influence*, London, 2003.

Trethewey, Ken, Ancient Lighthouses – —6: Lighthouses After the Pharos, unpublished paper, https://www.researchgate.net/publication/333557637_Ancient_Lighthouses_-_6_LighthousesAfterThePharos/citation/download

Truschke, Audrey, *Aurangzeb: The Life and Legacy of India's Most Controversial King*, Stanford, CA, 2017.

Tuohy, Thomas, *Herculean Ferrara: Ercole d'Este (1471–1505) and the Invention of a Ducal Capital*, Cambridge, 2002.

Turner, Nicholas, *Federico Barocci*, Paris, 2000.

Ugolini, Federico, 'A New Insight into the Iconography of the Civitas Classis Mosaic at Sant'Apollinare Nuovo, Ravenna', *Journal of Mosaic Research*, 11, 2018, pp. 287–95.

Valcanover, Francesco, *Jacopo Tintoretto and the Scuola Grande of San Rocco*, Venice, 2001.

'Valcimarra', *Turismo Marche*, https://www.turismo.marche.it/en-us/what-to-see/tourist-attractions/Valcimarra-The-fortified-village-of-Valcimarra/11151

Valenti, Anna Marie Megale, *Le carte de Antonio Cocchi*, Milan, 1990.

Van Creveld, Martin, 'Technology and War I: to 1945', in Charles Townsend, ed., *The Oxford History of Modern War*, Oxford, 2005, pp. 201–23.

Vasari, Giorgio, *Lives of the Most Eminent Painters, Sculptors, and Architects*, ed. Mrs Jonathan Foster, 6 vols, London, 1892.

Vatican Collections: The Papacy and Art, The, New York, 1982.

Vatican Library, The, DVL Digivatlib, https://digi.vatlib.it/

Vatican: Spirit and Art of Christian Rome, The, New York, 1982.

Vélez, Karin, *The Miraculous Flying House of Loreto: Spreading Catholicism in the Early Modern World*, Princeton, NJ, 2019.

'Venetian Diplomatic Agents in England', in *Calendar of State Papers Relating to English Affairs in the Archives of Venice, Volume 1, 1202–1509*, ed. Rawdon Brown, ed., London, 1864, pp. cxxii–cxxix. *British History Online*, http://www.british-history.ac.uk/cal-state-papers/ venice/vol1/cxxii-cxxix

Venetian Protestant Cemetery, https://www.veniceprotestantcemetary.com

'Venice and its Lagoon', Media Center for Art History, Art Atlas, Colombia University, New York, https://mcid.mcah.columbia.edu/art-atlas/venice-and-its-lagoon/biblioteca-nazionale-marciana

Venice in Peril, https://www.veniceinperil.org

Vermeulen, Frank, Verhoeven, Geert J., and Hay, Sophie, 'Potentia: An Integrated Survey of a Roman Colony on the Adriatic Coast', *Papers of the British School of Rome*, 74, 2006, pp. 203–36.

Verstegen, Ian, 'Federico Barocci', *Renaissance and Reformation,* Oxford Bibliographies, https://www.oxfordbibliographies.com/view/document/obo-9780195399301/obo-9780195399301-0442.xml

Veryard, E., *An account of divers choice remarks, as Well Geographical, as Historical, Political, Mathematical, Physical, And Moral; taken in a journey through the Low-Countries, France, Italy, and part of Spain*, London, 1701.

Viale, Mattia, 'Consumer Behaviour and Material Living Standards in a Transition Economy: Venice 1650–1800', PhD dissertation, University of Antwerp and Universita degli Studi di Verona, 2017.

Victoria and Albert Museum, London Collections, https://www.vam.ac.uk/collections

Villa Dolfin dal Martello known as La Mincana, https://villevenetetour.com/en/venetian-villas/villa-dolfin-dal-martello-known-as-la-mincana

Villa Pisani Vescovana, https://www.visitabanomontegrotto.com/en/venetian-villas/villa-pisani-vescovana-en/

Villa Veneto Tour, https://villevenetetour.com

Villani, Stefano, 'Between Anatomy and Politics: John Finch and Italy, 1649–1671', in Margaret Pelling and Scott Mandelbrote, eds, *The Practice of Reform in Health, Medicine, and Science, 1500–2000: Essays for Charles Webster*, New York, 2017, pp. 151–67.

Vine, Angus, *In Defiance of Time: Antiquarian Writing in Early Modern England*, Oxford, 2010.

Virgil, *Aeneid: Books 7–12, Appendix Vergiliana*, trans. H. R. Fairclough, rev. G. P. Goold, Cambridge, 2001.

Visit Cesenatico, https://visitcesenatico.it

Vistoli, Fabrizio, 'La Via Flaminia antica più vicina a Roma: stato dell'arte e prospettive future', in Fabrizio Vistoli, ed., *La riscoperta della Via Flaminia più vicinia a Roma: Storia, luoghi, personaggi. Atti dell'Incontro di studio*, Rome, 2010, pp. 17–26.

Vivi Terni: Portale Turistico del Comune di Terni, https://turismo.comune.terni.it

Vivi Tolentino, https://www.vivitolentino.com

Vivian, Francis, *Il Console Smith mercante e collezionista*, Venice, 1971.

—, *The Consul Smith Collection: Masterpieces of Italian Drawing from the Royal Library, Windsor Castle – Raphael to Canaletto*, Munich, 1989.

Walsh, Kirsten, 'Newton's Epistemic Triad', PhD dissertation, University of Otago, 2014.

Ward, Adolphus William and Waller, A. R., eds, *Cambridge History of English Literature, vol. 8: The Age of Dryden*, Cambridge, 1932.

Water Museum of Venice, https://www.watermuseumofvenice.com

Wegmann, Karl W. and Pazzaglia, Frank J., 'Late Quaternary Fluvial Terraces of the Romagna and Marche Apennines, Italy: Climatic, Lithologic, and Tectonic Controls on Terrace Genesis in an Active Orogen', *Quaternary Science Reviews*, 28, 2009, pp. 137–65.

Weil, Mark S., *The History and Decoration of the Ponte S. Angelo*, University Park, PA, and London, 1974.

Weinberg, Alan M., *Shelley's Italian Experience*, New York, 2016.

Weinshenker, Anne Betty, *Falconet, His Writings and His Friend Diderot*, Geneva, 1966.

Westphall, Allan F., 'A Borrowing from Eadmer of Canterbury's *Liber de Excellentia Virginis Mariae* in a Middle English Translation of the *Meditationes Vitae Christi*', *Notes and Queries*, 70, 1, 2003, pp. 4–6.

White, Eric Marshall and Needham, Paul, *Peter Schoeffer: Printer of Mainz: A Quincentenary Exhibition at Bridwell Library*, Dallas, TX, 2003.

Wilk, Sarah, 'Titian's Padua Experience and Its Influence on His Style', *The Art Bulletin*, 65, 1, 1983, pp. 51–61.

Williamson, Tom, *The Archaeology of the Landscape Park: Garden Design in Norfolk, England, c. 1680–1840*, Oxford, 1998.

Witcombe, Christopher L. C. Ewart, 'The Vatican Apartment of Cinzio Aldobrandini: Notes and Documents', *Archivium Historiae Pontificiae*, 19, 1981, pp. 173–90.

Wood, Carolyn H., 'The Indian Summer of Bolognese Painting: Gregory XV (1621–23) and Ludovisi Art Patronage in Rome', PhD dissertation, University of North Carolina, Chapel Hill, NC, 1988.

Woolf, David, *The Social Circulation of the Past: English Historical Culture, 1500–1730*, Oxford, 2003.

Worsleely, Giles, *Inigo Jones and the European Classicist Tradition*, New Haven, CT, 2007.

Wright, Edward, *Some Observations made inthrough travelling through France, Italy, etc, in the years 1720, 1721 and 1722*, 2 vols, London, 1730.

Wynne, Michael, 'Members from Great Britain and Ireland of the Florentine Accademia del Disegno, 1700–1855', *The Burlington Magazine*, 132, 1049, August 1990, pp. 535–8.

Zanetti, Antonio Maria, *Descrizione di tutte le pubbliche pitture della città di Venezia e isole circonvicine*, Venice, 1733.

—, *Græca D. Marci Bibliotheca codicum manu scriptorum per titulos digesta*, Venice, 1740.

—, *Delle antiche statue greche e romane, che nell'antisala della Libreria di San Marco, e in altri luoghi pubblici di Venezia si trovano*, 2 vols, Venice, 1740–43.

—, *Latina et italica D. Marci Bibliotheca codicum manu scriptorum per titulos digesta*, Venice, 1741.

Zanetti, Vincenzo, *Guida di Murano e delle celebri sue fornaci vetrarie*, Venice, 1866.

Zeri, Federico, *La Galleria Pallavicini in Roma: catalogo dei dipinti*, Florence, 1959.

—, and Gardner, Elizabeth E., eds, *Italian Paintings: Venetian School: A Catalogue of the Collection of the Metropolitan Museum of Art*, New York, 1973.

Zwierlein, Cornel, *Imperial Unknowns: French and British in the Mediterranean, 1650–1750*, Cambridge, 2016.

Zwingle, Erla, 'The World's Greatest Goldbeater', *Craftmanship Quarterly*, Spring 2021, https://craftsmanship.net/the-worlds-greatest-goldbeater

INDEX

Individual artists, artworks, basilicas and cathedrals, and churches are listed under collective headings. All dates are AD unless specified.